SAHARAN DUST

SCOPE 14

C. Morales

International Council of Scientific Unions (ICSU)
Scientific Committee on Problems of the Environment (SCOPE)

SCOPE is one of a number of committees established by a non-governmental group of scientific organizations, The International Council of Scientific Unions (ICSU). The membership of ICSU includes representatives from 68 National Academies of Science, 18 International Unions, and 12 other bodies called Scientific Associates. To cover multidisciplinary activities which include the interests of several unions, ICSU has established 10 scientific committees, of which SCOPE, founded in 1969, is one. Currently, representatives of 34 member countries and 15 Unions and Scientific Committees participate in the work of SCOPE, which directs particular attention to the needs of developing countries.

The mandate of SCOPE is to assemble, review, and assess the information available on man-made environmental changes and the effects of these changes on man; to assess and evaluate the methodologies of measurement of environmental parameters; to provide an intelligence service on current research; and by the recruitment of the best available scientific information and constructive thinking to establish itself as a corpus of informed advice for the benefit of centres of fundamental research and of organizations and agencies operationally engaged in studies of the environment.

SCOPE is governed by a General Assembly, which meets every three years. Between such meetings its activities are directed by the Executive Committee.

Frank Fenner
Editor-in-Chief
SCOPE Publications

Executive Secretary: Dr V. Smirnyagin

Secretariat: 51 Bld de Montmorency
75016 PARIS

SCOPE 14

Saharan Dust

Mobilization, Transport, Deposition

SCOPE 14

Saharan Dust

Mobilization, Transport, Deposition

Edited by
Christer Morales
Swedish Natural Science Research Council
Ecological Research Committee
Stockholm, Sweden

Papers and Recommendations from a Workshop held in Gothenburg, Sweden, 25—28 April 1977

Published on behalf of
Scientific Committee on Problems of the Environment (SCOPE)
of the
International Council of Scientific Unions (ICSU)
by

JOHN WILEY & SONS

Chichester · New York · Brisbane · Toronto

Library of Congress Cataloging in Publication Data:

Main entry under title:

Saharan dust.

(SCOPE [report]; 14)
'Contains the review and recommendations . . as well as the major part of the papers presented at the workshop.'
1. Dust–Sahara–Congresses. I. Morales, Christer. II. Scientific Committee on Problems of the Environment. III. Series: Scientific Committee on Problems of the Environment. SCOPE report; 14.
QC882.S23 551.3'7'30966 78–8687

ISBN 0 471 99680 7

Typeset in IBM Press Roman by Preface Ltd, Salisbury, Wilts.
Printed in Great Britain by Unwin Brothers Ltd, The Gresham Press, Old Woking, Surrey.

Foreword

The Scientific Committee on Problems of the Environment (SCOPE) was established by the International Council of Scientific Unions (ICSU) in 1969 to advance knowledge of the influence of the human race upon its environment, as well as the effects of those alterations upon human health and welfare. It was intended to give particular attention to those effects which are either global or shared in common by several nations. It serves as a non-governmental, interdisciplinary and international council of scientists, and as a source of advice for governments and intergovernmental agencies with respect to environmental problems.

SCOPE seeks to synthesize environmental information from diverse scientific fields, identifying knowledge gaps, and disseminating the results. During the last several years, the main emphasis has been on the following topics:

1. Biogeochemical cycles;
2. Dynamic changes and evolution of ecosystems;
3. Environmental aspects of human settlements;
4. Ecotoxicology;
5. Simulation modelling of environmental systems;
6. Environmental monitoring;
7. Communication of environmental information and societal assessment and response.

A number of publications have resulted, including *SCOPE 10: Environmental Issues* which provides an overview of the environmental challenges of the next decade.

Few problems illustrate more concretely and visibly the international and multidisciplinary character of these issues than does the question of the origin, transport, and deposition of Saharan dust. The eolian particles from the Sahara are seen and have their effects at distances of thousands of kilometres from the Mediterranean to the Caribbean. Their mobilization is influenced by a wide range of physical, biological, and social factors. Understanding what produces them and shapes their movement, and their influence upon the receiving environment calls for research in biology, chemistry, geography, geology, meteorology, oceanography, pedology, and related social science.

SCOPE was happy to join in the initiative taken by the Secretariat for International Ecology of Sweden in organizing the workshop from which this volume of papers is drawn. The officers of SCOPE are grateful to Dr. Christer Morales for his

leadership with the support of the Ecological Research Committee of the Swedish Natural Science Research Council, the Royal Swedish Academy of Sciences, the Monitoring and Assessment Research Centre, and UNESCO. We also acknowledge the assistance of Dr. A. J. Dyer, who ably reviewed the manuscript.

GILBERT F. WHITE
President of SCOPE

University of Colorado,
Boulder, Colorado, U.S.A.

Contents

(c) Transport

(d) Monitoring and Deposition

List of Figures

List of Tables

Preface

A Workshop on Saharan Dust was held in Gothenburg, Sweden, from 25th to 28th April, 1977. The initiative to the Workshop was taken by the Secretariat for International Ecology, Sweden (SIES). It was hosted by the Ecological Research Committee of the Swedish Natural Science Research Council and the Swedish National SCOPE Committee of the Royal Swedish Academy of Sciences. Financial support for the meeting was mainly received from the Monitoring and Assessment Research Centre (MARC) of SCOPE and UNESCO. SCOPE is the acronym of the Scientific Committee on Problems of the Environment of ICSU. Professor Cyrill Brosset and his institution (the Swedish Water and Air Pollution Research Laboratory in Gothenburg) hospitably placed their localities at the Workshop's disposal.

The outline of the Workshop was suggested during the WMO/UNEP meeting on Wet Deposition, which was held in Gothenburg in October 1976.

This document contains the Review and Recommendations from the Workshop as well as the major part of the papers presented at the Workshop. It has been compiled by the Ecological Research Committee in cooperation with SCOPE.

CHRISTER MORALES
Scientific editor

SECTION 1

Introduction and Aims of the Workshop

Wind-blown dust from the Sahara has been observed and reported from ships in the equatorial Atlantic for centuries (Ehrenberg, 1862). This area was known as the 'Dark Sea' because of the remarkable reduction of horizontal visibility. A long geological record of deposition of eolian dust from the Sahara is provided by loess deposits particularly in areas near the northern desert fringes and by eolian strata in marine sediments of the sea bottom in the Atlantic and the Mediterranean.

In recent years, and particularly during the Sahelian drought and famine catastrophe, a question has been raised concerning the quantity, origin and ecological impact of the soil dust which is being blown out of the Sahara and surrounding areas. Which processes are responsible for this transport of dust and what does it mean in terms of the loss of productive soil by wind erosion in Africa, of the pollution of the air and possible impact on climate, and of the supply of nutrients and soil particles to oceans and land areas under the dust trajectories? It is particularly appropriate to address these questions to a group of scientists of different disciplines now, in the year of the United Nations Conference on Desertification, because better knowledge of dryland dust storms and their ecological impact may contribute to more efficient counteraction of desertification.

The workshop on Saharan dust was arranged by the Ecological Research Committee of the Swedish National Science Research Council and funded by the Scientific Committee on Problems of the Environment (SCOPE) and by the Monitoring and Research Assessment Center (MARC) in London, as well as by UNESCO in Paris. The outline of the workshop was suggested during the WMO/UNEP meeting on wet deposition, which was held in Gothenburg in October, 1976. The workshop is also a follow-up of earlier work done by the Secretariat for International Ecology, Sweden (SIES), during 1973–1976.

The workshop in Gothenburg was attended by 26 scientists representing different disciplines in natural sciences, scientists coming from Africa, Europe and North America (cf. list of participants ending Section 1).

As a basis for the discussions during the workshop about 20 lectures were presented by international experts covering the wide variety of interdisciplinary topics which are relevant to the field.

The aim of the workshop was to review the present state of the art and to give recommendations for future research and monitoring concerning mobilization, transport and deposition of air-borne Saharan soil dust and connected ecological implications.

The processes of dust formation and transport, as well as the details of the

ecological impact, primarily in the Saharan area, are indeed very poorly understood. This is partly due to the fact that only in recent years has the possibility of large scale processes induced by man become of general concern, and partly because the Saharan area has, despite many individual efforts, remained a largely unknown area in many respects.

The workshop, therefore, had to start with a critical review of the present state of our knowledge about these problems, and to point out the most important gaps in our knowledge.

At the same time, it had become clear that the problems of impact of Saharan dust production, transport and deposition were in need of considerable clarification, and the discussions during the workshop about these aspects were intended to be another important aim of the workshop.

The lectures and the ensuing discussions of the workshop should identify those areas where research is urgently needed, and follow up with recommendations for measuring and monitoring those variables which may be of key importance for a better understanding of the problem.

The Three Important Processes Involved

During the later part of the workshop the group was divided into three working parties, each concentrating on one of the three processes:

Mobilization
Transport, and
Deposition

These three sections of this report are based on the findings of the three working parties. It is inevitable, therefore, that there is a certain amount of overlapping between the contents of these three sections.

1 MOBILIZATION

1.1 Introduction

The generation of dust in the Sahara has not been studied extensively. Yaalon and Ganor (1979), in their description of dust emission in Israel, state that desert dust originates from wind erosion of alluvial deposits in desert mountain valleys. Thus fine dust material would be ultimately produced by chemical weathering of primary minerals in mountains. On the other hand, Junge (1979) suggested that some of the desert dust may be produced by crystalline breakage of saltating grains. This latter possibility is supported by evidence of Krinsely and Doornkamp (1973). The generation of fine air-borne particles from semi-arid soils has been described by Gillette (1979) who showed that the ratio of vertical flux of fine particles to horizontal flux of all air-borne particles (roughly equivalent to the saltation flux) is variable with respect to soil mineralogy and wind speed but that in general it was much less than 1%. The production of dust was related to variables of surface soil texture, wind speed, vegetation, vegetative residue, surface roughness, soil aggregate size distribution and soil moisture.

To determine larger scale patterns of dust production in the Sahara, a compilation of Dubief (1952, 1979) on the frequency of occurrence of dust storms in different areas provides a valuable foundation for future studies. Discussions by Dubief (1979) and Kalu (1979) established that there are different kinds of disturbances affecting different parts of the Saharan area during different seasons. In the winter they are connected with the Mediterranean polar front and/or with strong upper tropospheric troughs. In summer they are related to the easterly waves associated with the Intertropical Convergence Zone.

The impact of human action can considerably increase wind erosion through: cultivation (for subsistence and market economy), improper grazing, unsuitable trampling by animals, cutting wood for fuel or other use, certain types of tourism, etc.

There are also many means for human action to limit wind erosion: exclusion from grazing (seasonal or pluriannual), scattering of numerous wells with small water discharge fitted to the grazing capacity of the land, limitation of cultivation to cover nutritional needs (only in areas with high erodibility), preferably limitation of settlement to a reasonable level, watershed management, and appropriate soil conservation techniques.

1.2 Research aspects

A. Surface Conditions

Aside from the aerodynamic conditions which are partially determined by the surface, the generation of dust is determined by surface conditions. The conditions listed below are the primary factors effecting erodibility for a given surface wind stress and the threshold velocity for raising dust.

Disintegration of soil aggregates and discrete particles into suspendible dust particles. Experiments should be undertaken to determine the amount of crystalline breakage and production of fine particles from coarse particles. This study, along with quantitative estimates of generation of fine particles from dunes and from alluvial deposits with fine material available will be helpful in estimating the relative importance of these processes for the production of fine particles.

Soil. The chief condition determining erodibility is the size distribution of surface soil aggregates. These aggregates, which may be in the form of a surface crust, large clods, or small pellets of soil, largely determine the threshold velocity and intensity of erosion. The stability of the aggregates determines future erodibility if more soil moisture is not forthcoming. Thus we should know the relationship of the size distribution of soil aggregates with physical properties, surface texture, modulus of rupture, soil moisture and compaction, and with chemical properties such as soil mineralogy and organic content. Organic studies should include organic films generated by soil microorganisms which tend to stabilize soil aggregates.

Vegetation. Vegetation exerts an influence in various ways: by the stabilization of soil by roots, by the absorbtion of some momentum flux to the soil surface, by alteration of the moisture and heat exchange to the soil, and by decay, which adds organic material to the soil. Soil erosion produces a negative effect on vegetation through the sand blasting effect of saltating sand grains on the plant leaves and stems. Studies should be undertaken to develop species of plants that are more resistant to this hazard. Studies should further be undertaken using a systematic approach to determine vegetation-desert soil interactions.

Plant litter. This material protects the surface by covering the soil. Upon decomposition, humus is added to the soil. Studies of plants that provide the best protective litter should be undertaken.

Rocks and boulders. These elements provide the surface with non-erodible elements which absorb part of the wind stress. They also disturb the wind field in general and can concentrate the wind in certain regions. This alteration of the wind should be studied.

Microscale roughness. Roughness of the soil traps sand grains and inhibits saltation. The distribution and cause of this surface roughness should be surveyed.

B. Geomorphological and Topographical Surface Conditions

Wind velocity is greatly modified either by single topographical components such as hills, mountains or large watersheds (in which valleys exert an influence), the (even slight) degree of slope of embankments, the cliffs (leeward or windward), pediments and alluvial fans. For this reason typical erosion and deposition patterns may be associated with geomorphological forms. Study methods should be based upon:

- statistical study of the frequency of 'dust-winds' in meteorological stations located at the pediment base;
- information provided by the indigenous population and by local geographical names;
- aerial photographs and ground checks;
- satellite images.

C. Wind Conditions

Measurements of wind profiles indicate wind stress which along with such factors as soil, vegetation, plant litter, rocks and boulders and microscale roughness, permits estimates of erodibility in most circumstances. Studies of how surfaces affect the wind flow, concentrating the wind in certain regions, should be carried out. Studies should also be made of threshold values for wind velocities with regard to dust mobilization under different surface conditions.

D. Large Scale Weather Mechanisms

The logical way to study such mechanisms is to make synoptic meteorological case studies of typical dust-storm conditions at the surface and upper levels of typical disturbances during different seasons. Satellite images of dust-storms are highly desirable for these studies. The weather maps should encompass a sufficiently large area to permit studies of the relation of disturbances to the general weather situation, and to permit investigations of dust transport from the Saharan area into other areas, particularly westward over the Atlantic, northward over the Mediterranean or eastward across the Red Sea.

*E. Statistical Studies of Air-borne Dust Over the Saharan Area Based on SYNOP and METAR Reports.**

The SYNOP and METAR reports contain valuable information on air-borne dust, permitting a study of frequency distribution in space and time. Such studies may reveal sources and sinks of desert dust.

An important factor in relation to air-borne dust is the critical wind speed for the raising of dust. Apart from wind-tunnel experiments and similar research, these SYNOP and METAR reports could be utilized for studying this problem, particularly through the information they provide concerning wind speed, present weather and visibility. Such studies should, however, be done only after a careful evaluation of the reliability of the reports in different regions of the Saharan area.

Several studies have indicated that there is a relationship between the visibility and the mass concentration of dust in certain particle size ranges even if this relation is fairly vague. A statistical study over the Saharan area of the visibility, as included in the SYNOP and METAR reports, could contribute to the knowledge of the distribution of air-borne dust in this area. Such a study should be supplemented by a systematic comparison of visibility and dust load at a few selected sites.

1.3 Monitoring aspects

As regards monitoring of air-borne dust in Africa, the group concluded that it is essential that some immediate steps be taken. The following action is proposed:

- initiating further studies of the best methods for observation of dust deposition, including comparisons between different methods;
- adding regular observations by high volume samplers of suspended particulate matters at all regional stations and at a selection of climatological stations in Africa north of the Equator. Analysis of content of suitable mineral and organic trace elements and of particle size distribution;
- adding wind observations as necessary at suitably distributed stations over Africa north of the Equator for establishment of the relationship with air-borne dust frequency;
- encouraging the establishment of further WMO regional air pollution stations in the area with the specific purpose of monitoring particulate matter.

2 LONG RANGE TRANSPORT

2.1 Introduction

The problem of the long-range transport of Saharan dust consists of two major parts:

The documentation of dust in the air, both for the area within and outside the

*SYNOP and METAR are coded routine reports on weather observations.

Sahara, and the possible impact of this dust on both the ecology of the adjacent areas and the regional and global climate.

To evaluate any possible effect of Saharan dust on the ecology and on climate not only within but also beyond the area of the Sahara, it is necessary that the phenomenon of this dust in the atmosphere, its transport, and its short and long term variations be much better documented and understood.

2.2 Dust in the Sahara

Essential for any understanding of the phenomenon of Saharan dust in the atmosphere is an adequate knowledge of the source areas, their location and strength and the seasonal and more long-term variations. Such information will enable more reliable estimates to be made of the fluxes leaving the Sahara, their further fate in the atmosphere and their composition. A suitable starting point to obtain such data is provided by the statistics compiled by Dubief (1952, 1979), which cover a period of over 20 years from 1929 to 1950. This information should be supplemented by a thorough statistical evaluation of available weather observations which have accumulated in the meantime. For the purpose of studying long-range transport, a spatial resolution of $5^\circ \times 5^\circ$ of sources of air-borne dust in the Saharan area and its surroundings is considered sufficient. The time resolution should be in accordance with the variability of the observed dust concentration, but for most global considerations monthly mean values would seem sufficient.

As mentioned in Section 1.2 E, it would be desirable to make a statistical investigation on weather data based on SYNOP and METAR reports.

These simple statistics should be supplemented by synoptic studies based on weather maps. Such investigations should cover a period of a few years in order to obtain sufficient representative information on the origin and behaviour of the dust storms and the resulting direction of the dust transport. This would certainly be a major undertaking and it could best be carried out by a synoptic meteorologist visiting the various meteorological services in the affected areas, and making use of the locally available material and expertise. Such a study should be facilitated by the cooperation of WMO.

Two-dimensional horizontal flow patterns, based on mean monthly winds over Africa, are available and were discussed during the workshop (Newell and Kidson, 1979). There are significant differences in the flow patterns between the 1958–63 moist period and the 1969–73 dry period and there are corresponding differences in the global general circulation. It is more difficult to appraise the vertical motion field through the continuity equation and the uncertainties make deduced year-to-year variations unreliable.

While the question of the dust transport by the mean circulation versus that by the transient features was not resolved at the workshop it is highly likely that the two aspects of the circulation are inter-related.

Another aspect of investigation of the Saharan dust phenomenon is the chemical

and mineralogical composition of the soil material as a function of particle size, as well as the same information for the aerosols during dust-storms at the same locations. Such information is at present almost completely lacking for the source areas. It would also give important information about fractionation processes between soil and aerosol as a function of environmental conditions and particle size.

The analysis of the dust samples should be made according to both elemental and mineral composition because only this double analysis will permit understanding of fractionation with particle size. It is very desirable also to identify major biological components in the samples, since they may serve as useful tracers.

In addition, it is considered desirable to have aircraft data on the vertical distribution of dust over the Sahara.

The sample collection should cover all areas of the Sahara that may function as dust sources and should be made in such a way as to constitute a representative and unique data bank. It is essential that in all cases sufficient reserve samples should be stored for future research and follow-up studies. It is most essential that the sampling be done by scientifically trained personnel and that it should be supplemented by mapping expeditions carried out by a team of soil scientists, geologists and mineralogists. Again, this program should be organized in such a way as to obtain maximum support from the appropriate international organizations.

For study of the possible ecological and climatological impact, it is essential to have information on the long-term trends of the strength, location and composition of the Saharan dust sources. From the available evidence it is clear that considerable changes have occurred on all time-scales, including the ice ages. It is most likely that both natural changes of the climate and man-induced modifications of the ecology are involved, but these processes are very difficult to separate. Any attempt to clarify these questions should involve the most complete compilation of all available evidence from geology, sedimentology, hydrology, archeology, etc. Some of this evidence is already available in meticulous studies, such as those by Nicholson (1976) and Street and Grove (1976), but further studies of this type should be made in other fields of research.

Changes in source strength over periods of, say 10–20 years may also be significant. There have been suggestions that an observed three-fold increase in the dust concentration in Barbados between 1965 and 1976 might be due to an increase in source strength in the Sahara, possibly associated with the drought in the Sahel (Prospero and Nees, 1977). However, the observed increase could also be due to changes in the circulation pattern and the real cause has yet to be determined. Furthermore, some measurements (Jaenicke and Schütz, 1977) close to the Sahara indicate that no major change in source strength occurred in recent years.

2.3 Dust outside the Saharan area

There are several well-documented cases where an atmospheric transport of Saharan dust over distances of up to several thousand kilometres has been evident.

Probably the best known such situations refer to the transport westwards across the Atlantic at a latitude of about 10°N to 25°N (Carlson and Prospero, 1972), and the transport north-east to Israel (Yaalon and Ganor, 1979). These transports have been documented by measurements of turbidity and of dust concentration in the air (Prospero *et al.*, 1979) and have also been confirmed by satellite images.

There is also no doubt that a considerable transport of dust takes place from an area southwest of the Tibesti region of the central Sahara into the countries along the northern coast of the Gulf of Guinea (Kalu, 1979). Occasional intrusions of Saharan dust into Europe as far north as Scandinavia have also been documented. Our knowledge about possible transports of Saharan dust towards the east and south-east is very limited. However, satellite images have shown several cases of dust-storms moving in an ESE direction over the Red Sea. No significant transport of dust into the southern hemisphere has been observed and there are good reasons to believe that such a transport is effectively hindered by the precipitation scavenging associated with the intertropical convergence zone (ITCZ).

Regarding the meteorological processes that give rise to the mobilization and long-range transport of dust it is important to know whether they are more or less continuous processes or if they take place only under very special circumstances. In the former situation it would be pertinent to concentrate the study on the mean circulation pattern. In the latter case, we should of course concentrate our study on those particular occasions. From the information available it seems that there is a rather continuous transport of smaller dust particles (less than 1 μm) out over the Atlantic but that the transport of larger particles is more discontinuous. This is reflected in the relatively small variation of turbidity out over this area in contrast to the relatively larger day to day variations of total particle load (Prospero *et al.*, 1979; Jaenicke and Schütz, 1977). The transport in the north-easterly direction is probably a less continuous process. The conclusion is that we need to study both the mean circulation and the typical circulation patterns for situations when the transport is augmented. By getting to know the weather patterns that are conducive to long-range transport and by studying the variations of their frequency of occurrence, we might be able to establish whether changes in the observed dust concentrations can be explained by such variations.

2.4 Possible climatic effects of Saharan dust

It has been estimated that the contribution of Saharan dust to the total burden of tropospheric aerosols is about 60–200 million tons per year (Junge, 1979). This should be compared with other sources of mineral dust in the world estimated at roughly 200 million tons per year. It is clear that the Sahara is a very important source and that appreciable variations in its strength could have significant implications on the regional and global burden of mineral dust. This implies that changes in the dust output from the Sahara (be they natural or man-made) is of potential significance for the regional and global climate. However, quantitative estimates of such impacts have yet to be made. Among specific mechanisms of such impact on

climate, one may mention:

- radiation effects in cloud-free air;
- effects on cloud albedo and
- effects on the formation of clouds and raindrops (condensation and ice nuclei).

A better understanding of these possible impacts does not depend so much upon the collection of more observations of the aerosols as upon a better understanding of the physical processes involved. However, it is imperative that reliable vertical profiles of the aerosol concentration are available.

2.5 Measurement and monitoring aspects

Appropriate techniques

There are two principal methods for measuring soil dust concentrations in the atmosphere which can be recommended at the present time. For areas with relatively high concentrations of soil dust (within the Sahara, or in dust plumes outside the Sahara) the desert derived insoluble fraction dominates and can be determined with high volume samplers by weighing filter samples. If the concentrations are less there may be significant fractions of aerosol components other than mineral dust present. In this case the total aerosol no longer represents soil dust and it is recommended to use more sensitive and specific methods, such as to analyse filter samples for silicon, iron, aluminium, or other elements which originate exclusively from crustal sources and to subsequently correct for average crustal composition. If the samples contain considerable fractions of particles larger than 10 μm radius, care should be taken to make the sampling isokinetic. The techniques mentioned have been used successfully in a variety of research projects and are well documented in the literature.

Monitoring of soil derived dust in the WMO network

Since rather large variations of the global production of soil dust occur and can be expected to continue in the future, it is recommended to monitor the soil dust component of the aerosol at the baseline and regional stations with an extended program of the WMO network of monitoring stations all over the globe.

For long range monitoring, weekly or monthly sampling is generally satisfactory but daily observations would be desirable for certain specialized research programs, for example, when the relation to transport trajectories is investigated. The samples should be analysed by a central laboratory to ensure comparability and high quality. The most suitable technique of choice is the elemental analysis of filter samples discussed above.

3 DEPOSITION

3.1 Introduction

Loess and other eolian deposits in soils of the desert fringe area present an excellent record of past rates and variations in dust deposition and could help in understanding natural trends or the influence of man on deposition. Saharan dust is the source for loess deposits in places such as Israel, the Nile Valley and the Cape Verde Islands (Matznetter, 1960). It has even been suggested that it also may be the origin of some soils on islands in the western Atlantic such as Bermuda, the Bahamas and the West Indies (Bricker and Prospero, 1969).

Studies, like those carried out in Israel, should be extended to circum-Saharan regions, with particular attention to the distribution of loess and to the eolian contribution to quaternary soils. As the rate of dust deposition seems to be a function of the distance and nature of the desert source, such studies would indicate past cycles of extension or recession of the desert. Detailed stratigraphic studies on carefully selected sites – e.g. from Tunisia and Israel – should be correlated with similar studies of deep sea cores off the West African coast, Barbados, the Mediterranean and other places where a fairly complete stratigraphic record is available. Recent studies have shown that many soils in desert fringe regions, in addition to the obviously loessial ones, have received considerable supplements of desert dust. Each case requires a specific analysis because of the slow but steady deposition the dust becomes totally assimilated in the soil forming environment. It seems desirable to carry out studies of dust contribution to soils, e.g., in northern Nigeria, the delta area of Egypt, the Sudan and Tunisia. Such studies would reveal not only the past extent of dust deposition but would also assess the posssible contribution of dust to the fertility of the soils and hence its ecological function. The studies should be made in conjunction with monitoring of the present rate and nature of dust deposition in desert fringe areas.

The ecological impact of dust transport in desert fringe areas may have been beneficial in the past; however, this process could easily be reversed by inappropriate land use, causing the desert fringe region to become a dust source rather than a dust accumulator. Careful attention must be given to dust generation processes during development projects whether agricultural or industrial in such areas, – particularly since the effects of the dust on human health are unknown.

3.2 Possible ecological effects of Saharan dust

There appears to be no reliable data, collected systematically over an adequate number of years, on the quantity, periodicity and chemical composition of dust received by soils of natural and agricultural ecosystems in oases, desert fringes and more distant peripheral vegetation. Such deposition is of particular importance with respect to the possibility of dry-flushing effects of dust with macro-nutrients (e.g. K, P, Ca, Mg, S) and with micro-nutrients (e.g. B, Mn, Co, Cu, Zn, Mo) in an oasis.

Studies are needed on the nutrient regimes of such soils and the dynamics of nutrient inputs via dust in order to determine whether air-borne deposition represents a significant percentage of the 'total', 'plant-available' and 'supplying-power' for nutrients in different soil-types.

It is also important to investigate whether the dust deposition (including any organic materials) enhances the physical soil structure for plant growth, possibly in conjunction with soil micro-biological investigations designed to relate dust receipt to micro-organism activity. Consideration should also be given to similar studies in different vegetation types in peripheral areas of deserts. Such studies would necessitate an experimental review of available inexpensive deposition gauges.

The ecological impact of dust deposition on marine life off near-shore and coastal deserts also requires attention and investigation. The deposition of nutrients associated with Saharan dust into ocean waters may under certain conditions contribute significantly to the productivity of the surface waters, for example, off the coast of West Africa (Lepple, 1975; Lundholm, 1979). Since phosphorus is probably the most important element in this connection we recommend that measurements of this element be made in different particle size ranges. This will enable the deposition to be quantitatively estimated. The importance of deposition of phosphorus on terrestrial ecosystems also needs to be investigated.

There is little doubt that, in the tropical and subtropical Atlantic east of the mid-Atlantic ridge, a major part of the non-biogenic sediment is of Saharan origin.

Saharan dust most likely also gives significant contributions to the sediments in the Mediterranean and in the western parts of the equatorial Atlantic. Study of these sediments may provide useful information about the past climate of the desert regions (Parkin and Shackleton, 1973).

There is also evidence that the dusty environment with its different components may affect human health. Attention should be drawn to the transport of potentially toxic elements from surface-mining and other human activities in the desert situation and to their possible bioaccumulation in food-chains, bearing in mind the ease with which a deposited pollutant may be resuspended under arid conditions. This latter item is of special importance for the Sahara, which is surrounded by grazing lands.

3.3 Measurement aspects

In order to understand the importance of atmospheric dust deposition and its effect on soils and organisms at different ecological sites, we recommend the collection and analysis of soil deposition from deserts and surrounding dry-land fringes.

A generally accepted standard method for measuring dry fall-out or dust deposition does not exist (cf. conclusions from the Report of the WMO Expert Meeting on Dry Deposition, Gothenburg, April 18–22, 1977) (WMO, 1977). However, for specific research purposes where only relative measurements are needed, several methods can give acceptable results.

In order to obtain as valuable data as possible with deposition gauges it is necessary that:

- the terrain be characterized with respect to surface roughness and 'filtration efficiency' of vegetation cover;
- the aerosol be characterized with respect to its particle size distribution and mean horizontal velocity;
- calibration studies be carried out where the deposition of an aerosol to a gauge in different types of terrain is correlated to the deposition on the ground. The calibration curve so obtained could then be used to estimate actual deposition so long as there is no major change in the aerosol or terrain type.

Several types of dust deposition collectors which are in use elsewhere are recommended for testing in desert and semi-arid areas, namely:

- dust collectors with beads of glass or plastic for collection of dry and wet fall-out (Yaalon and Ganor, 1979);
- buckets with water, as used in dust pollution studies;
- moss bags or other retentive surfaces, used until now mainly in highly polluted areas, may prove useful also for the analysis of the chemical composition of desert dust.

3.4 Research aspects

Because of the close inter-relationships between meteorological, pedological and biological studies in these proposals, it is desirable to develop a carefully co-ordinated integrated programme at each of the field sites under investigation.

– A careful study should be undertaken of all loess deposits and their stratigraphy on the northern and southern fringes of the Saharan desert. Stratigraphic analysis and dating of the sedimentary and pedogenic cycles should be made, using all of the modern techniques of radioactive, paleomagnetic and archaeologic dating.

– Specific pedologic studies should be undertaken to identify dust contribution in the soils of northern Nigeria, the Nile delta and Gezira, using geochemical, mineralogical and biological methods. These studies should be made in conjunction with monitoring the present rate and nature of dust deposition for purposes of comparison.

– To identify more accurately the actual sources of long distance transport of dust from the desert, a field study of suspected desert areas should be undertaken and samples collected from representative surfaces which characterize the source area. As clay minerals seem to be particularly sensitive to climatic influences, special attention should be given to these components in dust, soils, oceanic cores etc.

– A comprehensive nutrient budget of a selected typical oasis should be carried out over a period of several years.

– Methods of dust collection used in various places should be tested for their

applicability on the desert and desert fringes and made more widely known. Particular efforts should be made to standardize methods and thus improve comparisons.

– Historical records of dust deposition have been provided by analysing the content of eolian dust in cores from glaciers in the Caucasus. This method could also be tried for suitable glaciers in other areas, e.g. the Alps for the dust plumes from North Africa.

Conclusions

It is clear that the mobilization of dust in the Sahara and its subsequent transport through the atmosphere well beyond the Sahara is a natural process that has been going on for a long period of time. This has led to the formation of eolian sediments in the Mediterranean and the Atlantic and to the formation of loess soils in many adjacent regions and possibly as far away as the islands of the West Indies. Possible short-term effects of the dust deposition on terrestrial and aquatic ecosystems are less well understood.

During recent years, several cases of a substantial long-range transport (up to several thousand kilometres) have been documented by direct measurements. Rough estimates show that the Saharan area may contribute annually 60 to 200 million tons of soil dust to the troposphere. This is a substantial part – perhaps about 50%– of the total soil dust emissions into the troposphere and may have some influence on the regional and the global climate.

It is not at present possible to estimate what effect human activities such as agricultural practices, livestock grazing, wood cutting, etc., may have on the large scale mobilization and transport of dust in this region. Before such an assessment can be made, we must obtain much more knowledge of the processes that contribute to erosion and to the transport and deposition of the dust locally as well as at greater distances. The required research and measurement work is of an interdisciplinary nature and will have to include aspects of at least mineralogy, pedology, ecology and meteorology.

Among the main areas of research the following may be mentioned:

– Comprehensive studies to understand the production of fine particulate material by weathering and disintegration processes as a first and important step in dust production.

– A thorough documentation of arid soil composition in terms of mineralogy, particle size distribution and chemical composition, etc., for the whole Saharan area.

– A comprehensive study of Saharan meteorology on all scales to understand the rise and transport process of mineral dust within the Saharan area and out into the surrounding areas.

– Studies to understand the positive and negative ecological effects of dust deposition along the fringes of the desert areas, especially the relation to productivity of soils and coastal waters.

We make no proposal for specific institutional arrangements but note the desirability of having the investigations closely tied to institutions in the countries

surrounding the Sahara. An important task for UNEP and other international organizations is to strengthen such institutions by supporting research and measurement programmes.

References

Bricker, O. P., and Prospero, J. M. (1969). Air-borne dust on the Bermuda Islands and Barbados. *EOS, Trans. Am. Geophys. Union*, **50**, p. 176.

Carlson, T. N., and Prospero, J. M. (1972). The large-scale movement of Saharan air outbreaks over the northern equatorial Atlantic. *J. Appl. Meteorol.*, **11**, 283–297.

Dubief, J. (1952). Le vent et le déplacement du sable du Sahara. *Trav. Inst. Rech. Sahariennes, Alger*, **8**, 123–164.

Dubief, J. (1979). *Review of the North African Climate with Particular Emphasis on the Production of Eolian Dust in the Sahel Zone and in the Sahara*. In this report.

Ehrenberg, C. G. (1862). Erläuterungen eines neuen wirklichen Passatstaubes aus dem atlantischen Dunkelmeere vom 29. Okt. 1861. *Monatsber. Kgl. Preuss. Akad. Wiss.*, Berlin, 202–224.

Gillette, D. A. (1979). *Environmental Factors affecting Dust Emission by Wind Erosion*. In this report.

Jaenicke, R., and Schütz, L. (1977). A comprehensive study of physical and chemical properties of the surface aerosols in the Cap Verde Islands region. – Submitted to *J. Geophys. Res.*

Junge, C. (1979). *The Importance of Mineral Dust as an Atmospheric Constituent.* In this report.

Kalu, A. E. (1979). *The African Dust Plume: Its Characteristics and Propagation across West Africa in Winter.* In this report.

Krinsley, D., and Doornkamp, J. (1973). *Atlas of Quartz Sand Surface Textures.* Cambridge Univ. Press, Cambridge, 91 pp.

Lepple, F. K. (1975). *Aeolian Dust over the North Atlantic Ocean.* Ph.D. thesis, Univ. Delaware, Delaware, USA, 270 pp.

Lundholm, B. (1979). *Ecology and Dust Transport*. In this report.

Matznetter, J. (1960). Die kapverdischen Inseln. *Mitt. Oesterr. Geogr. Ges.*, **102**, 1–40.

Newell, R. E., and Kidson, J. W. (1979). *The Tropospheric Circulation over Africa and its Relation to the Global Tropospheric Circulation.* In this report.

Nicholson, S. E. (1976). *A Climate Chronology for Africa: Synthesis of Geological, Historical and Meteorological Information and Data.* Ph.D. thesis, Univ. Wisconsin, U.S.A., 323 pp.

Parkin, D. W., and Schackleton, N. J. (1973). Trade wind and temperature correlations down a deep-sea core off the Saharan coast. *Nature*, **245**, 455–457.

Prospero, J. M., and Nees, R. T. (1977). Dust concentration in the atmosphere of the equatorial North Atlantic: Possible relationship to the Sahelian drought. *Science*, **196**, 1196–1198.

Prospero, J. M., Savoie, D. L., Carlson, T. N., and Nees, R. T. (1979). *Monitoring Saharan Aerosol Transport by Means of Atmospheric Turbidity Measurements.* In this report.

Street, F. A., and Grove, A. T. (1976). Environmental and climatic implications of late quaternary lake-level fluctuations in Africa. *Nature*, **261**, 385–390.

WMO. (1977). *Report of the Expert Meeting on Dry Deposition* (Gothenburg, 18–22 April, 1977), 8 pp.

Yaalon, D. H., and Ganor, E. (1979). *East Mediterranean Trajectories of Dust Carrying Storms from the Sahara and Sinai.* In this report.

Participants in the Workshop

Research Ass. Mr. Randolph D. Borys
Department of Atmospheric Science
Colorado State University
Ft. Collins, Colorado 80521
USA

Professor Cyrill Brosset
Swedish Water and Air Pollution Research
Laboratory
P.O. Box 5207
S-40224 Gothenburg
Sweden

Professor Jean J. Dubief
150 rue de l'Université
F-75007 Paris
France

Professor K. Gösta Eriksson
Dept. of Geology
Chalmers University of Technology and
University of Gothenburg
S-40220 Gothenburg
Sweden

Dr. Dale A. Gillette
National Center for Atmospheric Research
P.O. Box 3000
Boulder, Colo. 80307
USA

Professor Gordon Goodman
Monitoring and Assessment Research Centre
Chelsea College
University of London
The Octagon Building
459a Fulham Road
London SW 10 OQX
UK

Dr. Michael Inskip
Monitoring and Assessment Research Centre
Chelsea College
University of London
The Octagon Building
459a Fulham Road
London SW 10 OQX
UK

Dr. Ruprecht Jaenicke
Max-Planck-Institut für Chemie
(Otto-Hahn-Institut)
Abt. Luftchemie
Postfach 3060
D-6500 Mainz
W-Germany

Dr. G. Leonard Johnson
Code 461
Office of Naval Research
Arlington, Va 22217
USA

Professor Dr. Christian E. Junge
Max-Planck-Institut für Chemie
(Otto-Hahn-Institut)
Abt. Luftchemie
Postfach 3060
D-6500 Mainz
W-Germany

Mr. A. E. Kalu
Research and Training Institute
Meteorological Department
Oshodi, Lagos
Nigeria

Dr. Bernhard Koopmann
Geol. Inst.
Univ. Kiel
Olshansenstr. 40–60
D-2300 Kiel (1)
W-Germany

Professor Gösta H. Liljequist
Dept. of Meteorology
University of Uppsala
Box 516
S-75120 Uppsala
Sweden

Dr. Bengt Lundholm
Ecological Research Committee
Swedish Natural Science Research Council
Sveavägen 166, 15th floor
S-113 46 Stockholm
Sweden

Dr. Christer Morales
Ecological Research Committee
Swedish Natural Science Research Council
Sveavägen 166, 15th floor
S-113 46 Stockholm
Sweden

Professor Reginald E. Newell
Dept. of Meteorology
Massachusetts Institute of Technology 54–1520
Cambridge, Mass. 02139
USA

Dr. Georges Novikoff
Desert Biome Project
c/o Smithsonian Projects
Angle Rue Dr. Skalli et Rue Ali Belhaouane
Khereddine
Tunisia

Professor Joseph M. Prospero
School of Marine and Atmospheric Science
University of Miami
4600 Rickenbacker Causeway
Miami, Fla. 33149
USA

Research Ass. Dr. Kenneth A. Rahn
Graduate School of Oceanography
University of Rhode Island
Kingston, R. I. 02881
USA

Professor Anders Rapp
Dept. of Physical Geography
University of Lund
Sölvegatan 13
S-223 62 Lund
Sweden

Dr. Henning Rodhe
Dept. of Meteorology
University of Stockholm
Arrhenius Laboratory
S-10691 Stockholm
Sweden

Dr. Thomas Rosswall
Swedish SCOPE Committee
The Royal Swedish Academy of Sciences
S-10405 Stockholm
Sweden

Dr. Bengt Steen
Swedish Water and Air Pollution Research
Laboratory
P.O. Box 5207
S-40224 Gothenburg
Sweden

Dr. Carl Christian Wallen
UNEP/GEMS c/o WMO
Case postale No 5
CH-1211 Genève 20
Switzerland

Dr. Douglas M. Whelpdale
Atmospheric Environment Service
4905 Dufferin Street
Downsview
Ontario M3H 5T4
Canada

Professor Dan H. Yaalon
Dept. of Geology
The Hebrew University
Jerusalem
Israel

SECTION 2

(a) General Description and Ecology

CHAPTER 1

Review of the North African Climate with Particular Emphasis on the Production of Eolian Dust in the Sahel Zone and in the Sahara

J. DUBIEF

ABSTRACT

The aridity of the Sahara, the high temperatures reached by its soil and the strong diurnal thermic turbulence resulting from these, are all factors which favour the production of dust and its suspension in the atmosphere. The directions of the prevailing winds account for the transport of this dust towards the Sudan and towards the Atlantic off Mauretania and Senegal, but do not explain its movements towards the continent of Europe. In order to understand the latter movements one must take into account the effects of depressions crossing the Sahara and the distribution of the air currents in its medium and upper atmosphere, which from autumn to spring is in the sphere of influence of the polar front of the medium latitudes.

The extent of the areas affected by the Saharan dust, at the present time, is large in the southern and western Sahara as well as in the Middle East and is not negligible, though infrequent, over the European continent. It is likely that this dust has played an important role during the last ice ages, particularly in the melting of the glaciers.

The Sahara, with its axis along the Tropic of Cancer, is a hot desert, perhaps a typical example of a hot desert. It is therefore above all a region where rain is extremely rare. One day of rain in a hundred, and then only for a few hours, as at In Salah in the centre of this 'attenuated desert' which forms the western half of the Sahara, two days in a thousand in the other half, the 'absolute desert' (cf. Gautier, 1928). Drought is therefore a normal phenomenon. Its average duration in the middle of the Sahara is five or six months in the west and two years in the east. There is, however, a difference between the northern and southern edges. In the northern Sahara the drought occurs mainly during the summer and is broken up into short periods, while in the south and the Sudanese Sahel a long desert period alternates each year with a short rainy period. The first is centred on the cold

season and extends well beyond it. The second is limited to the middle of the summer and decreases rapidly with increasing latitude. Of course, these periods without precipitation become less pronounced with increasing altitude in mountainous areas.

To this lack of rain is added the great evaporating power of the Saharan atmosphere. If this is not a necessary factor in the creation of deserts – which can be proved by the existence of coastal deserts – it nevertheless remains a factor which enhances this. It depends mainly on the great saturation deficit of the Saharan air, i.e. the great difference between the actual water content of the air, which during the winter season is of the same magnitude as in France, and the theoretical water content at saturation at the actual temperature. This results in an extremely low relative humidity favouring the development of a field of static electricity around all isolated bodies like dust particles. The relative humidity is about 10% in the central Sahara and can sink to 2% in extreme cases. This saturation deficit depends partly on the high air temperature in the Sahara, characteristic for the latitude, partly on its continentality and partly, as will be shown below, on the fact that the air masses crossing the desert normally move from north to south – from the Mediterranean towards the Sudan – i.e. towards increasingly warmer areas, and this without being able to absorb any water vapour. To this might be added the capture by giant ions and dust particles of the water vapour in the atmosphere.

The lack of rain and the strong potential evaporation result in a thinning out and, later, disappearance of the vegetation cover, which in its turn facilitates an eolian soil drift.

Owing to the small angle of inclination of the sun's rays, the length of the day, not varying much during the year, the apparent annual migration of the sun in this area giving short winters and long summers, and the pronounced dryness of the atmosphere, the ground receives great quantities of heat from the sun. Notwithstanding the always great albedo of the Saharan soil and its own radiation, the surface layer of the ground always reaches high temperatures in the middle of the day: at least 30°C in wintertime in the northern part of the area, between 60°C and 65°C during the summer all over the flat parts of the Sahara and sometimes exceeding 70°C. This means that the air in contact with the ground is rapidly heated from sunrise on, which explains the high shade temperatures observed, particularly in the desert or semi-desert regions in the south, where, moreover, the thermic equator is situated. This heating of the gound leads during the day to a strong thermic turbulence, which reaches quite high altitudes: usually more than 2,000 m and sometimes over 4,000 m during the hot season. Even though this turbulence disappears at the end of the day, giving way at night to a great stability of the lower layers of the atmosphere caused by a strong temperature inversion close to the ground surface, due to the great cooling of this surface, the fine dust particles in suspension in the air do not reach the ground during the night, but stay in the atmosphere for weeks or even months. These particles give rise to the

so-called 'dry haze'. This haze is particularly dense and frequent in the southern Sahara and the Sudanese Sahel at all seasons and in the western Sahara in summer. It forms an almost inexhaustible reservoir of air-borne dust particles. When these move at low altitude with the prevailing winds they will only disappear after a rain or after the intrusion of a cold and stable air mass, provided they are not transported out of the desert area. Later we will see how they are renewed.

The normal pattern of the winds in the Sahara is dependent upon the yearly oscillation of a zone of low pressure, called the 'Intertropical Convergency Zone', which follows the apparent movement of the sun with a time lag of six weeks to two months. This is not peculiar to the Sahara, but is a part of a global phenomenon since this low pressure zone girdles the whole globe. Actually this zone consists of a string of low pressure centres, the position, extension and depth of which vary geographically and with the season. As far as we are concerned the zone is situated during the winter along the latitutde of the Gulf of Guinea in the west and at somewhat lower latitudes in the east. During the summer it comes in contact with the whole Saharan desert. In actual fact this convergency zone is restricted in Africa to a front towards which the winds converge and which is called the Intertropical Front (ITF). This front demarcates to the south a maritime equatorial air mass (the Sudanese monsoon, or more exactly the 'Sudanese summer monsoon') which is warm and moist with a high dew point (21°C to 22°C), from an air mass north of the front arriving from the northeast or east, which in wintertime is cool in the Sahara but warm in the southern Sudan, and during the summer always burningly hot. The most characteristic properties of this air mass are its dryness, a strong thermic turbulence and its great content of air-borne dust particles. In winter it comes from the belt of high pressure extending from the Azores to southern Asia via the northern Sahara and Arabia. The relevant wind is weak northerly at its source but turns gradually towards the east and increases with decreasing latitude.

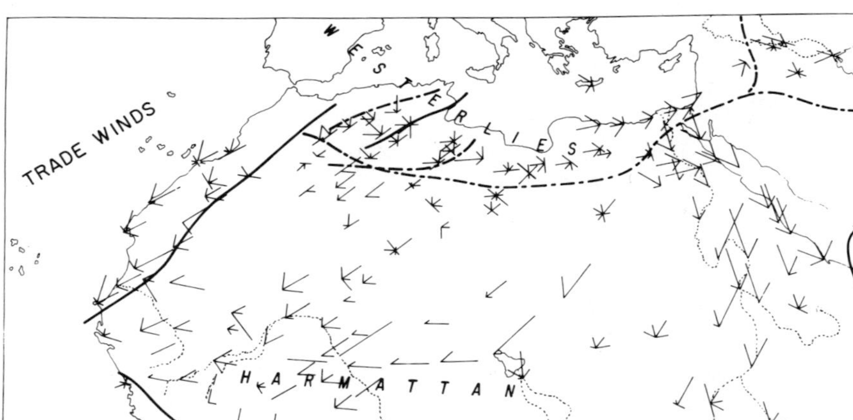

Figure 1.1 Winds in January

In the Sudan it is called the 'harmattan' (Figure 1.1). During the first months of the year the earlier barrier of high pressure breaks up and moves towards higher latitudes while the ITF moves northwards. During the summer it is only the Azorean high pressure which survives. This expands over part of the European continent and gives rise to northerly winds on its southeastern flank. These winds penetrate to the Sahara and the Middle East after having blown over the eastern Mediterranean. They were called 'the Etesian winds' during Classical times. They reach Africa and the Middle East having the character of a maritime trade wind, are rapidly heated over the continents, are dried out, turn towards northeast, later east and finally show the same characteristics as the harmattan (Figure 1.2). The ITF is then situated close to or just south of the Tropic of Cancer. It is these winds which cause the summer drought in northeastern Africa and the Middle East. One should add that from autumn to spring the harmattan may be more or less mixed with, if not replaced by, polar air, after the passage of a 'family of polar front depressions' crossing southern Europe and the Mediterranean. We will see that these circumstances are of some importance in relation to the present problem. Further west, over the Mauretanian areas and over the eastern Atlantic the true trade winds prevail. We are speaking of maritime tropical air originating from the Azorean high pressure. These cool and moist trade winds blow from a northerly direction the whole year round. They can be surmounted by the harmattan to the south, sometimes even to the north. Like the latter they may be more or less mixed with polar air. The harmattan and the trade winds increase rapidly in vertical extent when approaching the Sudanese area. They surmount the northward-moving monsoon and reach higher altitudes as the latitude decreases. Finally they join at high altitude the equatorial easterlies. Note that as there is a very small angle between the Sudanese monsoon and the horizontal plane its vertical extent is never great: 2,500 to 3,000 m in the southern parts at a maximum; nevertheless it can rise above

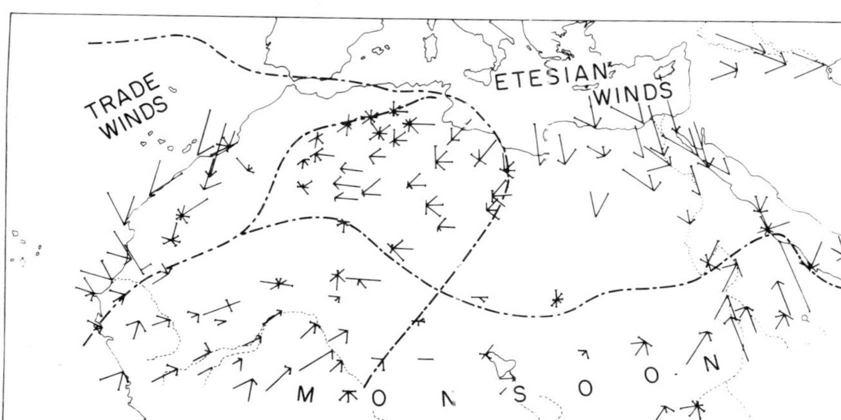

Figure 1.2 Winds in August

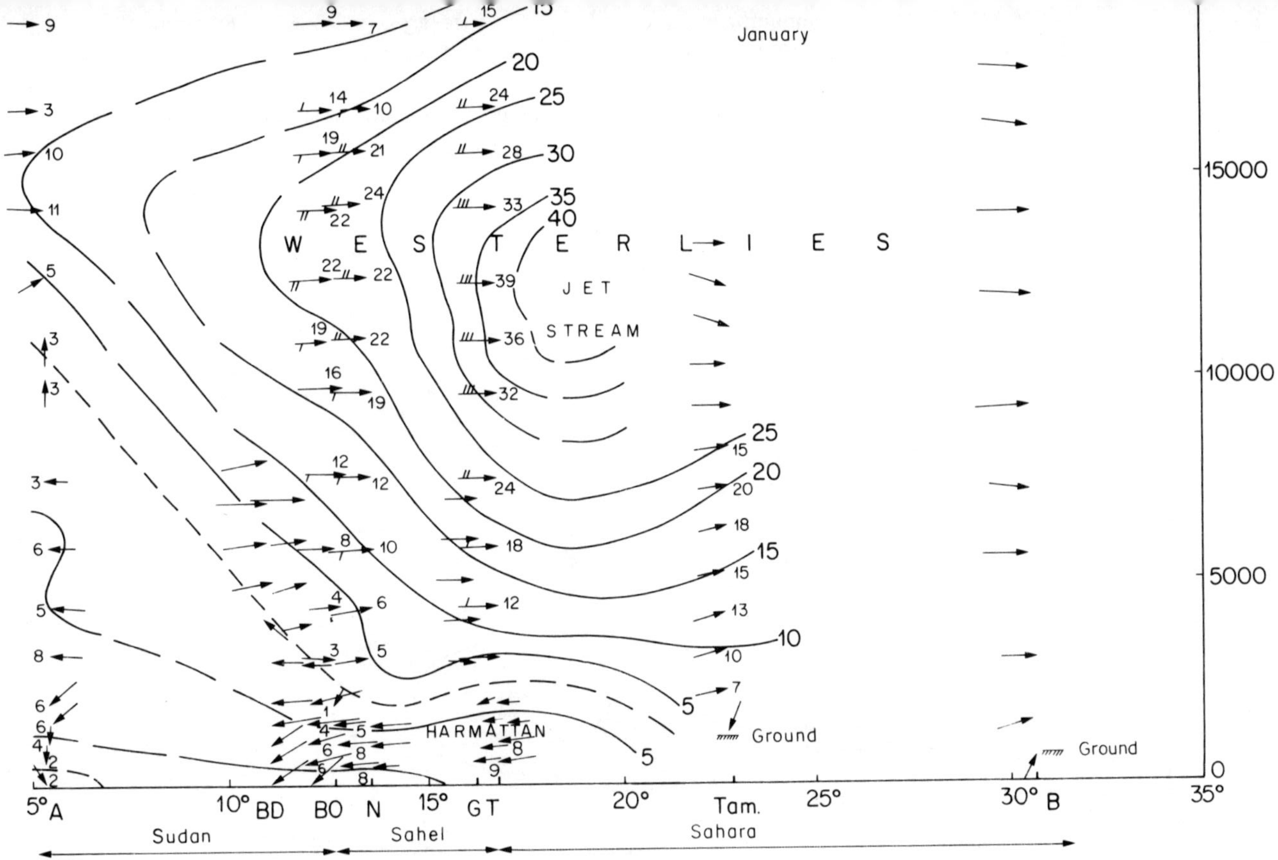

Figure 1.3 Predominant wind directions and mean wind speeds in upper levels at 0° longitude in January

A = Abidjan
BD = Bobo Dioulasso
O = Ouagadougou
B = Bamako
N = Niamey
G = Goa
T = Tombouctou
Tam = Tamanrasset
B = Béchar

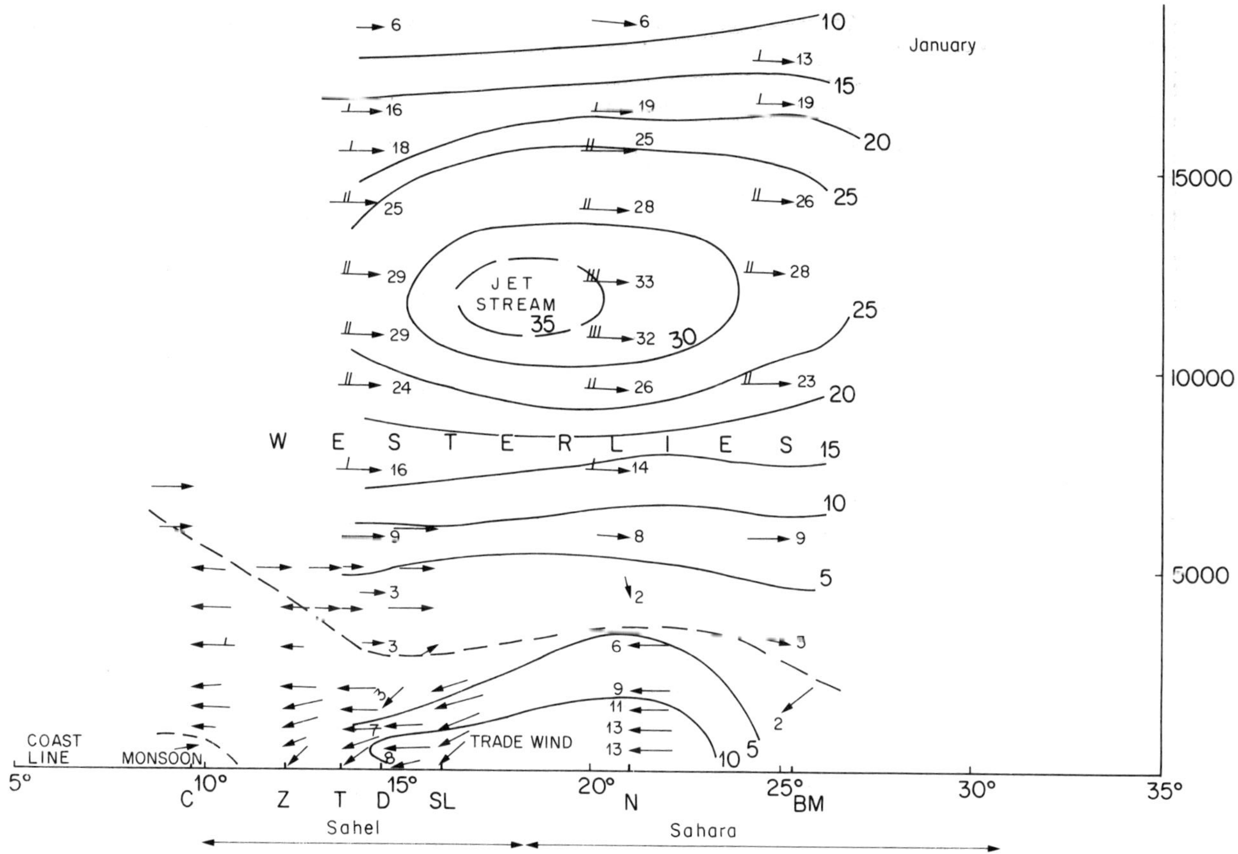

Figure 1.4 Predominant wind directions and mean wind speeds at upper levels at the African Atlantic coast in January

C = Conakry	T = Tabacounda	SL = Saint Louis	BM = Bir Moghreim
Z = Ziguinchor	D = Dakar	N = Nouadhibou	

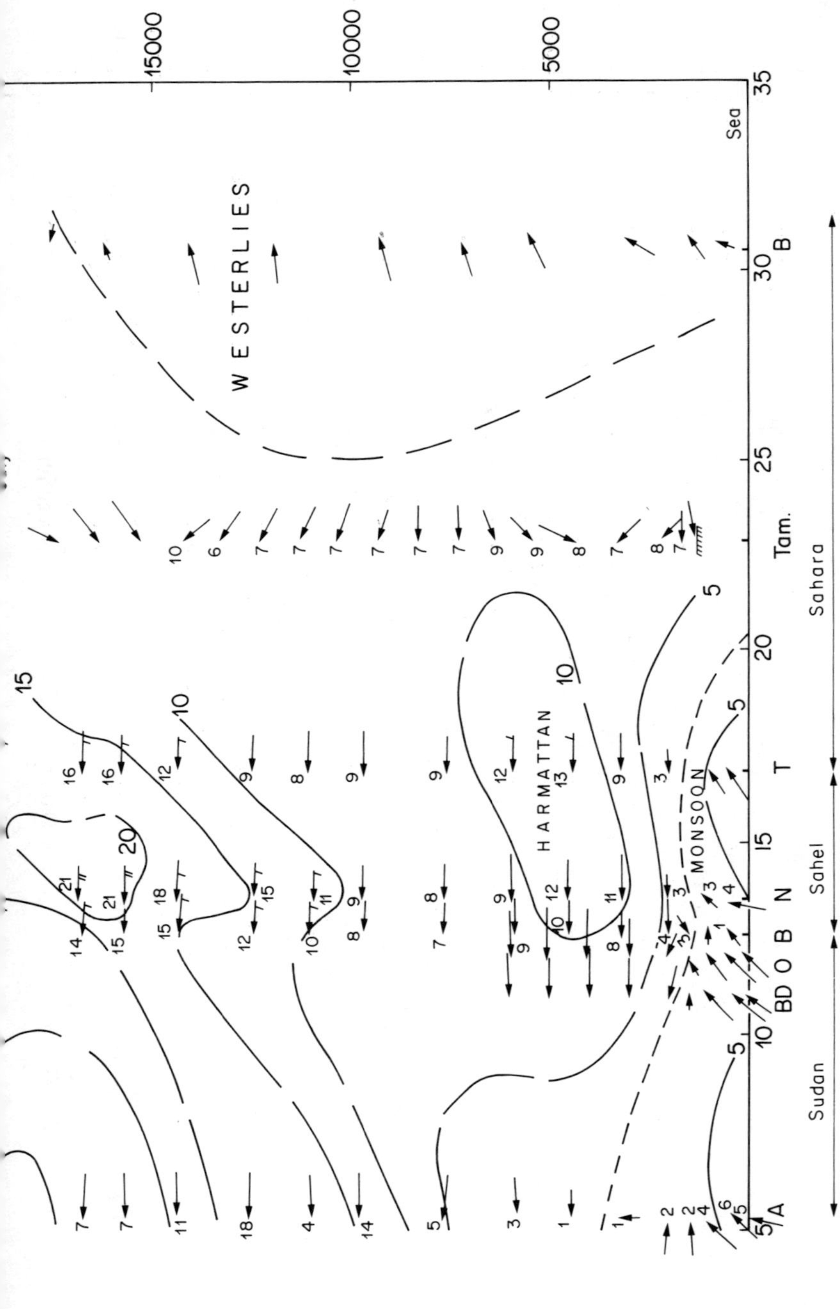

Figure 1.5 Predominant wind directions and mean wind speeds in upper levels at 0° longitude in July

A = Abidjan O = Ouagadougou N = Niamey Tam = Tamanrasset
BD = Bobo Dioulasso B = Bamako T = Tombouctou B = Béchar

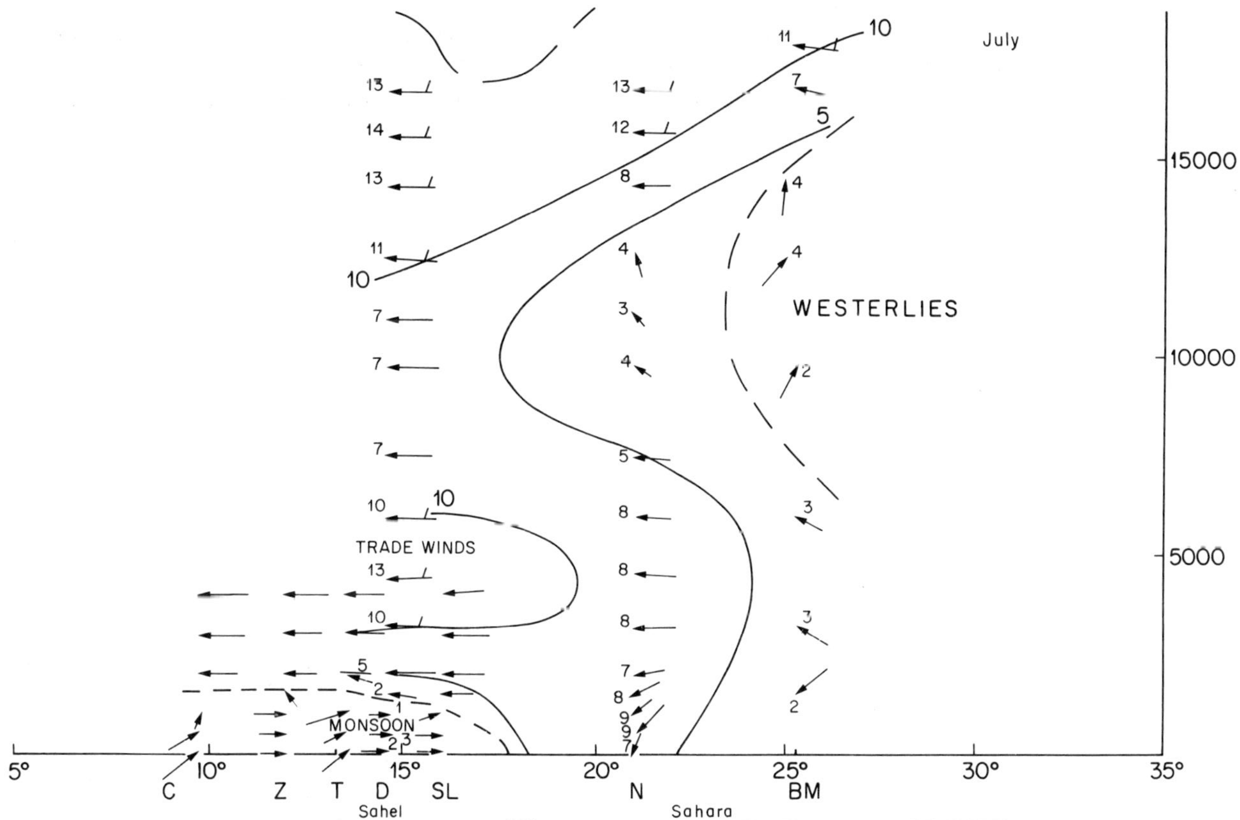

Figure 1.6 Predominant wind directions and mean wind speeds in upper levels at the African Atlantic coast in July

C = Konakry
Z = Ziguinchor
T = Tabacounda
D = Dakar
SL = Saint-Louis
N = Nouadhibou
BM = Bir Moghreim

4,000 m in more mountainous areas as is the case in summer at Mount Cameroun, which it literally deluges. Of course the dust particles in the harmattan air are transported above this monsoon towards very low latitudes until they are brought down to the ground by thunder showers.

The harmattan and the maritime trade winds are bounded on the north and surmounted largely on the south by the westerlies, which – and this is very important – are coupled to the low pressures of the medium latitudes. If this were not so the transport of dust towards Europe would not be possible. As the latitude decreases the westerlies are found at higher and higher altitudes, while their speeds increase. In winter a very strong westerly air current may be found at 8,000 to 10,000 m above the area from southern Mauretania to the loop of the Niger: the 'jet-stream' (Figures 1.3 and 1.4). Further east its course is from southwest to northeast as far as the north of Arabia, then frequently being marked by very characteristic high clouds. In summer the jet stream moves back towards North Africa and its speed decreases. To the south of this great air current the west winds gradually disappear and give way to the equatorial easterlies (Figures 1.5 and 1.6).

The above pattern is well illustrated by a graph of the winds at Tamanrasset, situated in the middle of the Sahara 100 km south of the tropic and 1,300 m above sea level. In winter the westerly winds prevail from about a thousand metres above the ground up to the highest altitudes. They have very high speeds, particularly at around 10,000 metres (Figure 1.7). This system is maintained until the end of May, but with diminishing wind strengths. In June it changes completely in the lower layers where easterly winds set in, from ground level up to 7,000 metres, above which altitude the westerly winds persist. In July and August the easterlies prevail everywhere, wind speeds are low. From September on, westerly winds reappear at high altitude. They draw successively closer to the ground until the end of October, at which time the winter situation is resumed. To sum up, from autumn to spring practically the whole of the high Saharan and north Sudanese troposphere forms part of the great stream of westerly winds governed by the low pressure systems of the medium latitudes. These winds draw progressively nearer the ground with increasing latitude. In summer, on the other hand, the whole of the troposphere of the southern Sahara, and with greater reason of the Sudan, above the monsoon, is in the sphere of influence of the easterly winds – harmattan or trades – surmounted by the equatorial easterlies, that is to say of the equatorial system. This distribution of the winds in altitude enables one to understand the long-distance transport of Saharan dust.

It has been established that in winter and at ground level all the Sudanese regions are under the influence of the Saharan desert climate. Swept by the harmattan, drought sets in everywhere, ponds disappear and their alluvial surfaces become the prey of wind erosion. Thermic turbulence, due to high near-ground temperatures, is considerable in the daytime and keeps in suspension the dust torn away from the soil of the southern Sahara and the Sudanese Sahel during dust storms. This constitutes what has been aptly named the 'harmattan haze' or 'dry haze' or again

'dust haze'. It is particularly dense in the neighbourhood of the ITF, where there is also a maximum amount of freezing nuclei, strongly concentrated in clay and especially in kaolinite (Bertrand, 1974). This haze is a striking phenomenon for aeroplane passengers as it entirely conceals the countryside, in the same way as low clouds (See Figure 1.8). Accompanying the ITF in its movements, it is to be found

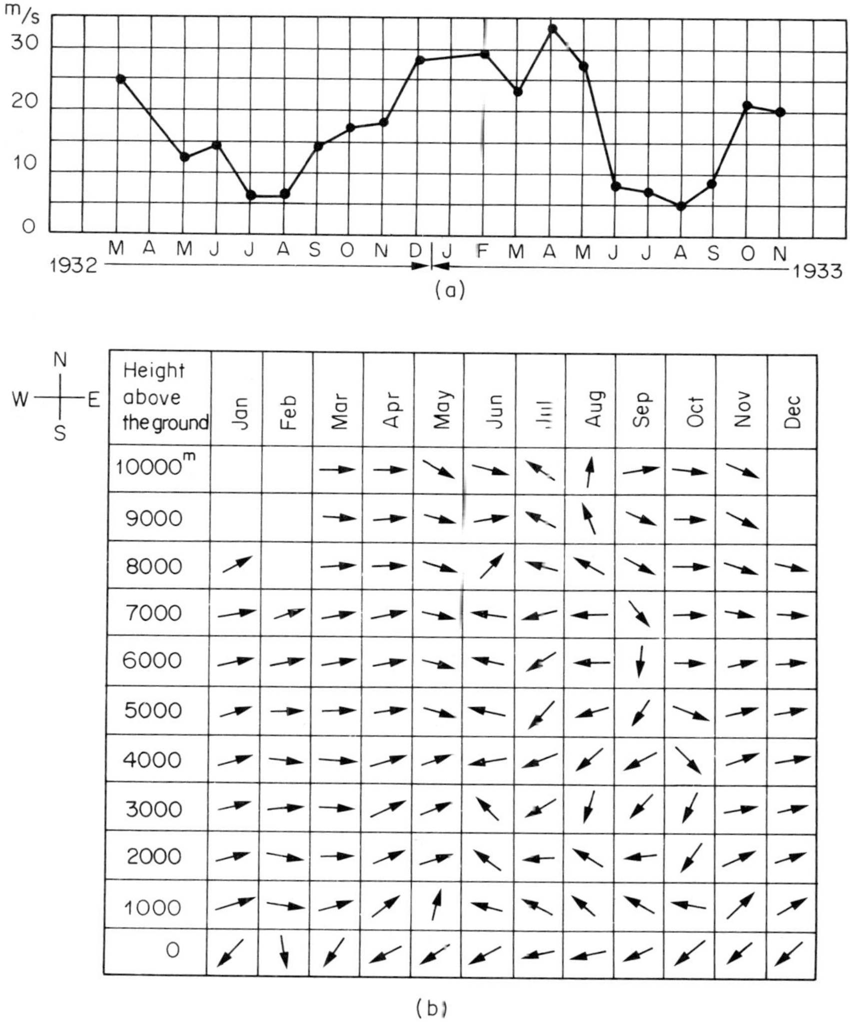

Figure 1.7 (a) Mean wind speed at an altitude of 10,000 m (above the ground) at the station Tamanrasset.

(b) Mean wind direction at the station Tamanrasset (the mean for the period 1932–1933; wind measurements made between 07[h] and 08[h])

Figure 1.8 Cirrus clouds above a layer of dense dry haze as seen from a Caravelle flying at a height of 8000 m between Douala in Cameroon and Ndjamena (former Fort Lamy) in Tchad. The photo was taken 26 January 1969

in spring, more or less mingled with fine ash from bush fires, from south of the Tropic to the regions where it is precipitated to earth in rainstorms, namely between 200 and 500 km south of the ITF. That is to say in regions where the layer of the monsoon is sufficient to maintain the cumulus through the dry layer of the harmattan. This pattern is, however, rather oversimplified. Indeed, it is difficult to distinguish between haze due to dust in suspension and that caused by humidity in the air, when the two are not combined. Their effect upon the soil is, however, not the same. A. Cerf (1974) has shown, for instance, that in the tropical humid zone the aerosols manifested themselves by an increase in ground temperature for the whole 24 hours during the period of the harmattan, while in the dry zone where the humidity is low the warming up and greenhouse effect is limited to the daytime. The greenhouse effect is cancelled at night by the increased radiative cooling of the dust due to the absence of water in the atmosphere. In summer and particularly in August, the ITF is in its most northerly position. It may quite often be observed in the Hoggar and the Tibesti areas and it sometimes reaches the plateau of Tademaït, in the western Sahara. In autumn, following the southwards withdrawal of the monsoon, the Saharan desert climate quickly replaces the equatorial climate in the Sudan.

The above describes average movements of the ITF, which are not literally true: in fact the latter oscillates widely on both sides of these theoretical average positions. Neither does it show the straightness of outline of the average position, in

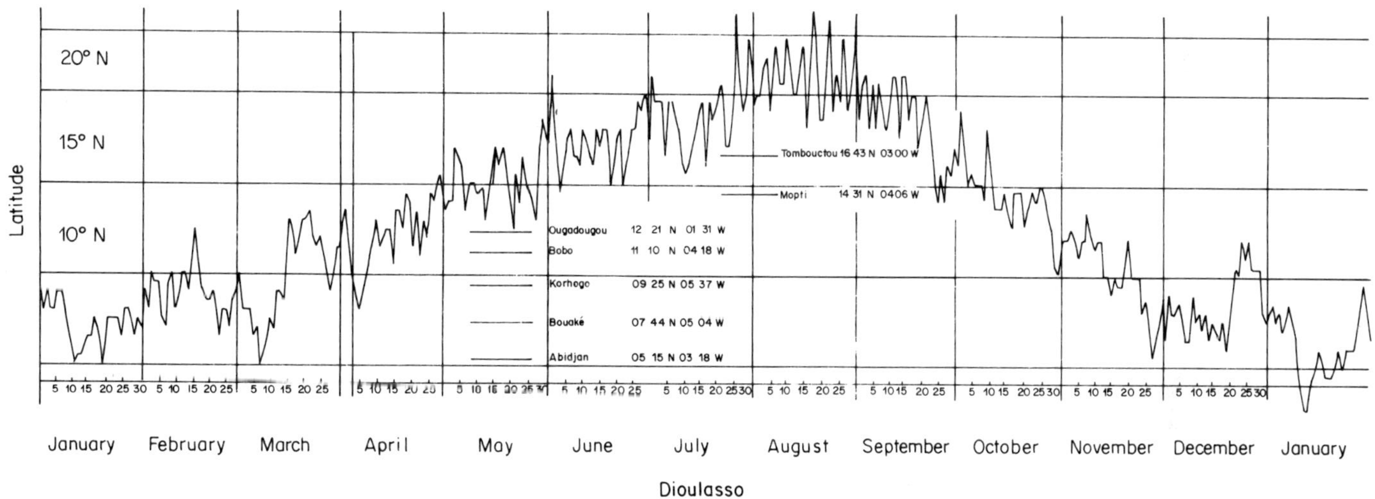

Figure 1.9 Position of the Intertropical Front (ITF) at surface at the longitude 5°W during 1973, in accordance with the meteorological charts prepared at Abidjan

Figure 1.10 The tops of cumuliform clouds emerging from a layer of dense dry haze. It is possible to distinguish within this layer much more voluminous cloud masses which are partly invisible from an observer in an aircraft. The photo was taken from a DC 9, north of Lagos, Nigeria on 14 January, 1969, just before landing

fact it curves back and forth across this, that is why these advances and withdrawals from the front of the monsoon may be compared to the waves of the sea, rolling different distances up the beach, while still ebbing or flowing with the tide (Figure 1.9).

This distribution of prevailing winds during the year appears to be confirmed by the traces of eolian erosion which may be observed in photographs of the Sahara taken from satellites. However, a careful study of aerial photographs taken from a plane leads to much less conclusive proofs, at least in the northern part of the desert, and, in order to understand the orientation of some dune formations in relation to the above distribution, one has to resort to some rather rash hypotheses as in Mainguet and Canon (1976) (Figure 1.10). The study of sand and dust winds in this Saharan region shows, finally and above all, that their most frequent directions are totally different from those of the prevailing winds. As proof of this observations may be taken in the sirocco in the Maghreb, the ghibli in Libya and the khamsin in Egypt. The above conclusion has led us to study the frequencies of

winds from the various points of the compass in relation to their strength (Dubief, 1952). We soon see that the prevailing directions at low speeds are very often different from those observed at high speeds, when they are not directly opposite (Figure 1.11). Thus, the frequency of winds between southwest and northwest is seen to increase with their strength, until they finally represent almost all the violent winds, that is to say those which lift and transport most of the dust and sand. Apart from the summer season (Figure 1.12) and the southern regions of the Sahara, easterly to northerly winds do indeed prevail but are light. This is why, in the northern part of the country, the northern and continental air masses cannot contribute to the morphology of the ergs and the production of dust, with the possible exception of Egypt and the Atlantic coast of the Sahara, but there the problem is more complex. To this may be objected at first that sand and dust winds are sometimes encountered with relatively low wind speeds. But in this case their origin has another cause. A lifting of sand and dust occurs when the dunes are approached by 'contrary' winds, if I may use the expression. That is to say winds in the opposite direction to the equilibrium previously established by winds in the contrary direction. One could also say that moderate or strong winds are too infrequent to account for the eolian transport phenomena. This is to forget that these are accidental phenomena, just like rain, and thus it is not surprising that they should be produced by exceptional causes such as high velocity winds: exceptional causes produce exceptional phenomena. Remember, too, that the force of the wind is not alone in causing these phenomena. To it should be added its turbulence,

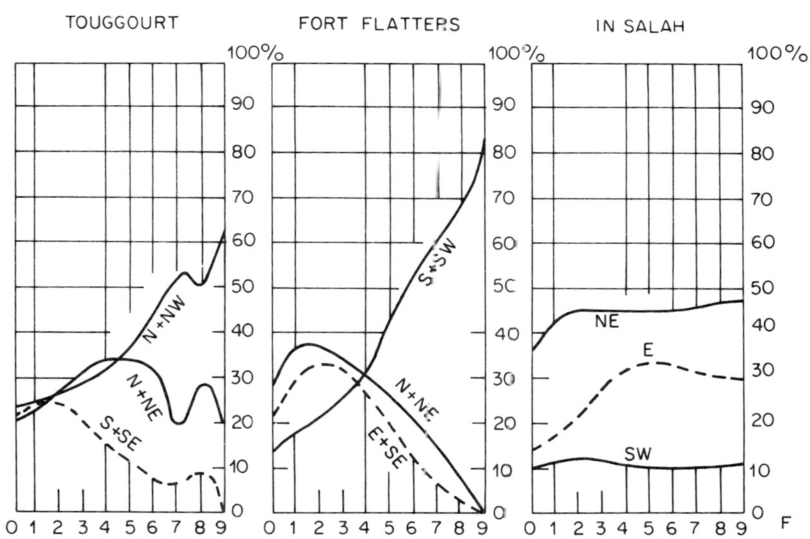

Figure 1.11 Windforce frequency at different wind directions (F: Windforce according to the Beaufort scale)

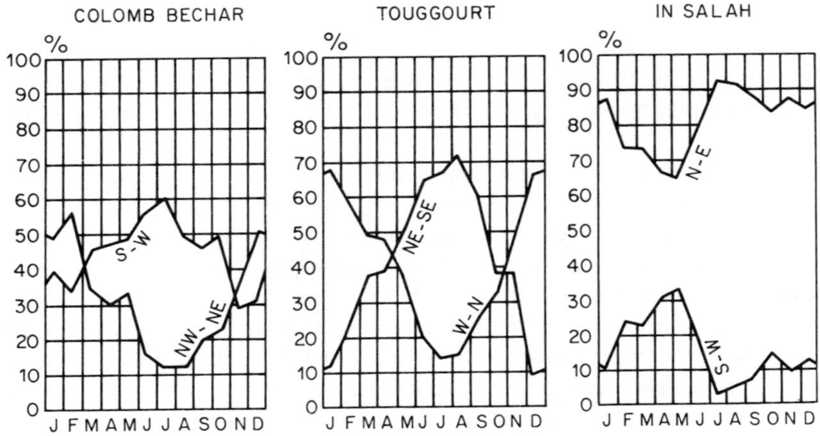

Figure 1.12 Yearly variations of predominant wind sectors (only wind forces above 3 Beaufort are considered)

thermic, orographic or otherwise, in order to explain eolian liftings and transports. A strong, steady, laminary wind lifts little sand and dust and, if it does so, to no great height (one or two metres only). This is why dust winds are often associated with sharp rises in temperature and are frequently encountered in the warm sectors of depressions. I would like to emphasize that often, not always, as sand winds are sometimes to be met with in the cold sector of a depression and also, especially in the southern Sahara, on the intrusion of cold winds, but here their turbulence is of orographic origin: this is the case, for instance in the southeast of the Tibesti.

These moderate to strong southwesterly winds and the accompanying movements of sand can usually be related to depressions involving the Sahara. The most important of these are the ones which arise locally in the warm sector of these perturbations. They may be seen at first as small rivulets of sand, very wavy, separate from each other and close to the ground. By degrees, as the speed and turbulence of the wind increase, these rivulets of sand run together, the air quickly darkens, while the depth of the air-borne dust increases, normally reaching around 2,000 metres. These hot winds full of sand mixed with dust increase slowly in intensity during the morning, reaching a maximum during the afternoon. Their turbulence is therefore mainly of thermic origin. When the depressions are on a large scale and allied to European perturbations, these winds of sand and dust may extend to the whole of the Sahara and the dust transported by the upper winds, from between southeast and southwest, may reach the Mediterranean basin and Europe. This is what Navlikine (1969) called 'intercontinental dust-storms'.

These depression have various origins. They may be:

(1) Mediterranean depressions skirting the Mediterranean coast of Africa. Their

action is reinforced when they halt for some days at certain spots on the coast or in the eastern Mediterranean.

(2) Atlantic depressions, connected, like the above, to polar front depressions when these approach the Sahara by the south of Morocco. This usually only occurs in winter.

(3) Secondary south-Moroccan depressions originating in the southern part of Morocco when the trajectory of the Atlantic depressions reaches a turning point at the level of the Iberian Peninsula and then moves towards the northeast. Paul Queney (1936) drew attention to these and formerly designated them as 'Saharan depressions'. Moving eastward through the north of the Sahara and the Maghreb, they reach Libya and the eastern Mediterranean basin. If they are linked to strong European polar front perturbations they enable the dust, raised south of the Atlas chain, to reach higher latitudes and the continent of Europe. It is probably these last two types of depression which have inspired the Meteorological Office in London to think that the majority of 'khamsin depressions' in Egypt originate from the Atlas Chain wind (Bejjani, 1975).

(4) Depressions arising along the ITF, thus in the Sudan area. With Paul Queney (Dubief and Queney, 1935) we at first called them Saharo–Sudanese depressions, then, with Robert Capot-Rey (1953), Sudano–Saharan depressions. They are in fact, and more simply, tropical depressions. They develop when the polar or pseudo-polar air mass, in conjunction with modifications of the upper atmosphere, meets a vast wave of the polar front. Jalu and his collaborators have recently studied them with greater resources than we formerly used (Jalu, 1965; Jalu *et al.*, 1965).* At first their trajectories stretch across the Sudan from east to west, like all the Sudanese perturbations, then, often after encountering a turning point, they move towards the north or northeast. Then, swept along by the westerly upper winds, they cross the Sahara and reach the Mediterranean coast, where they continue on their way, more or less linked to the polar front depressions (Figure 1.13). Their most frequent trajectories go from Senegal to Morocco in autumn, from the loop of the Niger to the Gulf of Sirte in spring and, to a lesser degree, in May–June, from the north of Nigeria to Lower Egypt. In the latter country they are known as 'khamsin depressions'. It is they which usually give rise to the Libyan 'ghibli' (Sutton, 1946). It should be noted that these 'khamsin depressions', after having caused strong sand and dust winds in Egypt, sometimes as far as Assouan, reach the Lebanon and Syria, bringing with them considerable clouds of dust (Bejjani, 1975). Contrary to our opinion, El Tantawy (1964) believes that they might be caused, either by a divergency in the upper atmosphere in the neighbourhood of the sub-tropical jet stream, or by a special development of this divergency zone.

*The terminology used by Jalu to designate the air masses is different from ours and may lead to confusion. To him the monsoon is a humid tropical air mass and the harmattan a northerly trade wind; whereas to us, as has been seen, the former is an equatorial maritime air mass and the latter a tropical continental one.

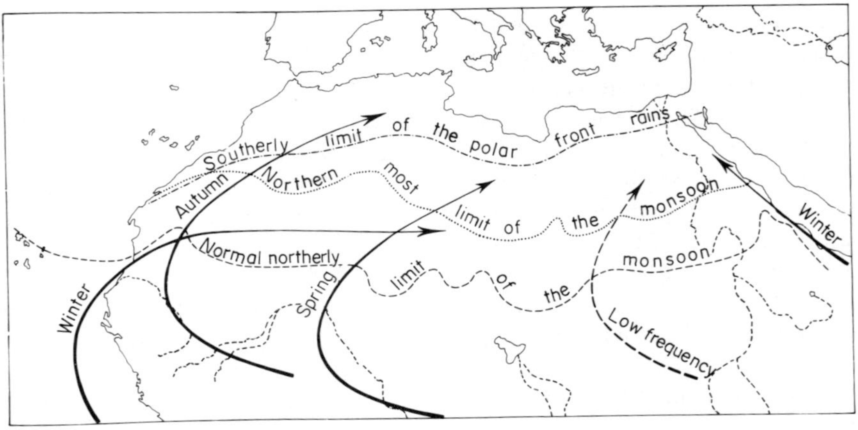

Figure 1.13 Trajectories of Sudano–Saharian depressions and limits of the polar front and monsoon rains

In order that the dust torn away from the Saharan soil may reach Europe, it is necessary, as we have said, for it to be linked to European polar front depressions and the 700 to 500 mb contour charts show a very marked trough (a "thalweg"), with a north-south axis and touching North Africa. This is shown on the charts given by Bücher and Lucas (1975) and by Champollion (1963). In this case the dust may attain altitudes of 5,000 metres and more and reach areas as far away as the north of the Federal Republic of Germany. Because the upper air currents take slightly different directions at different levels, dust from the same Saharan region may be spread across a vast area by the end of its journey, as shown on Champollion's map of April 1962 (Champollion, 1963). The duration of this transport, one or two days at most to reach France, and the distance travelled are determined by the meteorological circumstances: speed of the upper winds, zones of precipitation, calms, etc. In exceptional instances the Saharan dust can reach remote parts of Europe, as happened in March 1903, when falls were observed over the Atlantic, from the Canaries to the Azores, in Great Britain and Central Europe, while Spain, curiously enough, was spared (Hermann, 1903). Zamorsky has studied the movements of dust clouds on the 20 and 24 March 1964 which reached the Caucasus and then the northern part of the Caspian Sea and the Penza area, near the Volga, at latitude 53°N (Nalivkine, 1969).

It should be noted that these Saharan dust clouds may be augmented by other forms of dust, algae, pollen – from hazel for instance (Bücher and Lucas, 1972 and 1975), even alfa fibres (Champollion, 1963, page 80) lifted from the soil of the countries on the way and adjoining the desert. This is the case when the rising current persists for a long enough distance at the beginning of its course. They may

also remain in the upper air and be unnoticed from the ground, as happened in Spain in 1903, where all that was observed was a darkening of the atmosphere. This was also the case in the Bagnères area (Central Pyrenees) on the same date (Bücher and Lucas, 1972). Usually this dust is precipitated by rain (rains of mud or 'blood') (Combier *et al.*) or snowfall (Clement *et al.*). Nevertheless, it may gradually fall earthwards simply by means of gravity, owing to an exceptional calm in the atmosphere and perhaps, too, aided by a subsiding movement of the air. This is what occurred at Bagnères de Bigorre in February 1903 and, recently, in February 1972 (Bücher and Lucas, 1972).

In the rear of these Saharan depressions, the arrival of the cold front is sometimes accompanied by a raising of sand which gives the impression of a 'wall of sand and dust'. Two to three thousand metres high, it is spectacular in the extreme. This is what is depicted in paintings and photographs when wishing to illustrate sandstorms engulfing caravans (*sic*). Several hundred kilometres long and some tens of kilometres wide, they are of fairly short duration (from ½ to 2 hours) for they travel at rather high speed. While they are passing a particular spot the visibility is suddenly and very greatly reduced. Unlike the previously-mentioned liftings, the wall of sand is essentially migratory and may be easily followed on weather charts. Due to its origin, it is accompanied by a drop in temperature and a change of wind direction to the north.

When, in late spring and early summer, the whole atmosphere of the southern Sahara is occupied by easterly winds and a zone of relatively high pressure is found at 500 or 200 mb, the above-mentioned Sudano-Saharan depressions move only from east to west. If they are very active they give rise to strong sand and dust winds which involve the whole of the southern Sahara and the Sahel. This was the case between the 5 and 9 June 1967 (Figure 1.14). There were, in fact, two depressions rapidly following each other from the loop of the Niger to the sea area off Senegal. Their northern sectors comprised dry, burning, violent east to north-east winds. At Saint-Louis, for instance, maximum temperatures went from 28°4 on the 5th to 40°3 on the 7th and 42°8 on the 8th, dropping to 29° on the 9th, while an east-northeast wind blew, reaching a strength of 18 m sec^{-1} at midday on the 7th. It was evidently accompanied by a particularly intense dust haze which reached Dakar and remained there from the 7th to the 9th. Injected into the medium layers of the atmosphere, this haze pursued its course over the eastern Atlantic, the famous 'sea of shadows'. We have not been able to study the origins of the dust cloud which crossed the Atlantic from 29 June to 4 July 1969 and which was photographed by a satellite (Courrier de l'UNESCO August to September 1973).

Sand and dust winds may have quite a different origin to the above. It is then a question of so-called 'accelerating winds of sand and dust' (Berenger, 1963). These are relatively cold. Although they affect the desert on a large scale, they are less extensive and the reduction of visibility is less than in the above. Also the dust is raised to a lesser height. They are less turbulent and the turbulence is mainly from orographic causes and their speed is higher than in the preceding cases. They blow

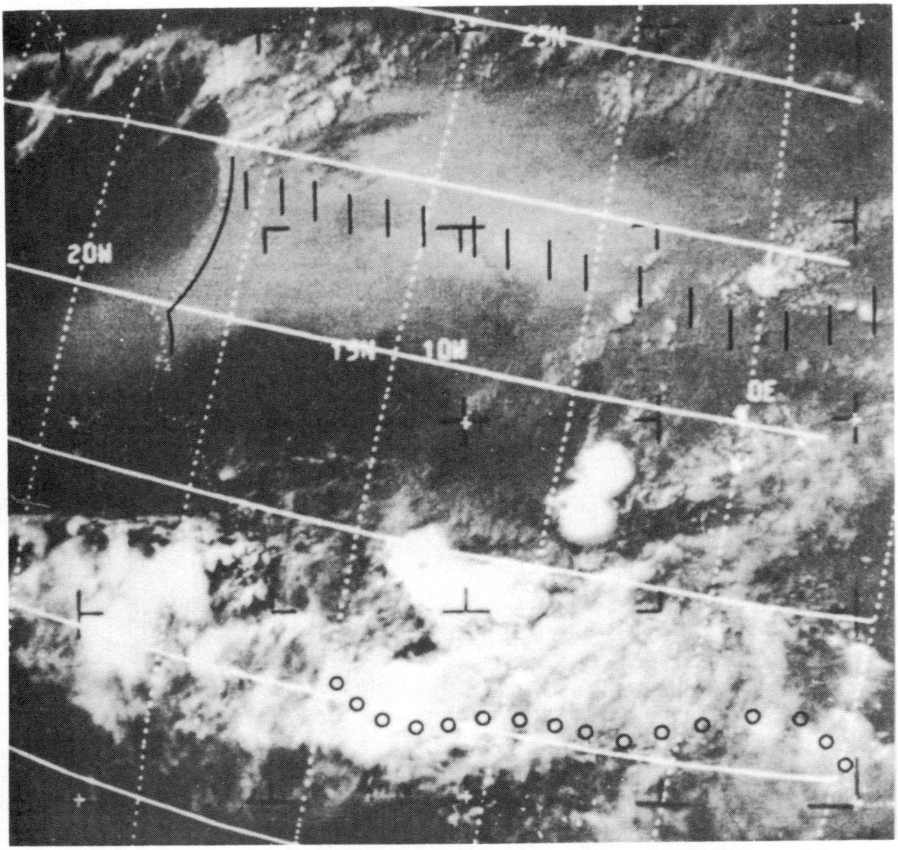

Figure 1.14 Sand- and dust-storm over western Sahara, photographed by ESSA on 5 June 1967. Driven by very warm and dry violent winds between north and east the dust cloud extends off the coasts of southern Mauritania and Senegal. The warm sector is situated north of the depression, the cause of the dust storm. The position of the depression was 14°N and 17°W on 5 June at 12^h GMT

mainly in the morning and not in the afternoon like the ones we have studied. They may be met with south of the Tidikelt plateau and southeast of the Tibesti, for instance, as well as in the Ténérés and the Tanezrouft of the southern Sahara. In the Tchad basin, these dust winds which follow the sand winds travel in a day or two and at a low altitude from Faya (Borkou) to Fort Lamy (Ndjamena), 800 km further south. They subsequently turn towards Nigeria and Upper Volta and in winter reach the coast of the Gulf of Guinea. Figure 1.15 shows an example of the progress of this kind of dust haze. They obviously feed the harmattan dry haze. Their precipitation over the Sudan area is considered beneficial by the Africans because of the fertilizing mineral elements they bring.

To these principal causes of dust clouds may be added secondary causes, giving

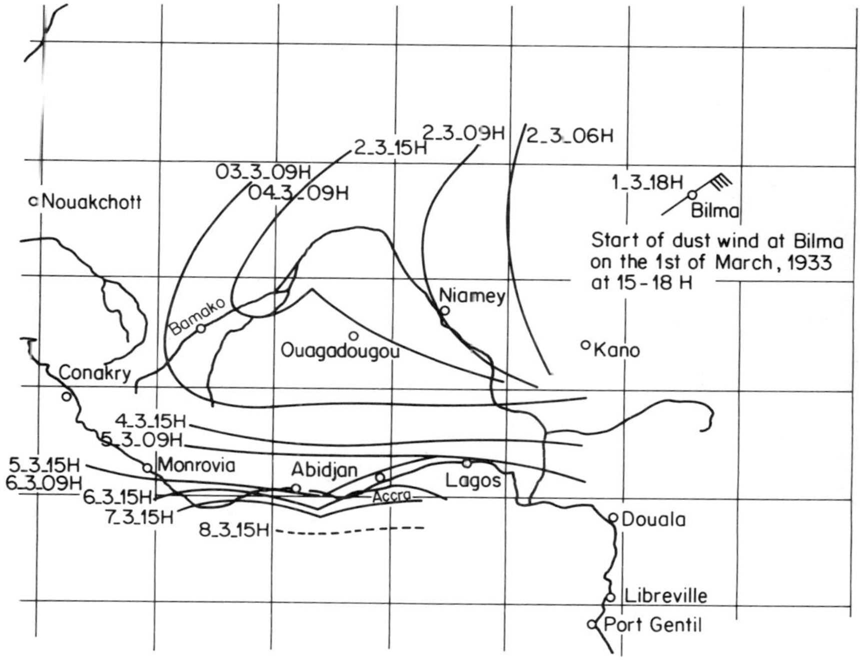

Figure 1.15 Migration of the front of dust haze above West Africa from the 2nd to the 8th March 1973 (after J. Bertrand, 1974)

rise to less spectacular phenomena. This is the case when particles of sand and dust are lifted in conjunction with thunderstorms. We meet this in the Sudanese region, mainly in the northern part and particularly in the Sudan itself. These are the 'haboobs' of the Khartoum area. They are walls of sand reaching up to the base of the Cumulo-nimbus and sometimes beyond these. They may be several kilometres long but of no great thickness. They are of variable, but usually short, duration, the average being half an hour. Similar but lesser phenomena may be seen at the onset of Saharan thunderstorms, especially in the summer, at night, in the northern part of the desert.

In addition, there are the small dynamic, thermic whirlwinds with vertical axes, the 'dust devils' which may be seen on the plain on calm and hot days. Some tens of metres in diameter at ground level, they may widen to several hundred metres at the top, thanks to their intense whirling movement and the violent rising current in the centre. As in the above, the amounts of dust which they can put in suspension in the air are minimal compared with the phenomena discussed earlier.

Dry haze may also become noticeable on the arrival of a humid air mass, by the absorbtion of water vapour by large ions in suspension in the air. We have observed this on various occasions at Tamanrasset, under the virgas of a rain cloud or on the arrival of a 'splash' of monsoon.

Lastly, dry haze linked to the harmattan may intrude very suddenly into the southern regions of the Sahara and particularly the central Saharan massif, following an exceptional northward movement of the ITF. It is then very dangerous as it drowns all the landmarks in dust and makes the most experienced guides lose their way. It causes even more mortal accidents than do the sand winds. When it appears in spring in the Hoggar, the Tuaregs believe it to mean rain in the Sudan.

To conclude, the Sahara, owing to its climate, appears to be the chosen territory of eolian erosion, this being particularly active on its northern and southern edges, where alluvia are still able to settle and where a dynamic balance in relation to the wind has not yet been reached. It is also a conspicuous centre for the dispersion of dust drawn up from the ground, and has been for thousands of years. If the normal course of this dispersion, in view of the prevailing winds, appears to be towards Black Africa and the Atlantic, it seems on the other hand to take a contrary course to these when moving towards the Mediterranean basin and Europe. This is only conceivable if one considers the effects of the various depressions involving the Sahara, in conjunction with those of the big air currents of the lower and middle atmosphere. For a large part of the year these currents are influenced by the low pressure systems of the medium latitudes. Strong injections of this dust into the upper altitudes permit it to reach the European continent, and, in exceptional cases enable it to arrive at some of the shores of America.

The big intrusions of cold air masses, more or less connected with the polar front, which from time to time overrun the Sahara from the north, reinforce the permanent action of the harmattan in winter and the 'Etesian winds' in feeding and displacing the mass of dust in suspension which remains in the neighbourhood of the intertropical convergency front and in the southern regions at all seasons, in summer in the centre of the western Sahara. These combined effects serve to compensate the losses to the mass of dust in suspension through tropical precipitation. The continuous action of the prevailing winds sweeps it towards the south and west of the Sahara. This dust haze gives to the Sudanese regions a climatic character very noticeable to travellers crossing or flying over them.

These dispersions of Saharan dust over Europe, while episodic at present are not without effect upon the climate there. Some years they certainly contribute to the melting of Alpine glaciers and it is very probable that their effects have been most important for thousands of years past.

REFERENCES

Bejjani, M.-C. (1975). Etude des perturbations synoptiques sahariennes associées à un soulèvement de sable; mise en évidence du rôle radiatif de ce sable. *Thèse Univ.*, Paris VI, 131 pp.

Berenger, M. (1963). Contribution à l'étude des lithométéores. *La Météorologie, Paris,* **72**, 347–374.

Bertrand, J. J. (1974). Mesure de la concentration en noyaux de congélation en Afrique Occidentale. In 'Reunions de Travail sur les Caractéristiques Physico-Chimiques et le Transport des Poussières d'Origine Africaine – 9–12 Septembre 1974'. *Inst. Observatoire Phys. Globe Puy-de-Dôme,* **26**, Clermont-Ferrand.

Bücher, A., and Lucas, G. (1972). Le nuage de poussière rouge du 7 février 1972. *Bull. Soc. Hist. Naturelle de Toulouse,* **108**, 437–455.

Bücher, A., and Lucas, G. (1975). Poussières Africaines sur l'Europe. *La Météorologie, Paris, Ser. 5,* **33**, 53–69. (English and French summaries).

Capot-Rey, R. (1953). *Le Sahara Français.* Presses universitaires de France, Paris. 564 pp.

Cerf, A. (1974). Action des aérosols sur le rayonnement. Incidence climatique. In 'Réunions de Travail sur les Caractéristiques Physico-Chimiques et le Transport des Poussières d'Origine Africaine – 9–12 Septembre 1974'. *Inst. Obser. Phys. Globe Puy-de-Dôme,* **26**, Clermont-Ferrand.

Champollion, M. (1963). Retombées de poussières et pluies colorées. *La Météorologie, Paris, Ser. 4,* **71**, 307–313.

Clément, R., Ricq de Bouard, M., and Thomas, A. (1972). La neige colorée du 9 mars 1972. *La Météorologie,* **24**, 65–83.

Combier, R. P. C., Gaubert, S. J. P., and Petitjean, L. (1937). Vents de sable et pluies de boue. *Mém. l'Office Nat. Météorol. France*, Paris, 135 pp.

Dubief, J. (1952). Le vent et le déplacement du sable du Sahara. *Trav. Inst. Rech. Sahariennes,* **8**, 123–164.

Dubief, J., and Queney, P. (1935). Les grands traits du climat du Sahara Algérien. *La Météorologie,* **119**, 80–91.

El Tantawy, A. (1964). The role of the jet stream in the formation of desert depressions in the Middle East. In 'High-level Forecasting for Turbine-engine Aircraft Operations over Africa and the Middle East'. *Proc. Joint ICAO/WMO Sem. Cairo–Nicosia*, 1961, WMO Tech. Note 64, (WMO – No. 159), 159–171.

Gautier, E.-F. (1928). *Le Sahara.* Payot, Paris, 232 pp.

Hermann, E. (1903). Die Staubfälle vom 19 bis 23 Februar 1903 über dem Nordatlantischen Ozean, Grossbrittanien und Mittel-Europa. *Ann. Hydrog.,* **31**, 425–438 and 475–483.

Jalu, R. (1965). Note sur le déclenchement des dépressions tropicales sahariennes. *La Météorologie,* **78**, 113–128.

Jalu, R., Bocquillon, M., and Bonnefous, M. (1965). Tempête de sable sur le Sahara. *La Météorologie,* **78**, 105–112.

Mainguet, M., and Canon, L. (1976). Vents et Paléovents du Sahara. Tentative d'Approche Paléoclimatique. *Rev. Geogr. Phys. Geol. Dyn.,* **18**, 241–250.

Nalivkine, D. V. (1969). Hurricanes, Storms and Well-developed Sand Whirls. *La Science*, Leningrad, 487 pp. (In Russian).

Queney, P. (1936). Recherches relatives à l'influence du relief sur les éléments métérologiques. *Thèse d'Etat. Fac. Sci. Univ. Paris. Soc. Météorol. France*, 97 pp.

Sutton, L. J. (1946). *The Climate of Egypt.* Schindler, Cairo, 48 pp.

CHAPTER 2

The Importance of Mineral Dust as an Atmospheric Constituent

C. JUNGE

ABSTRACT

The importance of mineral dust in the atmosphere is twofold: Dust deposition and possible effects on climate.

Depending on the conditions and dimensions of desert areas and meteorological transport processes, dust depositions along the fringe areas of the source region can be of geological, pedological and ecological importance. The loess deposits during the glacial age are the classical example. Present day deserts seem to be much less efficient sources for loess formation.

The mineral dust in the atmosphere affects both the radiation of cloud free air and the optical properties of the clouds. The interaction is complex but in general one can expect the mineral dust to affect primarily the absorption of aerosols and clouds for short wave radiation. Large scale natural or anthropogenic changes of dust production which have undoubtedly occurred should therefore be considered with respect to their effect on climate.

This workshop is concerned with the production and transport of dust from the area of the southern and western Sahara. This concern is based primarily on ecological reasons. But once the dust has become air-borne it is part of the atmospheric system and is itself of interest. I will discuss here the question of its importance for atmospheric sciences.

We have gaseous and particulate trace substances in the atmosphere. Both can be of local, regional and global importance. The aerosols influence cloud formation, radiation and air electricity. Changes in their concentration may induce variations of the climate by affecting the global radiation budget. There are aerosols from different sources and the question to be discussed here is the role of the dust component of the tropospheric aerosol in this context.

Aerosols in the troposphere have residence times of the order of a week (see Martell and Moore, 1974). This is long enough for considerable spread within both hemispheres with their strong zonal circulation, but it is too short for an exchange between the two hemispheres across the equator which is of the order of one year.

The aerosols of the two hemispheres remain, therefore, almost completely separated with some large scale mutual leakage resulting from the Indian Monsoon system. Since there is no substantial exchange between the hemispheres in the Atlantic sector, Saharan dust remains confined to the northern hemisphere for all practical purposes.

If we disregard the anthropogenic sources we can distinguish three dominant natural aerosol production mechanisms listed in the order of their quantitative importance:

1. Gasformed particles, produced within the atmosphere by oxidation and reaction of a variety of trace gases.
2. Sea spray.
3. Mineral dust.

Table 2.1 gives published estimates of their source strength, indicating the degree of uncertainty. The uncertainty is particularly large for mineral dust, because these figures were derived by rather dubious methods. For some strange reason, however, they seem to be in fairly good agreement with recent estimates based on fairly comprehensive data sets on atmospheric dust concentrations published over the last years. Covering essential parts of the world, these data became available by routine application of such analytical methods as neutron activation, absorption spectroscopy, and others, capable of determining simultaneously in clean air aerosol samples a large number of elements. The results show that a considerable number of these elements, in particular Fe and Al, occurred in ratios quite similar to those in crustal or soil material indicating that their dominant source was mineral dust. By applying appropriate factors it is, therefore, possible to derive fairly reliable figures for total mineral dust concentrations from these analyses. Figure 2.1 gives a survey based on such observations, with the most prominant authors indicated. We will come back to this Figure later. On the basis of such concentration data covering

TABLE 2.1 Global Estimates of the Major Natural Aerosol Sources in the Troposphere in 10^6 tons yr^{-1}

Source	Strength
Particles formed in the troposphere by oxidation of H_2S, NO_x, NH_3, HC and other trace gases	300–1100
Sea spray	1000
Mineral dust	100–500
Other sources (forest fires, volcanic, *etc.*)	3–150

Figures from the SMIC-Report (1971)(Table 8.1) and other sources.

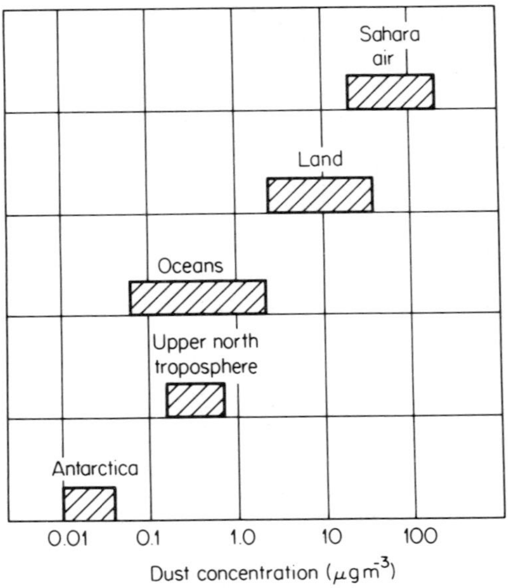

Figure 2.1 Survey of the concentration ranges of mineral dust in the troposphere based on numerous studies by – among others – the following authors: Blifford, Chesselet, Duce, Ferguson, Hoffman, Gillette, Goldberg, Griffin, Jaenicke, Prospero, Rahn, Schütz, Winchester, Zoller

most essential areas of the world and also of meteorological considerations about the vertical distribution within the troposphere, the global tropospheric dust burden can be estimated. With the knowledge of the average residence time for aerosols, we can then derive at figures for the total source strength. Table 2.2 gives such figures separate for both hemispheres for reasons indicated earlier together with our guess for reasonable uncertainty factors.

We think that the basis for these estimates is more reliable than that for the previous estimates. But it becomes equally clear that despite this fact the uncertainty is estimated to be even larger, which we feel is more realistic. We see that even without the Sahara source the northern hemisphere has a higher dust load than the southern one as was to be expected from the larger land areas. Included in Table 2.2 is the range of recent estimates of the Saharan dust production by Prospero and Carlson (1972) and Jaenicke and Schütz (1977). Clearly, the Sahara desert seems to play an important role in the northern hemispheric dust cycle, providing about half of it. The addition of the Saharan source results in a range for the global source strength of 260 to 400 x 10^6 tons yr^{-1} (disregarding the uncertainty of non Saharan production) which agrees well with the older estimates in

TABLE 2.2 Estimates of the Global Tropospheric Dust Cycle

Part of troposphere	Dust burden 10^5 tons	Source strength 10^6 tons yr^{-1}
Northern hemisphere[a]	3.0[b]	150[c]
Southern hemisphere[a]	1.0[b]	50[c]
Whole troposphere[a]	4.0[b]	200[c]
Sahara plume	1.2–4.0	60–200[d]
Total troposphere (plus Sahara)[e]	5.2–8.0	260–400
Total troposphere (plus Sahara)[f]	3.2–12.0	130–800

[a]Estimates disregarding any special production in deserts, particularly the Sahara area and its plume in the north Atlantic trade winds.
[b]Estimated uncertainty factor about ±2.
[c]The source strength was calculated from the dust burden assuming an aerosol residence time of 1 week. Estimated uncertainty factor about ± 3.
[d]Range given by Prospero and Carlson (1972) and Jaenicke and Schültz (1977)
[e]Without applying the uncertainty factors *b* and *c*.
[f]After application of the uncertainty factors *b* and *c*.

Table 2.1. Applying our uncertainty factors gives the possible range of 130 to 800×10^6 tons yr^{-1}.

But the actual uncertainty is still larger because there are other desert areas besides the Sahara which may act as huge individual sources. There are yet, at least to my knowledge, no data on the frequency and the magnitude of these sources. Figure 2.2 shows a map indicating areas covered with dunes which are potential sources for wind blown dust. Observation of dust haze over the oceans seems to indicate, however, that the transport of Saharan dust over the subtropical Atlantic is the most spectacular phenomenon of this type. But we should consider these facts to get a realistic feeling of the lack of reliable information in this field and for the uncertainty of the figures in Table 2.2.

It is perhaps of interest to compare the flux of the eolian transport of Sahara dust with other fluxes out of this desert. The transport of sand by moving dunes across the West coast of the Sahara by N to NE winds occurs apparently only along a front of about 80 km length south of Cape Blanc. Flux estimates give 0.15×10^6 tons yr^{-1}, with particle sizes peaking around 100 μm radius. Transport of mineral dust out of the Sahara by the river Niger is 60×10^6 tons yr^{-1}. These data are compiled in Table 2.3. It is clear that the long range eolian dust transport out of the West Sahara in Table 2.2 is the most important one, at least much more important than the other eolian transport mechanisms. This may of course be different for other arid zones and deserts.

Figure 2.2 shows also the known loess deposits. Loess is an eolian deposit and is

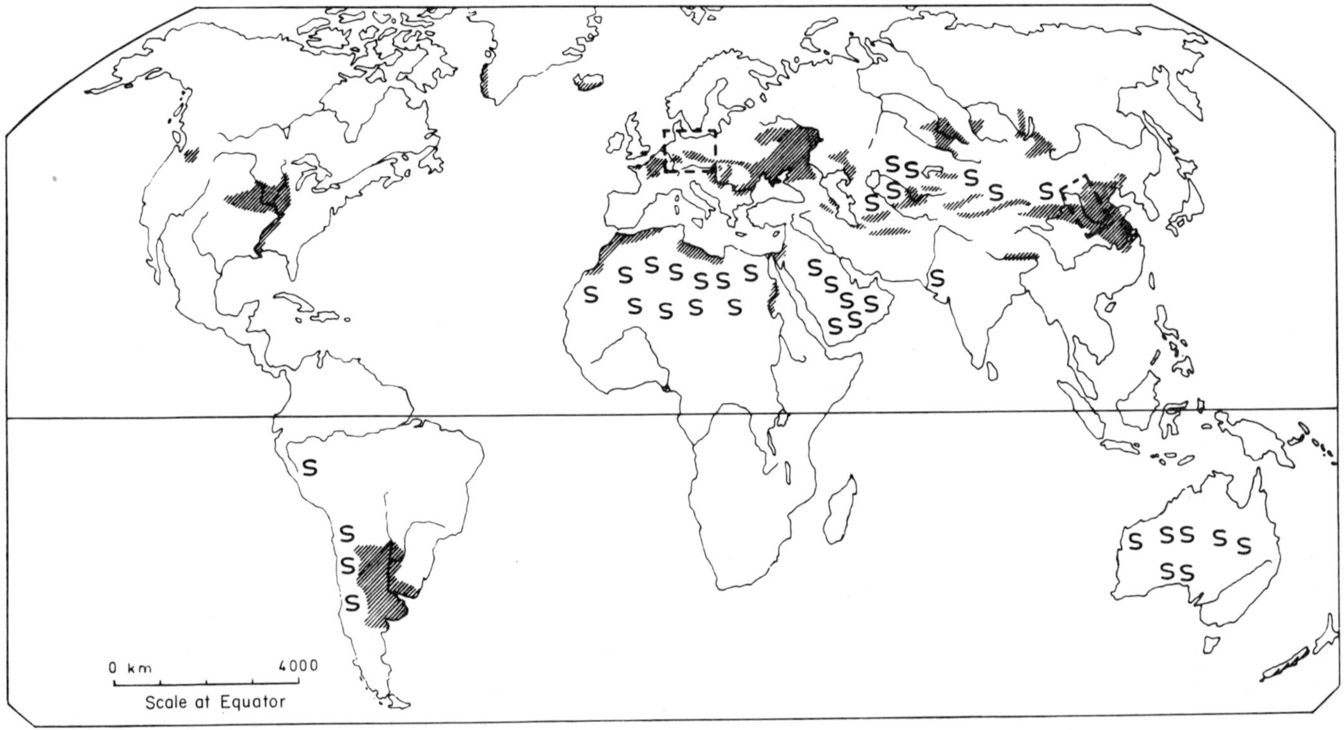

Figure 2.2 Distribution of loess on Earth (according to Scheidig, 1934) as well as desert areas with sand dunes (according Boyko, 1967). From Smalley and Vita-Finzi, 1968, Section of Central Europe, reproduced in Figure 2.6 indicated by a dotted square (after Füchtbauer and Müller, 1970)

Legend: Shaded areas = Proved loess deposits S = Deserts with sand dunes

TABLE 2.3 Comparison of various transport rates of Saharan material across the western coastline

Mode of transport	Flux in 10^6 tons yr^{-1}
Moving sand dunes	0.15^a
Saltating sand	$10–20^a$
River Niger	50^a
Eolian transport	$60–200^b$

[a] Sarntheim and Walger (1974).
[b] Table 2.2.

primarily made up of particles in the range 5 to 50 μm radius. Figure 2.3 shows cumulative mass distributions (size is expressed as diameter) of 8 different types of loess from various parts of the world showing a fairly uniform distribution. This is not too surprising because the fractionation process between loess and soil particles is always the same depending to some degree on the distance between source and deposition area. In desert dust storms even much coarser material becomes air-borne or is jumping along the surface but settles down again at much smaller distances. In

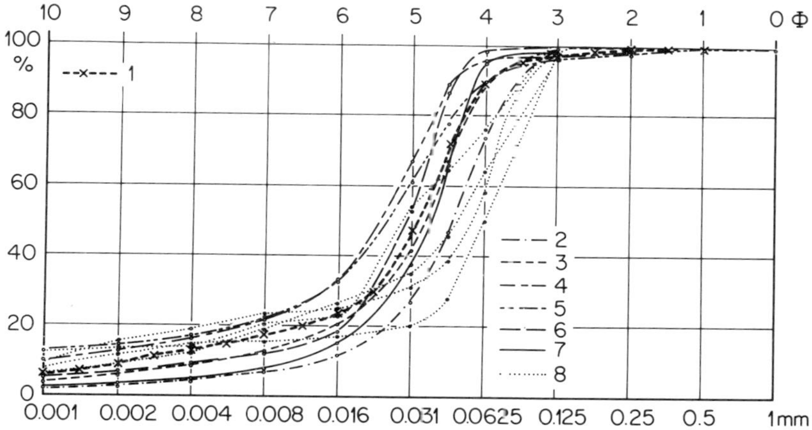

Figure 2.3 Particle size distributions (diameters!) of different kinds of loess (according to Swineford and Frye, 1955, and Teruggi, 1957) as well as recent eolian dust (according to Swineford and Frye, 1945) (after Füchtbauer and Müller, 1970)

Legend: 1. =Dust 1939, Meade, Kansas, U.S.A. 2. = Torino, Italy.
3. =St. Vallier, France. 4. = Achenheim, France.
5. =Wiesbaden, West Germany. 6. = Liege, Belgium.
7. =Thomas Co., Kansas, U.S.A. 8. = Loess from the Argentine

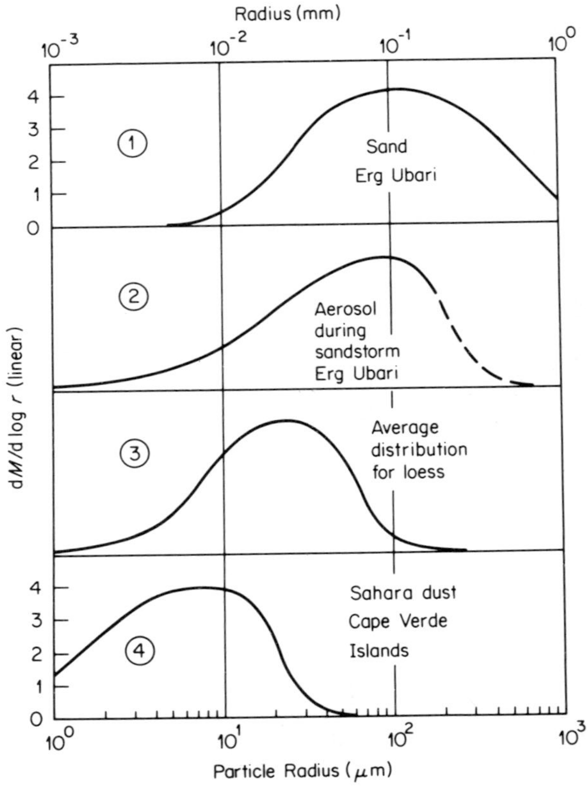

Figure 2.4 Comparison of different idealized mass distributions based on the following sources:

(1) Schütz and Jaenicke (1974), sand from the Libyan desert, Erg Ubari
(2) Same source and location as (1) but aerosol during sand storm
(3) Füchtbauer and Müller (1970), average of 8 loess distributions from various continents, according to Figure 2.3
(4) Jaenicke and Schütz (1977), average mass distribution over the Cape Verde Islands

Figure 2.4 we plotted linear mass distributions of sand and sand storm aerosols from the Libyan desert, loess (according to Figure 2.3) and air-borne dust in an idealized fashion to show the essential size ranges. Figure 2.5 demonstrates schematically how the major fractions are derived from the original source by successive stages of a winnowing process controlled by the interaction between air-borne transport and gravitational settling until Stage 4. For atmospheric travel-

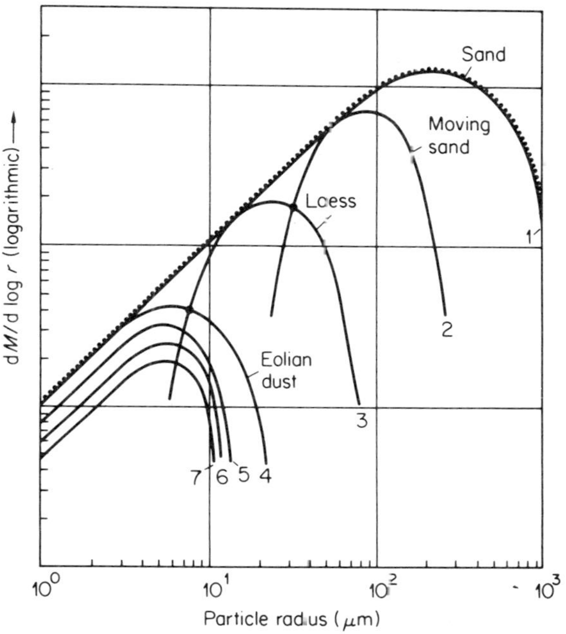

Figure 2.5 Sand fractionation processes by wind, schematic. The original sand distribution is fractionated into major fraction 2, 3, and 4 as a function of distance from the source. Curves 5, 6, and 7 depict the change in concentration due to both wet and dry removal from the atmosphere. The sum of curve 2, 3, and 4 should be equal to the original curve 1

ling times comparable to the average aerosol residence time in the troposphere removal by precipitation becomes important besides fallout by sedimentation. The size distribution of the individual fractions will, of course, vary with the environmental parameters, but Figure 2.3 suggests that the differences may not be very dramatic. In Figure 2.5 a smooth, continuous distribution was assumed for the original sand. In reality this may not be the case because of size dependent efficiencies of the various physical and chemical disintegration processes constantly active in the sand. In fact there are indications of bimodal distributions in the case of fraction 2, Figure 2.4, at some places, but in principle one should expect single mode distributions if the sand distributions are sufficiently smooth. Figures 2.4 and 2.5 demonstrate that there is and should be considerable overlapping in the tails of distribution curves.

It is clear that the loess fraction becomes air-borne only in fairly strong winds. The large majority of all loess deposits in Figure 2.2 were formed during or at the

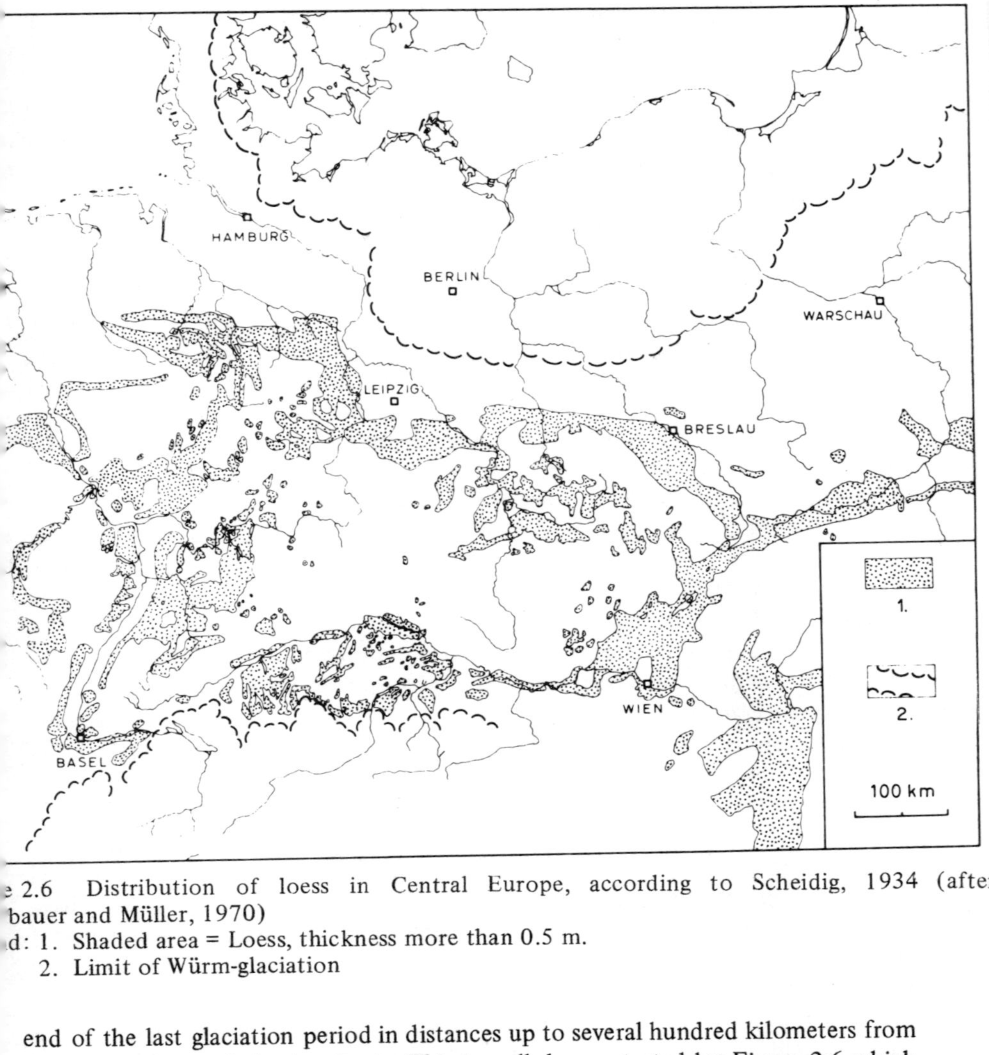

e 2.6 Distribution of loess in Central Europe, according to Scheidig, 1934 (after bauer and Müller, 1970)
d: 1. Shaded area = Loess, thickness more than 0.5 m.
2. Limit of Würm-glaciation

end of the last glaciation period in distances up to several hundred kilometers from the border lines of the ice sheets. This is well demonstrated by Figure 2.6 which shows that all the loess deposits in Central Europe are located between the ice shields of the last glaciation. The huge amounts of material of these deposits were most likely produced by physical weathering processes associated with glaciers. But there is sufficient evidence that deserts also can act as source areas for loess, though apparently at a much smaller scale. One example is the Negev desert in Israel (Smalley and Vita-Finzi, 1968). Apparently in most deserts the weathering rate just is not sufficient to produce deposits along their fringe areas.

The large areas covered with glacial loess deposits imply that at those times the

atmosphere in such latitudes must have carried a considerable dust burden, most likely due to stormy and dry weather conditions. This conclusion is in fact supported by high dust concentrations in Greenland ice layers during the end of the last glaciation. It is also implied by the actual absence of loess formation at present times in these latitudes. An intriguing question is if and to what degree such dusty atmospheres induced feed processes on the climate development during the ice age.

The processes by which dust becomes air-borne are complex. Structure, moisture content and other soil parameters determine a threshold wind speed at which the dust is picked up by the air. Once this threshold is passed the vertical flux of dust increases rapidly with wind speed (see e.g. Gillette, 1974). A considerable fraction of the material, however, remains near the surface and is soon redeposited again. It would, of course, be unreasonable to include the flux of such material in figures for the global dust cycle, as given e.g. in Table 2.2. Dust considered in this respect should remain sufficiently long in the air to become of importance for meteorology. It is suggested that a one day residence time would represent a reasonable lower limit for this purpose.

The dust component in atmospheric aerosols is highly variable as shown in Figure 2.1. The reasons for this are the rather short residence times of aerosols together with the pronounced variability of the dust sources with time and space. The data show that on a global scale the average dust concentrations at fixed locations usually fluctuate with time by about one order of magnitude. They also show that average dust concentrations range more than 4 orders of magnitude with values of 100 $\mu g/m^3$ in air coming from the Sahara and values as low as 10^{-2} $\mu g/m^3$ at the South Pole (for comparison: The normal range of sea salt concentrations over the oceans is 1 to 10 $\mu g/m^3$, i.e. only about one order of magnitude). Inside of desert dust storms the visibility can be less than 100 meters implying concentrations of the order of 10^4 $\mu g/m^3$.

We have shortly reviewed production, redeposition and atmospheric transport and distribution of mineral dust. Besides the geological, pedological and ecological effects of soil erosion and dust fallout the atmospheric dust may also have importance for the radiation budget of the atmosphere and hence for the climate if large scale changes of land surfaces have occurred or do occur, natural or man made. There are two ways in which the aerosol can influence the radiation budget: In a direct way by changing the radiation fluxes in a cloud free atmosphere, by scattering and absorption of radiation, and in an indirect way by modifying the optical properties, particularly the albedo of clouds (see SMIC Report, 1971).

The direct influence on the short wave radiation is primarily affected by the aerosol particles in the size range from about 0.1 to 2 μm radius. Most of the particles in this range are formed by gas reactions and under normal conditions the fraction of mineral particles in this size range is small. But we know that the mineral particles have, among other constituents, an iron content of about 5%, mostly present in the form of oxides which are good absorbers for radiation. In clean atmospheres the gasformed particles show little absorption, and here changes

in the concentration of dust may result in marked changes of the absorption characteristic of the aerosol even if the number concentration is not much affected.

The optical properties of clouds are influenced by the aerosol in two ways. The number concentration, particularly those larger than about 0.1 μm, controls the number concentration of cloud droplets during cloud formation because these particles act as cloud condensation nuclei. The number concentration of cloud droplets in turn affects the albedo of the clouds in the sense that more but smaller cloud droplets result in higher albedos. Due to multiple scatter of light on the cloud droplets within the cloud a certain fraction of light becomes absorbed in the cloud resulting in a decrease of the cloud albedo. The amount of absorbing material in cloud droplets therefore controls also the cloud albedo. Under normal atmospheric conditions the mineral dust will hardly influence the number concentration of cloud condensation nuclei and thus the cloud droplet concentration. But in clean air condition the mineral dust may be important for the absorption characteristic of the cloud droplets due to the content of iron oxides and other mineral compounds, just as it may be important for the absorption characteristics of the aerosols themselves.

Details of the radiation effects of aerosols are rather complex but in a simplifying way we may say that the gasformed particles affect primarily the aerosol number concentration whereas under clean air conditions the mineral dust component may affect the absorption characteristic of the visible light both of aerosols and clouds resulting in changes of the Earth albedo. Increase of the mineral dust content of the atmosphere should therefore under most conditions diminish the albedo and increase the amount of solar radiation absorbed by the Earth.

The discussion of loess formation clearly indicates that apparently considerable changes in atmospheric dust content have occurred during and after the ice age. There are also clear indications that man has increased the erosion rates of soil and modified the plant cover over large parts of the world due to agricultural activities over the last 5000 years or so, and is apparently still continuing to do so. This may be accompanied by large scale natural changes of the extent of the arid zones with corresponding changes of dust production. Turbidity data from meteorological observatories in the USSR, some of which are located in remote areas, apparently not affected up to the present time by industry and air pollution show, for instance, trends towards higher values over the last 30 years. It is possible that this trend is due to a large scale increase of dust production over Eurasia as a result of man's activity (Machta, 1972). It is for these reasons that mineral dust has in recent years attracted the interest of meteorologists and climatologists.

REFERENCES

Füchtbauer, H., and Müller, G. (1970). Sedimente und Sedimentgesteine. *Schweizerbartsche Verlagsbuchhandlung, Stuttgart.* Figures 4–8, p. 141.

Gillette, D. A. (1974). On the production of soil wind erosion aerosols having the potential for long-range transport. *J. Rech. Atmos.*, **8**, 735–744.

Jaenicke, R., and Schütz, L. (1977). A comprehensive study of physical and chemical properties of the surface aerosols in the Cape Verde Islands region. Submitted to *J. Geophys. Res.*

Machta, L. (1972). Mauna Loa and global trends in air quality. *Bull. Am. Meteorol. Soc.*, **53**, 402–407.

Martell, E. A., and Moore, H. E. (1974). Tropospheric aerosol residence times: A critical review. *J. Rech. Atmos.*, **8**, 903–910.

Prospero, J. M., and Carlson, T. N. (1972). Vertical and areal distribution of Saharan dust over the western equatorial North Atlantic Ocean. *J. Geophys. Res.*, **77**, 5255–5265.

Sarntheim, M., and Walger, E. (1974). Der äolische Sandstrom aus der W-Sahara zur Atlantik-Küste. *Geol. Rundsch.*, **63**, 1065–1073.

Schütz, L., and Jaenicke, R. (1974). Particle number and mass distributions above 10^{-4} cm radius in sand and aerosols of the Sahara desert. *J. Appl. Meteorol.*, **13**, 863–870.

Smalley, I. J., and Vita-Finzi, C. (1968). Formation of fine particles in sandy deserts and the nature of 'desert' loess. *J. Sediment. Petrol.*, **38**, 766–774.

SMIC (1971). Report of the Study of Man's Impact on Climate. *Inadvertent Climate Modification.* MIT Press, Cambridge, Mass., 308 pp.

CHAPTER 3

Ecology and Dust Transport

B. LUNDHOLM

ABSTRACT

Erosion is closely correlated with the vegetation cover. The boundary between areas with vegetation and areas without is always changing. Decreased vegetation cover gives increased erosion and dust transport. When human impacts damage vegetation cover the erosion and dust transport is accelerated.

Through erosion valuable nutrients concentrated in the surface are removed and transported away. The transported material with nutrients and organic material which is transported outside the desert areas is trapped in vegetation or in the oceans. This material will increase the productivity both in the terrestrial and marine ecosystems. However, our information about these effects is very limited and pilot research is needed.

The dust transported in the air is also of importance from human health point of view. Particles less than 2 μm are trapped in the lungs and may affect the physiology of the lung.

From pollution point of view arid areas are very different to humid areas. In humid areas the water sediments act as sinks for the pollutants. In arid areas pollutants in the dry fall out may be removed by soil erosion and re-enter the circulation.

In relation to monitoring of the desertification process it is necessary to include variables to follow changes in productivity.

The soils are protected from erosion by the vegetation cover. If this is broken or too thin the erosion can start. The thickness of the vegetation depends mostly on rainfall and temperature and, in tropical and subtropical areas, the rainfall is the determining factor. Is the precipitation too low there will be no vegetation at all. The boundary between areas with vegetation and areas without is always changing. We have seasonal changes, annual changes and in some cases periodical changes or long term changes. The vegetation is well adapted to the seasonal changes. There are also adaptations not only in the vegetation but in the whole ecosystem to the annual changes. Erosion is closely correlated to these changes. I will here elaborate on the effects of these annual changes on the erosion.

The rainfall in arid areas shows a large variation both in time and in space. The

meteorologists have studied the time variation intensively, but they have mostly neglected the variation in space. This space variation with its patchiness gives the same pattern to the vegetation. The wild animals and the nomads have adapted themselves to this patchiness – for this they need a high degree of mobility.

Figure 3.1 shows the conditions in an 'undisturbed' area, that is before modern technology and so-called aid appeared. The upper part of the diagram shows a year with good rains and the lower a dry year. All the time there is moderate grazing, that is not destroying the roots and underground reserves of the plants. During the next rainy year the vegetation will recover. During the dry year there is a moderate increase in soil erosion compared with the rainy year.

Figure 3.2 shows the conditions in an area with overexploitation. During the rainy year there is overgrazing. The grass-roots are damaged and the nutrient reserve is taken away. Trampling is damaging to the vegetation cover and to the soil structure. Agricultural activities may completely remove the natural vegetation cover. Fuel collecting might remove the woody plants. All these activities will result in increased soil erosion, even during the wet year. When the first dry year arrives, there will be a strong increase in soil erosion. At the same time the pressure on the remaining vegetation cover will be harder. A second dry year will complete the catastrophe.

The plants depend not only on moisture, but also on the structure and nutrient content of the soil. In arid areas the moisture transport is towards the surface resulting in a concentration of nutrients in the surface layer. Is this taken away by erosion, the productivity decreases. The most critical nutrient is nitrogen, as the supply is limited. The higher plants get nitrogen from two main sources, from

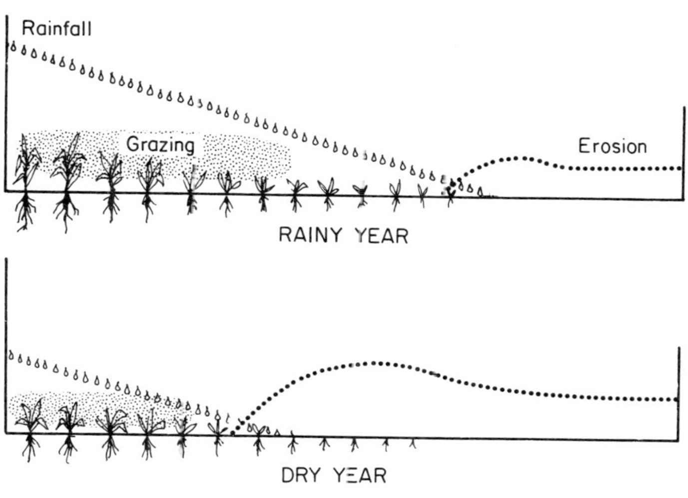

Figure 3.1

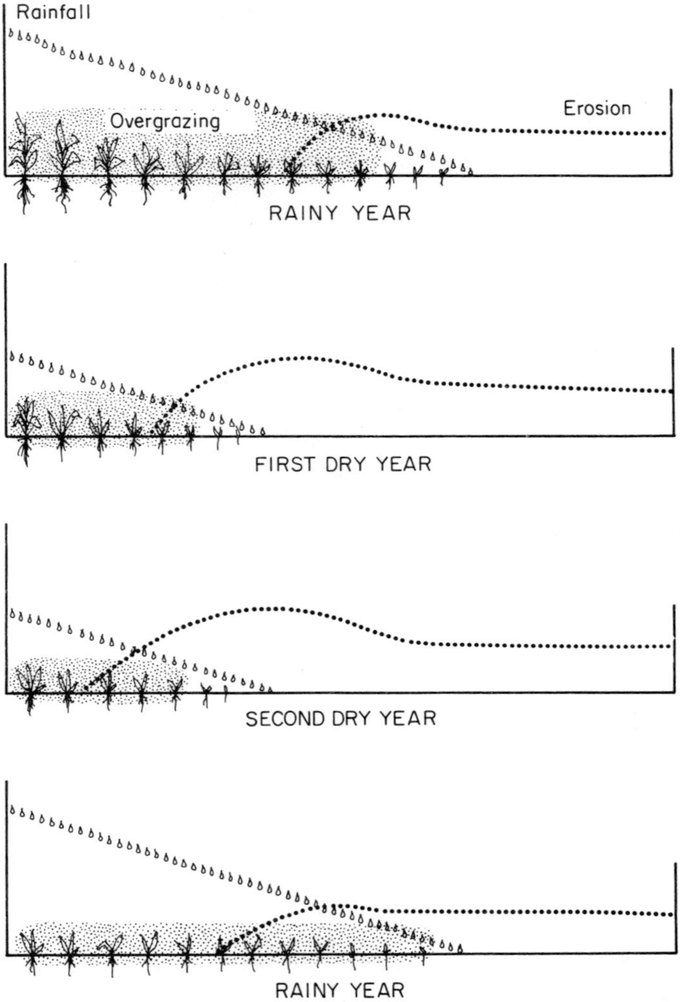

Figure 3.2

mineralization of organic matter and from nitrogen-fixing organisms. In both cases microorganisms are active. Nitrogen fixation in arid areas is done both by bacteria and algae. These organisms form a surface matting. If these algae mats are broken and fragmented and blown away, the nutrient situation is deteriorating. The organic matter in the soil is not only an important source for plant nutrients as nitrogen and phosphorus, but also of importance for the soil structure. Removal of the surface soil with its nutrients and microorganisms is a very serious event.

Also the organic material on the soil surface, wilted plants attached to the soil

with roots are important. If these parts are fragmented by trampling and get loose they are blown away. The result is decreased infiltration ability for rain water and increased erodibility. To sum up, surface erosion is not only removal of the valuable top soil, but it may also have very important consequences for the whole ecosystem.

Figure 3.2 shows also what happens when the good rains come back. We get a vegetation cover, but not so good as earlier and with different species composition. Earlier the perennials dominated now the annuals have taken over. The grazing is not so good any longer. The result depends, of course, on how far the destruction went. The question is if these changes are reversible or irreversible. The answer to the question depends on three factors. The first is the time factor. Within which time frame should the ecosystem recover? Given sufficient time – hundreds or thousands of years – a new similar ecosystem will appear. It is, however, remarkable how quickly eroded areas will recover. The second factor of importance is that no species are lost. If that is the case, a complete recovery is impossible. The dry grazing systems have evolved with large herbivorous mammals as important factors. If these large grazers are taken away, the whole system and its productivity changes. Also other important species may disappear and that may prevent complete recovery. The third consideration relates to the top soil. If this is gone, with its valuable content of nutrients and microorganisms, it takes such a long time to restore the conditions, that the changes might from man's short perspective be regarded as irreversible.

If we consider dust again some material is air-borne and transported over long distances. Figure 3.3 shows this outblow from Africa. Some of the particles settle in the Atlantic. Figure 3.4 shows dust fall over the Atlantic. This dust fall may be important from an ecological point of view. The nutrients to the sea may promote the growth of phytoplankton and contribute to the Atlantic fishing results. Figure 3.5 shows the plankton distribution in the Atlantic. This area of the Atlantic was earlier called 'the dark sea'. The name may have something to do with the African Savanna burning, which nowadays forms a black belt across the southern border of the Sahara, according to satellite images.

Over the continents the dust particles may settle on vegetation and get thus incorporated in the ecosystems. The mechanisms of this incorporation and its ecological importance is not well known or rather unknown. In Europe we are now dealing with a similar problem in relation to the so-called acid rain. We hope to be able to use the ideas and the technique which now is emerging from these European studies and apply it to the African problems.

What is the ecological importance of this wind-blown material? Taking the geological perspective we know that some of the most fertile soils are the loess soils and they are originated from wind-borne material caught in suitable vegetation. During the ice age, wind-transported material from the naked moraines deposited in central Europe. In Asia the Gobi Desert is mother to the fertile loess deposits in China to take a few examples. Since classical times Greece has lost 95% of its fertile soils. Where has this ended up?

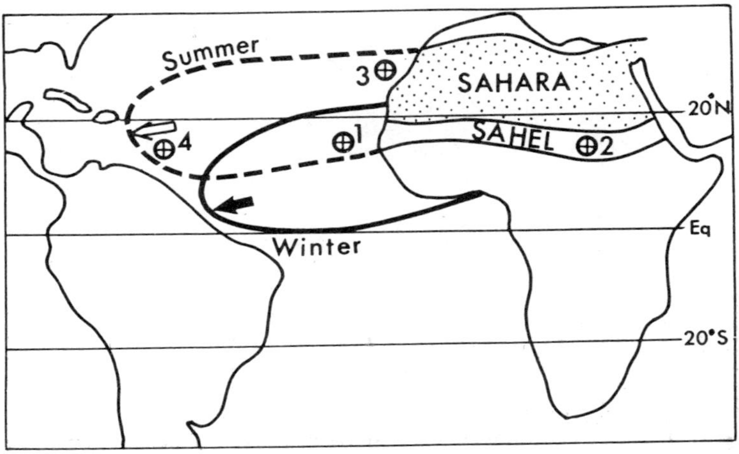

Figure 3.3 Dust transport over the Atlantic (after Rapp, 1974)
1 = Tenerife
2 = Jebel Marra Mountains, Sudan
3 = Cape Verde Islands
4 = Barbados

Most of these wind deposits are close to the origin. In our case we have to look for them in Africa at suitable places. A well-known Swedish traveller has recently told us in a book about a conversation on a Nile boat with an Egyptian. The Swede pointed out that the fertility in the Nile Valley ought to decrease as the fertile sediments are trapped in Lake Nasser (Nubia). The Egyptian replied that this was wrong as 'the Nile Valley is fed by the desert'. To me this reply opens up a new area of research.

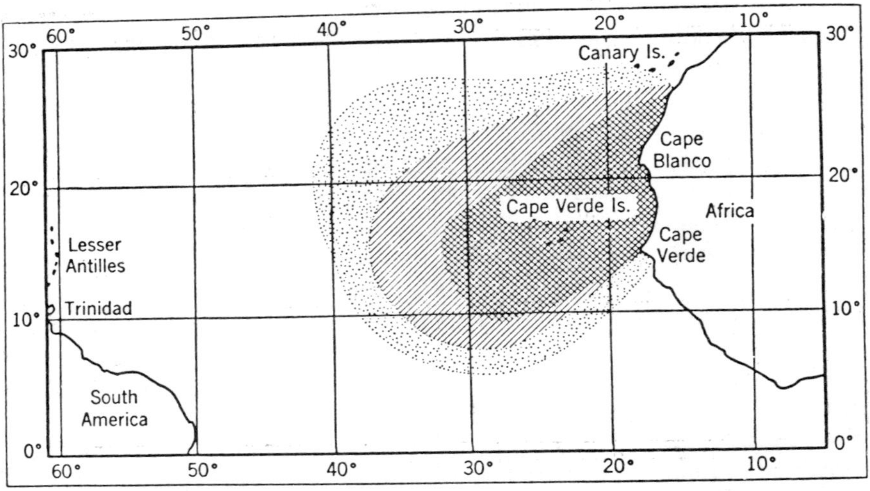

Figure 3.4 Dustfall over the Atlantic (after Kuenen, 1950)

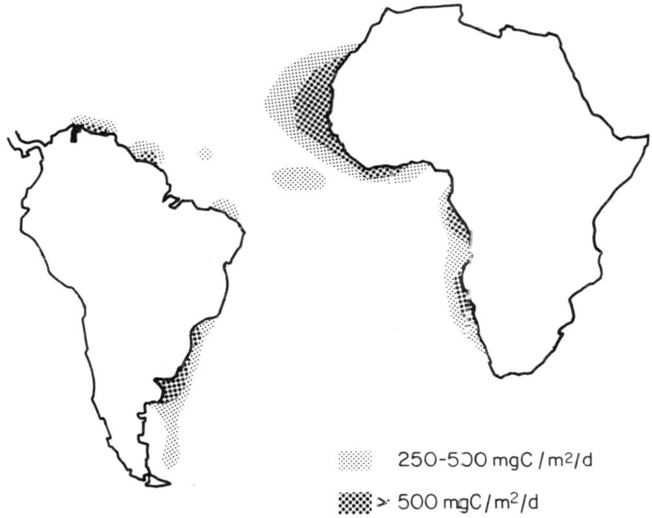

Figure 3.5 Phytoplankton production based on information from *Atlas of the Living Resources of the Sea* (FAO, Rome, 1972)

The vegetation outside and inside the Sahara is without any doubts trapping dust. We do not know to what extent and we can only speculate on its ecological importance. We know from air pollution research that vegetation and forests are acting as a filter for air-borne particles. We know that different types of vegetation have different ability to collect particles. When Professor Goodman worked out his method with moss bags (Goodman *et al.*, 1979) experiments in wind tunnels showed that different sizes of particles acted in a different way.

In the arid areas there are different traps for wind-transported material. Materials moving along the ground are deposited on the leeward side of solid objects such as plants.

The ideal dust traps are the date palms in the oasis. I have in vain tried to find information on the nutrient flow in an oasis. Such a research project, rather easy to carry out, would indicate the importance of dust-blown material. Considering the age of the oasis-culture without decreased productivity, it might be assumed that the nutrient inflow from the air is very important indeed.

It would also be of importance to study those problems when larger areas are involved. There is often an immense soil erosion during dry years from agricultural areas, where the soil has been laid bare. Some of this material is blown in overgrazing lands and might influence the productivity in a positive way. This is, of course, of special importance in Africa south of the Sahara where grasslands border the arid areas.

There is also another type of trap for erosion material. During the dry season soil

containing clay material has a tendency to crack. These cracks may be filled with material moving along the surface. These cracks improve the infiltration of rain water and the structure of the soil.

Dust in the air is also important from another point of view. Particles less than 2 μm are retained in the human lung. Particles more than 5 μm are filtered away in the nose, and particles between 5 and 2 μm are on the way down to the alveols. The airflow through the windpipe is about 150 cm per second but in the alveols it is zero. Here in the alveols the small particles are trapped by different mechanisms. This particle load may influence the physiology of the lung and may be a threat to health. I suppose that many of you have personal experiences of the effect of desert dust. If the particles carry materials or consist of material which is biologically active the impact on human health may be considerable.

As I pointed out earlier there is a tendency that water soluble material concentrates in or on the surface. In dry areas we find surface concentrations of all chemicals often forming salt pans. If this material is distributed and gets air-borne, we might expect a dust which is rather unhealthy. Very little is known about these effects. From the Kalahari area which partly is very overgrazed, a report states that Botswana has the highest death rate in lung diseases in the world.

There is a principal difference between dry and humid areas in relation to pollution in general and especially when air pollution is considered. In Figure 3.6 we have the situation in a humid area. Pollutants emitted by a factory are either washed out by rain or they are removed by dry fall out. These rather complicated mechanisms are now studied in relation to the 'acid rain'. The sink for these pollutants are the wet sediments, in these the pollutants are sealed off and withdrawn from the general circulation. During very special conditions this seal can be broken and the real final sinks are the marine sediments. Figure 3.7 shows emissions in a dry area. Here you have only a dry fallout and deposition in the vegetation. If the vegetation cover is disturbed the pollutants can return into the general circulation. The pollutant levels may thus be steadily increasing.

But are these speculations of any importance in Africa where we have so few industries? There might however be a very real danger in relation to mining

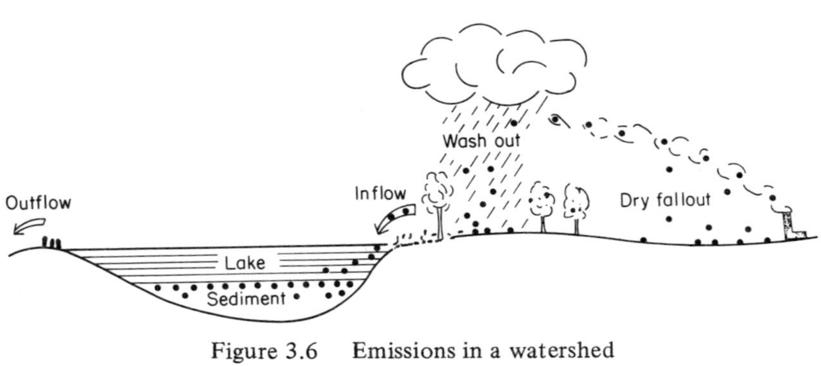

Figure 3.6 Emissions in a watershed

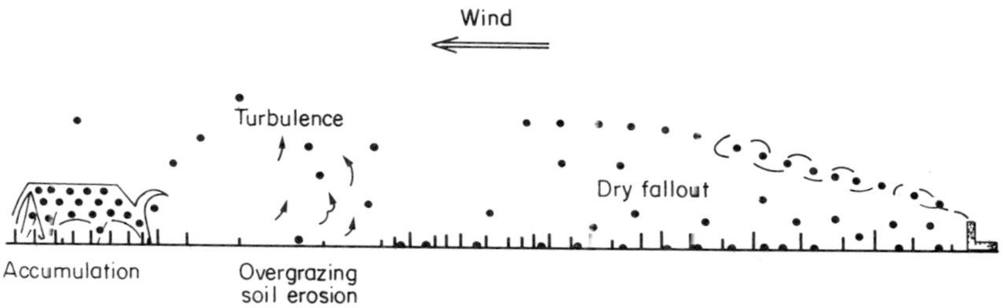

Figure 3.7 Emissions in an airshed

activities. In connection with copper mining there are emissions of both lead and cadmium. Both these metals have a very serious effect on human health. In the copper-mining areas in Zaire and Zambia as well as in the new mines at Selebi Pikwe in Botswana these problems with dry sedimentation in grasslands must be watched.

I am sorry that I have presented my case with so many question-marks and speculations. It is, however, strange that these problems have been so neglected. It would however be fairly easy to find out the fundamentals about the ecological effects. The first task should be to establish the quantity and quality of dry material trapped in the vegetation and its ecological effect. This might be called the *oasis-effect*. This project could also be described as a study of nutrient flow in an oasis.

There are several suggestions to monitor the desert encroachment. I think that it is important not only to measure the physical and vegetational changes, but also to include variables of relevance to the functioning of the whole ecosystem. Such variables are the soil structure (organic material), soil nutrients and micro-organisms.

REFERENCES

FAO Department of Fisheries (eds.) (1972). Atlas of the living resources of the seas. Map No. 1.1, *FAO*, Rome.

Goodman, G. T., Inskip, M. J., Smith, S., Parry, G. D. R. and Burton, M. A. S. (1979). The use of moss bags in aerosol monitoring. In this report.

Kuenen, H. (1950). *Marine Geology*. J. Wiley and Sons, Inc., New York.

Lundholm, B. (1975). Environmental monitoring needs in Africa. In Rickards, P. (ed.). African Environment. *Spec. Rep. 1: 9–24*, Int. Afr. Inst., High Holborn, London.

Raap, A. (1974). A review of desertization in Africa.—Water, Vegetation and Man. *Secr. Int. Ecology*, Rep. 1, Stockholm, Sweden, 77 pp.

SECTION 2

(b) Mobilization

CHAPTER 4

Environmental Factors Affecting Dust Emission by Wind Erosion

D. A. GILLETTE

ABSTRACT

Factors which influence the production of dust in the Sahara may be described in terms of aerodynamic and soil physical characteristics. Work on the threshold wind velocity for soil erosion, on total soil movement by wind, and on dust production is reviewed. An extremely important factor in desert environments is the aerodynamic partitioning of wind stress by nonerodible elements and erodible soil. The factors influencing dust production are presence of nonerodible elements such as vegetation and rocks, vegetative residue, surface roughness, aggregate structure of surface soil, soil moisture, soil mineralogy and texture, and wind stress.

4.1 INTRODUCTION

Dust production is affected by a wide variety of factors both natural and man-made. I will attempt to summarize some of the physical factors involved in dust emission but will not assign them to natural or man-made causes since in most cases they could be caused by either. I will also emphasize that part of the total transported soil material which can travel far from its place of origin. In order to travel long distances, dust must have a small settling velocity compared to the root mean square vertical velocity fluctuations of the supporting air. Fine dust, i.e. that having a potential for long-distance travel for almost all soil eroding winds, is generally smaller than about 20 μm (0.02 mm) in diameter.

Threshold velocity is one of the most important parameters of wind erosion since it is that parameter, along with the frequency distribution of wind speed, that determines the frequency of wind erosion of soil. The movement of soil by the wind is the other important characteristic of wind erosion that I will discuss. Soil movement is divided into the horizontal flux of sand (which describes most soil mass movement) and vertical flux of dust smaller than 0.02 mm (which describes a large part of permanent soil loss). I will discuss separately the aerodynamic and soil factors as they relate to threshold velocities and soil movement.

4.2 AERODYNAMIC FACTORS IN FINE DUST PRODUCTION

4.2.1 Threshold velocities for soil movement

In general, threshold velocities for soil movement are those velocities in which aerodynamic forces are sufficient to dislodge particles from the soil and initiate movement. Of course, this velocity is dependent on both the aerodynamic forces and the forces holding the particle in the soil. Theoretical studies have considered simple soil systems and idealized particles. Such studies typically equate moments due to lift, drag, and weight.

Experimental studies of threshold velocities for simple soil systems consisting of beds of loose, monodisperse, and similar particles are reported by Bagnold (1941), Ishihara and Iwagaki (1952), Chepil (1951), and Greeley *et al.* (1973). For the sake of illustration, the results* of Chepil (1951) are shown in Figure 4.1. It is to be noted that there is a minimum friction velocity that will produce motion in particles of diameter of about 100 μm. Particles larger than 100 μm require greater wind speed, presumably because of their greater weight, while smaller particles are lower in the boundary layer and require larger pressure fluctuations to initiate their

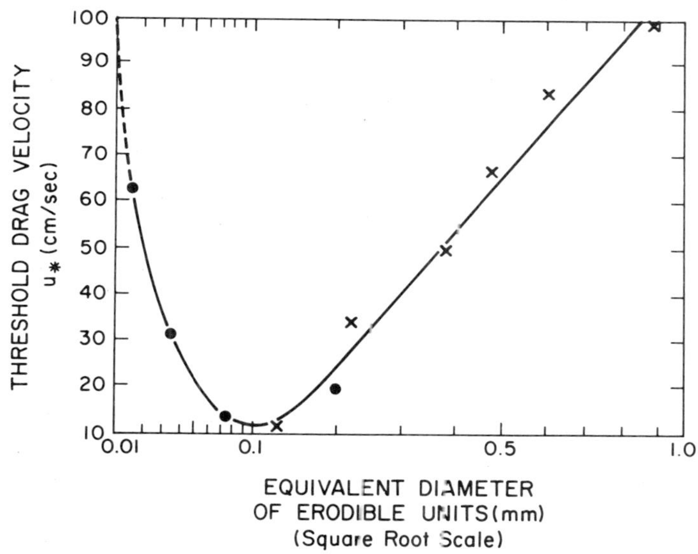

Figure 4.1 Threshold friction velocity *vs.* monodisperse particle size (after Chepil, 1951)

*Wind speeds are reported in terms of friction velocity $u_* = c_D^{1/2} u_1$ where u_1 is the mean wind at height 1, usually within a metre or two of the surface, and c_D is the drag coefficient for that height and that particular surface. For neutrally stratified air, the mean wind profile near the ground takes the form $u_2 - u_1 = (u_*/K) \ln (Z_2/Z_1)$, where u_2 and u_1 are mean wind speeds at heights Z_2 and Z_1 and K is von Kármán's constant.

movement. The main conclusion to be drawn from such data is that if loose particles are available in the soil, 100 μm particles require the lowest velocities for initiation of movement. Once particle movement has begun, momentum for downward soil movement is more effectively delivered by the soil particles colliding with the surface than by aerodynamic transfer. Natural soils, however, are rarely characterized by monodisperse particles and are rarely found in perfectly loose beds. Nonetheless, erosion is often initiated by a small quantity of loose 100 μm particles on a soil surface.

The effect of nonerodible roughness elements such as bushes or pebbles is to absorb some of the momentum being transported to the ground by the wind and thus to decrease the momentum felt by individual particles. Marshall (1971) and Lyles and Allison (1976) have made studies where the increase of threshold wind velocity due to nonerodible roughness could be calculated. Marshall measured partitioning of the momentum of roughness elements and the floor. This study was concerned mostly with the effects that the large-scale roughness elements such as bushes and boulders have in absorbing wind force. His roughness elements were cylinders having a diameter-to-height ratio of 0.5 to 5, as well as hemispheres. Lyles and Allison were concerned with the effect of standing stubble on threshold

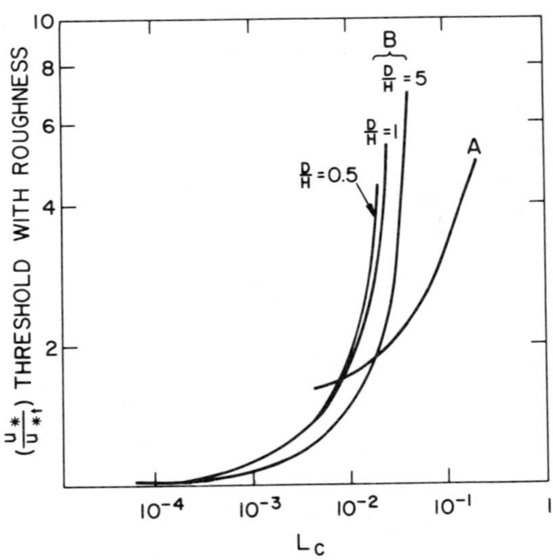

Figure 4.2 Ratio of threshold velocity with nonerodible elements present to that with none present. L_c is the ratio of silhouette area of one nonerodible element to the area of the floor divided by the number of nonerodible elements on the floor. A is the data of Lyles and Allison (1976), B is the Marshall data for elements with diameter to height ratios (D/H) of 0.5, 1, and 5

velocity and so in general were interested in much smaller diameter-to-height ratios. I have expressed the fractional increase of threshold velocity as a function of L_c which is a nondimensional ratio of silhouette area of a roughness element to the floor area occupied by one roughness element. The results are shown in Figure 4.2. In this figure u_*/u_{*t} is the ratio of threshold velocity with roughness elements to that without roughness elements, and D/H is the diameter-to-height ratio of the roughness elements. Both studies show that the presence of nonerodible elements is extremely important in the prevention of erosion.

4.2.2 The relationship of fine particles to coarse particles suspended by wind erosion

Since both fine and coarse particles are present in the soil, the emission rates of both size ranges of particles as well as the length of time they remain in the air must be considered. Figure 4.3 shows size distributions reported by Gillette and Walker (1977) of particulate mass in parent soil, air-borne particles in the first 1.3 cm above the ground, and at 1 m, for two soils at two different wind speeds. The size distribution of the particles in the first 1.3 cm for all cases is highly similar to the size distribution of the loose particles in the parent soil smaller than 0.4 mm. The size distribution very close to the ground thus reflects the availability of particles from the soil. On the linear scale used in these plots, the proportion of mass smaller than 0.02 mm is very small. At 1 m, soil I (a sand soil) shows that for both wind speeds the proportion of mass in particles greater than 0.04 mm is reduced so that a mode of particle size smaller than 0.02 mm (between 0.002 and 0.02 mm) is now important. Gillette and Goodwin (1974) showed that this change of proportion with height is due to the large sedimentation velocities of particles greater than 0.02 mm. Similarly for soil II (a loamy sand soil) the proportion of particles greater than 0.02 mm is greatly reduced from near the ground to 1.5 m, although in this case the proportion of fine to coarse particles increases with wind speed.

The enormous change in the proportion of fine to coarse particles reflects the much greater heights and distances that fine particles may be transported compared to coarse particles. A size distribution of air-borne dust, generated by wind erosion in Texas and western New Mexico and collected by aircraft in southwestern Missouri (Gillette, in preparation), is shown in Figure 4.4. The distance to the sampling location from the eroding areas, indicated in the figure, is about 800 km. The outline of the dust cloud from satellite photographs is shown, along with wind speed vectors. The size distribution, which also exhibits evidence of sulphur-containing particles ($d < 0.002$ mm), shows the dominance of particles smaller than 0.02 mm in dust carried great distances.

An explanation for the diminishing proportion of coarse to fine particles with increasing distance or height can be found in the probability distribution of the turbulent vertical air velocity. The distribution has a mean of zero, and near the ground is gaussian, with a standard deviation approximately equal to the friction velocity of the air (Lumley and Panofsky, 1964), a speed which is not referenced to

a specific height above ground. Since particles having a finite settling (downward) velocity must be supported by the upward fluctuations of the wind, a particle must have a favourable ratio of upward to downward motions in order to remain suspended.

By adding the settling velocity to the vertical air fluctuations, effective particle fluctuations result. The ratio of upward to downward motions of a particle carried in air having a normal vertical velocity distribution with zero mean and standard deviation u_* is shown in Figure 4.5. For reference, sedimentation velocities versus size (after Bagnold, 1941) are plotted in the figure. For a sedimentation velocity $V_{sed} = 0.4\ u_*$ the ratio of upward to downward motions is 0.5; in other words, for every upward motion there are two downward motions. The probability that such a particle will rise very high is thus rather small. Particles smaller than 0.02 mm are sufficiently small that their sedimentation velocities are less than 0.1 u_* for virtually all eroding winds (Gillette *et al.*, 1974).

4.2.3 Movement of soil particles as a function of wind speed

The flux of particles through a surface of unit width and infinite height that is mutually perpendicular to the ground and to the wind may be represented as 'q'. In simple terms we may think of q as the quantity of particles leaving an area of interest per unit width perpendicular to the wind. Several investigators have considered q as a function of wind speed greater than the threshold wind speed. Bagnold (1941) used dimensional arguments along with the assumption that all the momentum is transferred to the surface by collision of the air-borne sand grains (having positive momentum) with the surface in a jumping movement he called 'saltation'. His expression for q as a function of friction velocity of the wind is $q \propto u_*^3$. Other authors also give q which tends to be a function of u_* to the third power [Hsu, 1971; Kawamura, 1951; O'Brien and Rindlaub, 1936 (as reported by Horikawa and Shen, 1960); Belly, 1964]. A plot of the present authors' values for q *vs.* u_* for several different soils having a similar threshold shows a relationship $q \propto u_*^2\ (u_* - u_{*thresh})$ which tends to $q \propto u_*^3$ for large u_* (Gillette, 1974). The movement of air-borne soil particles is, however, dependent on soil type, since as seen in Section 4.2.1, non-erodible aggregates or roughness elements absorb varying fractions of the horizontal momentum delivered by the wind. In that case the dependence of q on u_* is altered greatly since only a fraction of the energy of the wind is being used to move soil particles. Also the agreement of soil movement as $q \propto u_*^3$ is probably due to the dominance in the size distributions of air-borne particles of sand-sized material. The derivation of Bagnold depends on air-borne sand grains striking the surface in the saltation mode of movement. If the bulk of the material did not move in saltation it is doubtful that the $q \propto u_*^3$ relationship would hold. Since, however, sand-sized grains have the lowest threshold velocity and make up the largest fraction of air-borne soil particles where erosion takes place, the approximation $q \propto u_*^3$ for large u_* seems to be good for loose soils having few nonerodible roughness elements.

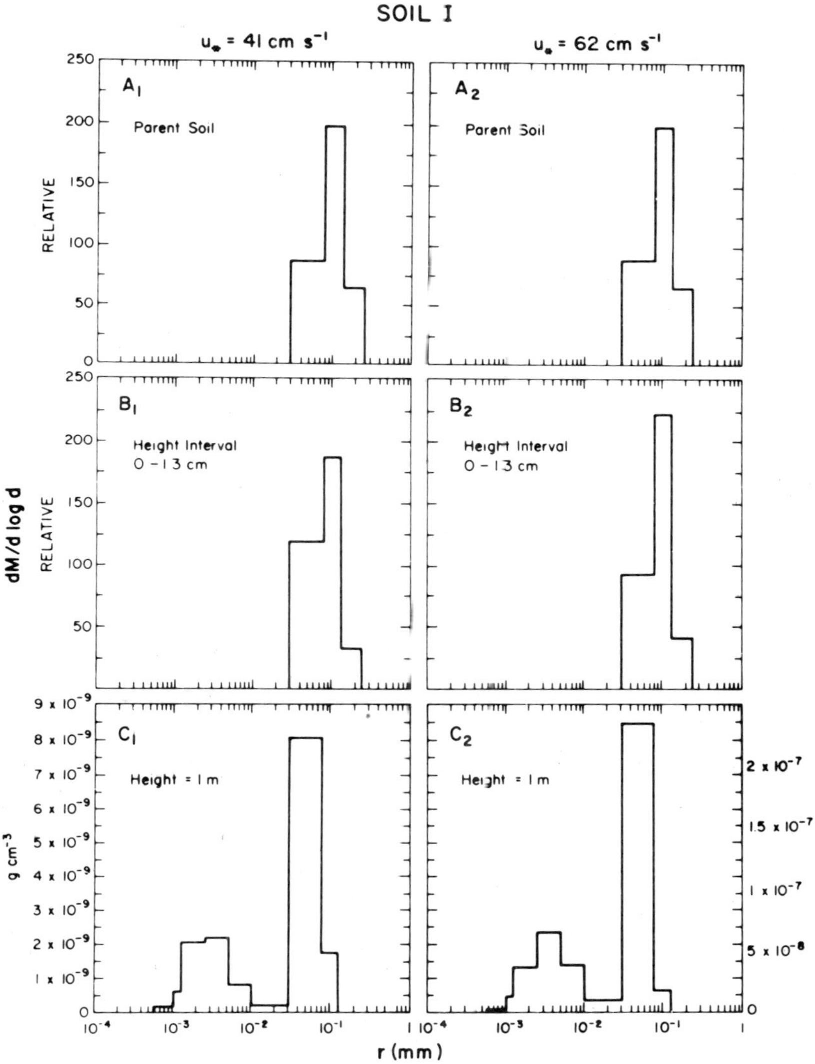

Figure 4.3 Size distributions of soil particles eroded from two soils. Soil I has a surface texture of fine sand and soil II has a surface texture of loamy fine sand. For each soil and each friction velocity 1 and 2, A is the relative size distribution of the dry parent soil, determined by wet sieving; B is the relative size distribution of

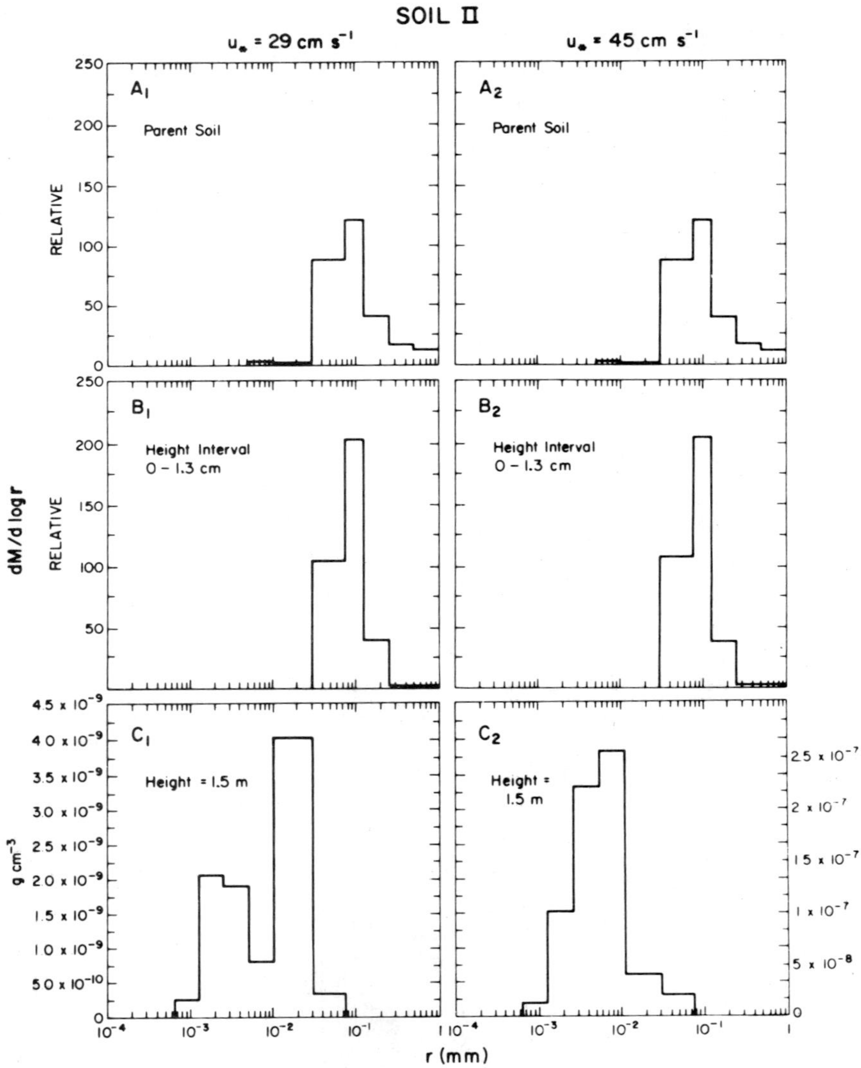

particles moving between the heights 0–1.3 cm above ground, and C is the mass size distribution for air-borne particles at 1 m above ground (after Gillette and Walker, 1977)

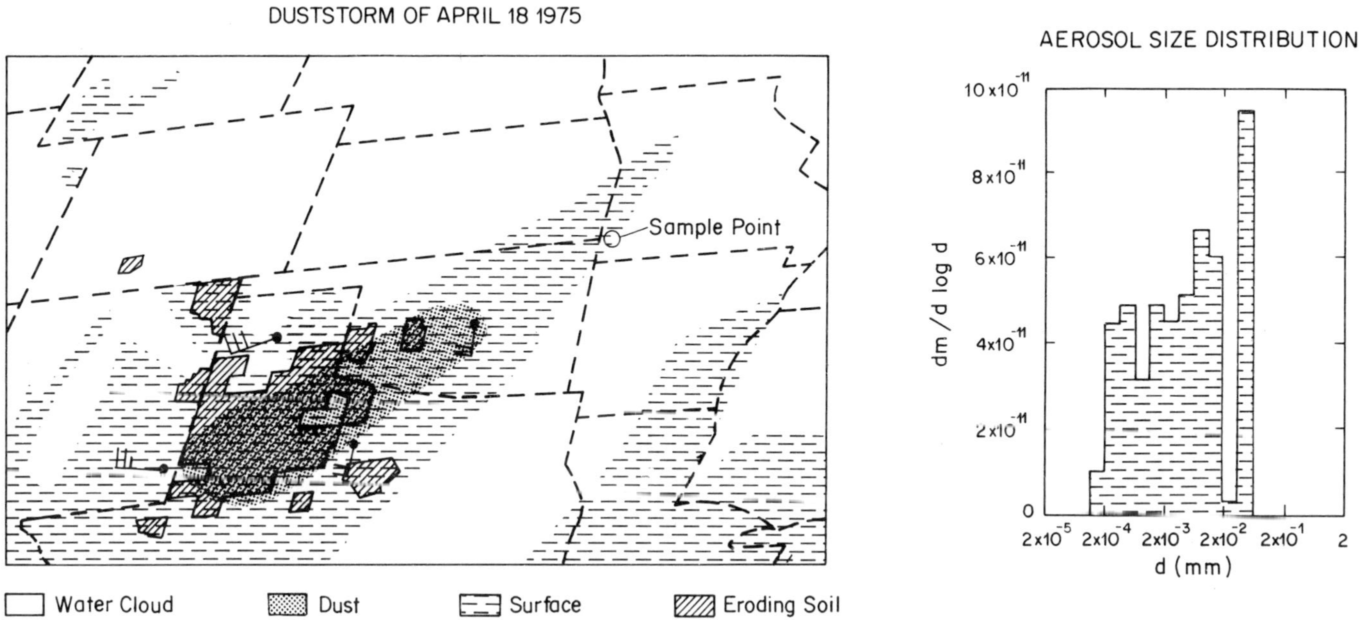

Figure 4.4 Size distribution of soil wind erosion particles carried a great distance. The dust was collected by aircraft. The production area of the dust is shown as the area of eroding soil in the figure (after Gillette, in preparation)

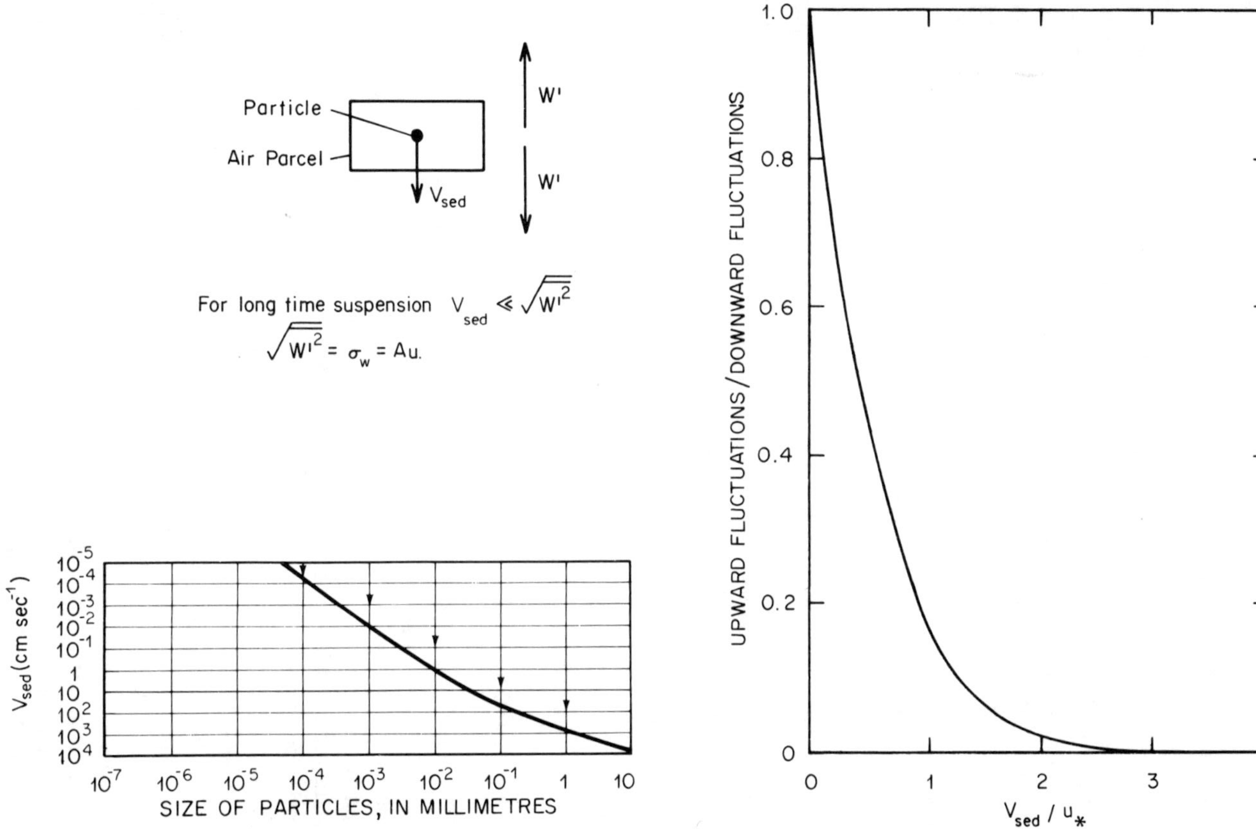

Figure 4.5 Sedimentation velocity V_{sed} compared to vertical velocity fluctuation w'; upward motions divided by downward motions for a particle having sedimentation velocity V_{sed} in air having vertical velocity fluctions with mean = 0, standard deviation u_*; sedimentation velocity *vs.* particle size (this part of figure after Bagnold, 1941); (the entire figure after Gillette, submitted to transactions off ASAE)

4.3 SOIL FACTORS IN DUST PRODUCTION

4.3.1 Soil factors affecting threshold velocity

Since threshold velocity is the minimum velocity at which the aerodynamic lift and drag is equal to the soil forces holding particles together, soil binding should be expected to have a major effect on threshold velocity. That is, a soil having greater coherence forces would require larger wind forces to set it in motion. Van der Waal's forces have been considered by Iversen *et al.* (1976) but little is known about the stronger effects of clay aggregation and water bonding. Chepil (1957) discussed erodibility in terms of soil moisture; he concluded that soil moistures near the wilting point of plants (i.e. approximately soil moisture content at 15 atmospheres of negative pressure) are ineffective in holding soil particles together. Chepil's correction for the presence of water will be given in the next section 4.3.2.1. Smalley (1970) stated that erodibility depends on cohesiveness of the soil, which may be measured in terms of its tensile strength. In Smalley's model of a simple soil, tensile strength is related to the packing density, the coordination number of the particles, and the interparticle bond strength. The tensile strength was shown in the simple soil system to be inversely proportional to the cube of the particle diameter, which suggests that very fine soils are less erodible than coarse soils; this is indeed observed.

Bisal and Ferguson (1970) investigated the effect of non-erodible aggregates on threshold wind velocity (in this case wind velocity at 30.5 cm above the soil surface). Their empirical relationship was

$$\ln V_T = 6.0438 + 0.02332\, C$$

where V_T is the threshold velocity in cm s^{-1} at 30.5 cm, and C is the percentage of soil mass in aggregates larger than 1 mm in a sample.

The classification by the US Department of Agriculture of soil texture classes into wind erosion groups also shows that soils having surface aggregates around 100 μm are more erodible than soils having aggregates of larger size. If the same wind erosion forces act on different soils, the more erodible soil will have a lower threshold velocity since the maximum soil transport at a given speed will be determined by aerodynamic factors. Thus more soil erosion implies a lower threshold velocity and more frequent erosion events. Table 4.1 (after Lyles, 1976, using data from Hayes, 1972) shows the magnitude of erosion for surface soil textural classes. Referring the textual classes to wind erosion, one notices that coarse soils are the most easily eroded, followed by clay soils (due to their tendency to form aggregates of sand size), calcareous loamy soils, and finally the soils of wind erosion groups 5, 6, and 7 (fine soils which form stable large aggregates). The erodibility of calcareous soils is quite interesting, showing chemical effects in the erodibility of soils.

TABLE 4.1 Descriptions of Wind Erodibility Groups (WEG)[a] (After Lyles, 1976)

WEG	Predominant soil textural class	Dry soil aggregates >0.84 mm	Soil erodibility 'I'
		Percent	Metric tons/ha/yr
1	Very fine, fine, and medium sands; dune sands	1	696
2	Loamy sands; loamy fine sands	10	301
3	Very fine sandy loams; fine sandy loams; sandy loams	25	193
4	Clays; silty clays; non-calcareous clay loams and silty clay loams with more than 35 per cent clay content	25	193
4L	Calcareous loams and silt loams; calcareous clay loams and silty clay loams with less than 35 per cent clay content	25	193
5	Noncalcareous loams and silty loams with less than 20 per cent clay content; sandy clay loams; sandy clay	40	126
6	Noncalcareous loams and silt loams with more than 20 per cent clay content; non-calcereous clay loams with less than 35 per cent clay content	45	108
7	Silts; noncalcareous silty clay loams with less than 35 per cent clay content	50	85

[a]Data from Hayes, 1972.

4.3.2 The effect of soil characteristics on the movement of soil

4.3.2.1 Total soil movement as affected by soil characteristics

Chepil and Woodruff (1959) related the total soil mass moved by a wind tunnel (χ) which produced winds of a given u_* (78 cm s^{-1}) to the average depth of furrows (the ridge roughness R), vegetative residue V, and an erodibility index I based on the percentage of soil mass in aggregates greater than 0.84 mm.

$$\chi = 400 \frac{I}{(RV)^{1.26}}$$

where R is measured in inches, V is measured in lb/acre, and I is given in Table 4.2. The present author modified a formula (Gillette, submitted to the ASAE) from Chepil (1957) to give the total soil flux 'q' through a surface of unit width that is mutually perpendicular to the wind and to the soil surface.

$$q = 0.16 \frac{LX}{16500} \frac{u_*^3}{78} \text{ (g cm}^{-1}\text{ s}^{-1}\text{)}$$

$$\text{for } \frac{LX}{16500} \leqslant 1$$

$$= 0.16 \frac{u_*^3}{78}$$

$$\text{for } \frac{LX}{16500} > 1$$

In these formulas L is the length of eroding soil parallel to the wind (in feet) and u_* is the corrected friction velocity (cm s^{-1}). X is defined above and the correction for the friction velocity for moisture content is

$$\rho u_*^2 = (\rho u_*^2)_{\text{observed}} - 6\left(\frac{W}{W'}\right)^2$$

where W is the amount of water held by the soil and W' is the amount of water held by the same soil at 15 atmospheres of negative pressure (Chepil, 1956).

The present author (submitted to ASAE) compared measurements of total soil flux to those calculated from the above formula. The calculated values were compared to observed fluxes in two categories: steady erosion and intermittent erosion. Since intermittent erosion does not meet the assumptions of the Bagnold (1941) formula (in which momentum is transported at the lower boundary by saltating sand grains), intermittent erosion was not expected to show agreement between observed and predicted mass fluxes. For steady erosion, agreement was good:

$$q_{\text{obs}} = 1.0005\, q_{\text{p}}^{1.08} \text{(corr. coeff.} = 0.84\text{)}$$

For intermittent erosion, however, agreement was poor:

$$q_{\text{obs}} = 0.58\, q_{\text{p}}^{1.49} \text{(corr. coeff.} = 0.70\text{)}$$

4.3.2.2 *The vertical flux of particles which may be transported great distances as affected by soil characteristics*

Since particles which may be carried great distances are a part of the total air-borne soil particulate suspension, we must consider that the factors affecting total soil movement also affect fine particulate emission. Consequently it is con-

TABLE 4.2 Soil Erodibility Index, I, Based on Percentage of Soil Fractions Greater than 0.84 mm in Diameter as Determined by Dry Sieving (after Chepil and Woodruff, 1959)

Percentage of soil fractions greater than 0.84 mm:	Units									
Tens	9	8	7	6	5	4	3	2	1	0
9										0.02
8	0.02	0.03	0.03	0.04	0.05	0.06	0.07	0.08	0.09	0.10
7	0.12	0.14	0.16	0.20	0.25	0.30	0.35	0.40	0.45	0.5
6	0.55	0.60	0.65	0.70	0.75	0.80	0.85	0.90	0.95	1.0
5	1.1	1.2	1.3	1.4	1.6	1.8	2.0	2.2	2.5	2.8
4	3.1	3.4	3.7	4.0	4.3	4.9	4.9	5.2	5.6	6.0
3	6.5	7.0	7.5	8.0	8.5	9.0	9.5	10.5	11.0	12.0
2	13	14	15	16	17	18	19	20	21	23
1	25	27	30	33	36	39	43	17	51	55
0	70	80	100	120	150	175	280	450	1000	–

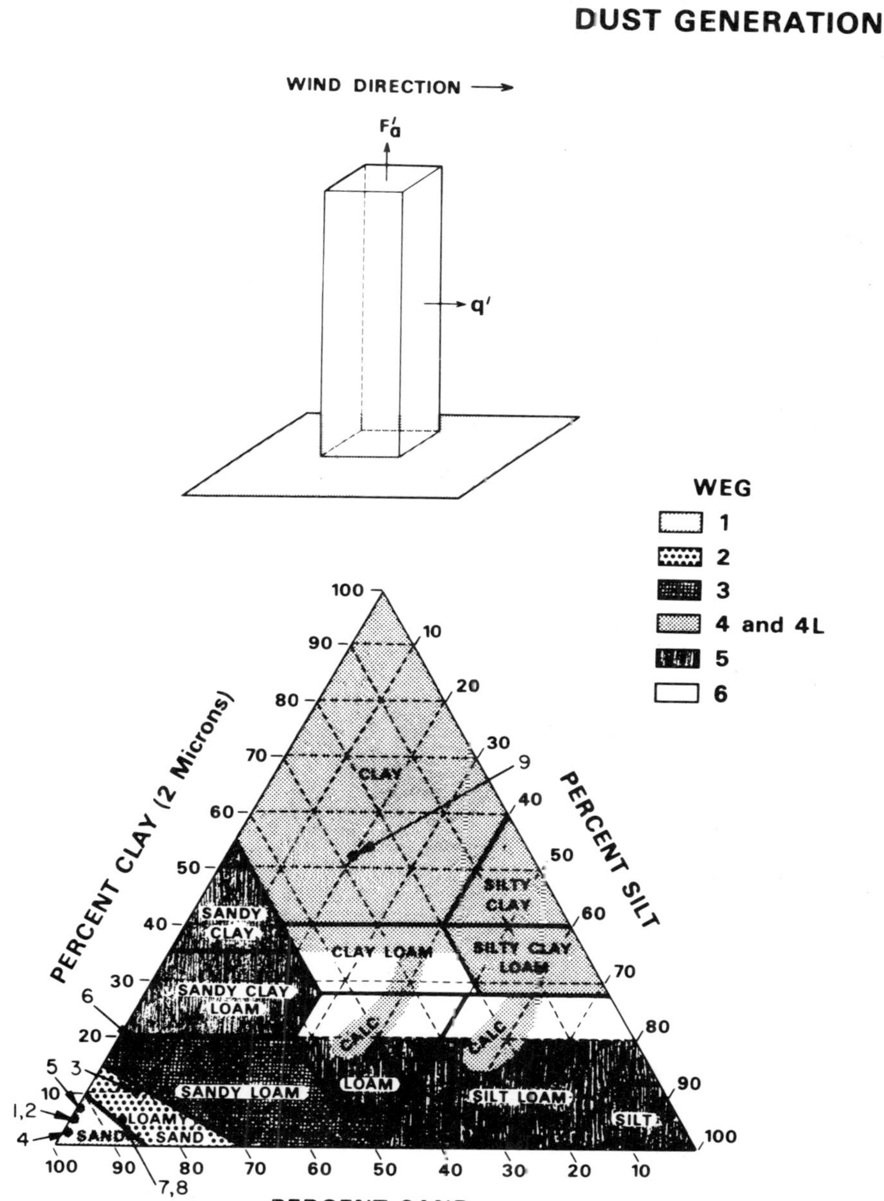

Figure 4.6 Textures of sampled soils; total soil movement *vs.* wind friction velocity; vertical flux of particles smaller than 0.02 mm *vs.* wind friction velocity; ratio of vertical flux of particles smaller than 0.02 mm to total soil movement per

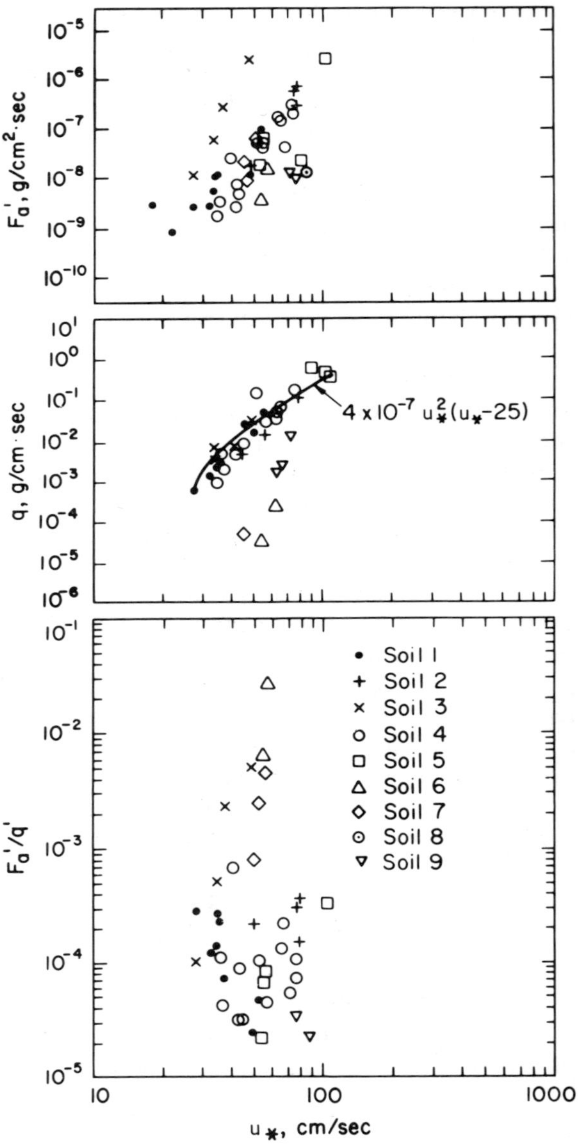

unit area per time versus wind friction velocity (after Gillette, submitted to transactions of ASAE)

venient to speak of the ratio of fine particles to total particles or of the ratio of the vertical flux of particles $d < 0.02$ mm (seen in a previous section to include most of the particles having the potential for long-range transport) to the total flux q. Ratios of the vertical flux of fine particles, F'_a to the horizontal flux of all air-borne soil particles*, q', were given by Gillette (submitted) and are shown in Figure 4.6. Nine different soils are shown in the soil textural diagram. Also included in the figure is a schematic diagram of F'_a and q' as well as F'_a *vs.* u_*, and q *vs.* u_*. Physical characteristics for the nine soils are given in Table 4.3. Soils 1, 2, 4, and 5 are sandy; soils 3, 6, 7, and 8 are loamy; and soil 9 is clay. The sandy soils show a fairly uniform trend of increasing F'_a with u_*, probably due to the uniformity of the dry aggregate structure of these soils. F'_a *vs.* u_* for the loamy soils shows a greater scatter, probably due to their widely different dry aggregate structures; the percentage of their mass in aggregates < 0.84 mm varied from 38% to 89%, which, in turn, affected q. Chepil and Woodruff (1963) showed that a greater percentage of mass in aggregates larger than 0.84 mm correlated with decreased wind erosion. Soil 9 was a clay soil whose eroding particles consisted of strongly layered pellets formed by shrinkage of a montmorillinitic clay upon drying from a mud state. These clay pellets were extremely resistant to breakage by hand when compared to the aggregates of the loamy soils and the sandy soils. This soil had a $u_{*thresh}$ of about 65 cm s^{-1} probably due to the large size of the aggregates (only 9.3% of the loose surface pellets were smaller than 0.84 mm). This high $u_{*thresh}$ and breakage-resistant aggregates help to explain the low values of F'_a for this clay soil. The largest values of F'_a for a given u_* are those for soil 3, a loamy soil with very little surface coherence. In general, the increase of fine particulate production (F'_a) with wind speed is greater than the linear increase of wind speed.

Values of F'_a/q' for sandy soils show great scatter but little evidence of trend with wind friction velocity u_*. For loamy soils there is a great increase with u_*, which is tantamount to an increase with wind speed in the proportion of fine ($d < 0.02$ mm) to total soil movement. The value of F'_a/q' for the clay soil is very low, showing the effect of the organized structure of the clay minerals on drying and shrinking in resisting breakage into fine particles. In general, soils having finer textures produced more fine dust for unit soil movement except for that fine-textured soil whose mineral components formed small aggregates hard enough to resist impact breakage at mean wind speeds up to 25 m s^{-1}.

4.3.3 Moisture and wind history of the soil

The moisture and wind history of the soil, including freezing and thawing and drought accompanied by frequent strong winds, influences the breakage of soil

*q' differs from q in that q' is the flux of particles into a surface 1 cm wide, perpendicular to the wind and ground with vertical dimensions from the surface to 76 cm. q is the flux into a surface 1 cm wide, mutually perpendicular to the wind and ground with vertical dimensions from the ground to infinity. Practically speaking, however, most air-borne soil mass is confined to heights less than 50 cm and the approximation $q' = q/76$ is good. The dimensions of q' are g/cm^2 sec whereas the dimensions of q are g/cm sec.

TABLE 4.3 Soil Parameters

Soil erosion parameter	Soil 1	Soil 2	Soil 3	Soil 4	Soil 5	Soil 6	Soil 7	Soil 8	Soil 9
Soil moisture, %	0.52 ± 0.61	0.99	1.29 ± 0.20	0.41 ± 0.10	0.52 ± 0.13	0.75	0.6	0.6	6.6
Cloddiness, % dry									
Aggregate <0.84 mm	95.0 ± 2.78	98.9	89.1 ± 5.0	95.9 ± 1.6	98.8 ± 0.6	36.7	60.0	53.0	9.3
Vegetative residue, g/m^2	26.67	8.25	3.67	91.6	19.1	3.5	161.0	39.0	2.9
Ridge roughness, cm	2.5	2.5	2.5	3.7	2.5	5.0	5.0	22.5	2.5
Erosion fetch, km	1.6	0.8	1.6	1.6	0.2	0.5	0.5	0.5	0.1
Soil Texture									
% Sand, $r > 25\ \mu m$	96.0	95.5	81.5	96.8	93.1	77.7	88.0	88.0	28.0
% Silt, $1 < r < 25\ \mu m$	0.5	1.0	8.5	1.4	1.0	3.3	3.2	3.2	20.0
% Clay, $r < 1\ \mu m$	3.5	3.5	10.0	1.8	5.9	19.0	8.8	8.8	52.0

aggregates into small erodible units which change the initial wind speed for soil movement ($u_{*thresh}$) as well as the horizontal soil flux (q) as a function of wind speed. Although it is not certain that the ratio F_a'/q' changes with weathering, it may change for certain soils.

4.4 CONCLUSIONS

Both aerodynamic and soil factors must be considered in studying dust emission by the wind. For the Sahara I must stress the importance of non-erodible elements and soil aggregation although other important factors have also been discussed. It is quite possible that human activities have a greater effect on soil surface condition and non-erodible elements than do other environmental factors.

The threshold wind velocity for wind erosion and movement of soil are affected by soil textures, wind speed, mineralogy, soil moisture, history, soil roughness, vegetative residue, and the presence of vegetation and other nonerodible elements such as rocks.

Fine particles are produced by the sandblasting effect of saltation, which acts to disaggregate fine particles on the surfaces of larger particles and by 'splashing' of the saltating particle into a reservoir of fine particles. Fine-textured soils tend to emit more fine particles upon unit soil movement unless the mineralogical structure of the soil aggregates is highly resistant to breakage. Total soil movement measurements show good agreement with values computed from formulae derived from ARS studies and this author's data.

4.5 ACKNOWLEDGEMENTS

This article was adapted from a longer review article 'Production of dust which may be carried great distances' by the same author, which was presented at the American Association for the Advancement of Science Annual 1977 meeting and which was submitted as a contribution to the Desert Dust Symposium volume.

REFERENCES

Bagnold, R. A. (1971). *The Physics of Blown Sand and Desert Dunes*. Methuen, London, 265 pp.

Belly, P. (1964). Sand Movement by Wind. *U.S. Army Tech. Memo.*, **1**, U.S. Army Corps of Engineers. 38 pp.

Bisal, F., and Ferguson, W. (1970). Effect of non-erodible aggregates and wheat stubble on initiation of soil drifting. *Can. J. Soil. Sci.*, **50**, 31–34.

Chepil, W. S. (1951). Properties of soil which influence wind erosion, 4, state of dry aggregate structure. *Soil Sci.*, **72**, 387–401.

Chepil, W. S. (1956). Influence of moisture on erodibility of soil by wind. *Soil Sci. Soc. Proc.*, **20**, 288–292.

Chepil, W. S. (1957). Width of Field Strips to Control Wind Erosion. *Tech. Bull.*, **92**, Kans. State Coll. Agric. Appl. Sci., 16 pp.

Chepil, W. S., and Woodruff, N. P. (1959). Estimations of Wind Erodibility of Farm Fields. *Prod. Res. Rep.*, **25**, U.S. Dep. Agric., 21 pp.

Chepil, W. S., and Woodruff, N. P. (1963). The Physics of Wind Erosion and its Control. *Adv. Agron.*, **15**, 1–301.

Gillette, D. A. (1974). On the production of soil wind erosion aerosols having the potential for long-range transport. *J. Rech. Atmos.*, **8**, 735–744.

Gillette, D. A., and Goodwin, P. A. (1974). Microscale transport of sand-sized soil aggregates eroded by wind. *J. Geophys. Res.*, **79**, 4080–4089.

Gillette, D. A., Blifford, I. H., Jr., and Fryrear, D. W. (1974). The influence of wind velocity on the size distributions of aerosols generated by the wind erosion of soils. *J. Geophs. Res.*, **79**, 4068–4079.

Gillette, D. A., and Walker, T. R. (1977). Characteristics of air-borne particles produced by wind erosion of sandy soil, high plains of west Texas. *Soil Sci.*, **123**, 97–110.

Greeley, R., Iverson, J. D., Pollak, J. B., Udovich, N., and White, B. (1974). Wind tunnel studies of Martian aeolian processes. *Proc. R. Soc. London, Ser. A.*, **341**, 331–360.

Hayes, W. (1972). Designing wind erosion control systems in the midwest region. *RT SCS-Agron. Tech. Note LI-9*, Soil Conserv. Serv., U.S.D.A., Lincoln, Nebr., 12 pp.

Horikawa, K., and Shen, H. W. (1960). Sand movement by wind action (on the characteristics of sand traps). *U.S. Army Corps of Engineers Tech. Mem.*, **119**, 51 pp.

Hsu, S. (1971). Wind stress criteria in eolian sand transport. *J. Geophys. Res.*, **76**, 8684–8686.

Ishihara, T., and Iwagaki, Y. (1952). On the effect of sand storm in controlling the mouth of the Kiku River. *Disaster Prevention Res. Inst.*, Kyoto Univ. Bull., **2**, 32 pp.

Iversen, J. D., Pollak, J. B., Greeleg, R., and White, B. R. (1976). Saltation threshold on Mars: The effect of interparticle force, surface roughness and low atmospheric density. *Icarus*, **29**, 381–393.

Kawamura, R. (1951). Study of sand movement by wind. *Rep. Inst. Sci. Technology, Univ. Tokyo*, **5**, (in Japanese) reported by Horikawa and Shen (1960).

Lumley, J. L., and Panovsky, H. A. (1964). *The Structure of Atmospheric Turbulence.* Wiley and Sons Inc., New York, 239 pp.

Lyles, L. (1976). Wind erosion: Processes and effect on soil productivity. *Paper 76–2016 presented at the Annu. Meet. of the ASAE, June 27, 1976*, Lincoln, Nebr., 10 pp.

Lyles, L., and Allison, B. (1976). Wind erosion: The protective role of simulated standing stubble. *Trans. ASAE*, **19**, 61–64.

Marshall, J. (1971). Drag measurements in roughness arrays of varying density and distribution. *Agr. Meteorol.*, **8**, 269–292.

O'Brien, M., and Rindlaub, B. (1936). The Transportation of Sand by Wind. *Civ. Eng.*, **6**, 325 pp.

Smalley, I. J. (1970). Cohesion of soil particles and the intrinsic resistance of simple soil systems to wind erosion. *J. Soil Sci.*, **21**, 154–161.

APPENDIX: OTHER DATA WHICH MAY BE OF INTEREST TO SAHARA DUST RESEARCH

I would like to quote two papers which might be of interest to the Sahara dust project.

1. Mass-visibility relationships

The first paper (Patterson and Gillette, 1977) deals with estimating dust concentration from visibility observations. Comparing our measurements with those of other workers, we found some discrepancies. Visibility, V (expressed in km), and mass concentration of dust, M (expressed in g m^{-3}), are related by

$$MV^{\gamma} = C$$

where C is a dimensional constant and γ is a nondimensional constant. The results of several authors for γ and C were given in Patterson and Gillette's Table 2, summarized below:

Study	C (g m^{-3} km for $\gamma = 1$)	γ
General Urban		
(Charlson, 1969)	1.8×10^{-3}	1
Bertrand *et al.* (1974)	1.4×10^{-3}	1.05
NCAR (Total mass)	2.3×10^{-2}	1.07
(SP $r < 20$ μm)	1.7×10^{-2}	1.08
(SP $r < 10$ μm)	1.3×10^{-2}	0.95
Chepil and Woodruff (1957)		
All data	5.6×10^{-2}	1.25
No local erosion	2.0×10^{-2}	—

The value of γ is reasonably close to 1 for all the studies except those of Chepil and Woodruff. The study concluded:

'Based on these data there is no single value that is generally appropriate for relating mass and visibility of soil-derived aerosols, and that can thus serve as a predictor of mass concentration when the visibility is known. In the presence of local erosion, our data and Chepil and Woodruff's data indicate that the value of C can vary between approximately 20×10^{-2} and 1.0×10^{-1} depending on soil conditions, visibility, and wind speed; the higher values for C are apparently characteristic of drought conditions. Under conditions of no local erosion, but still in the general source region for the dust generation, the value of $C = 2.0 \times 10^{-2}$ g m^{-3} km appears to be appropriate, while the Bertrand value of $C = 1.4 \times 10^{-3}$ g m^{-3} km may be appropriate for measurements made several thousand kilometers from the dust source.

Because of the variability in C, we feel that when the reduction in visibility is due to air-borne dust, mass concentration–visibility relations alone must be used carefully; if the fractional size distribution is known, however, the mass-visibility relations may be calculated using theoretical values of C.'

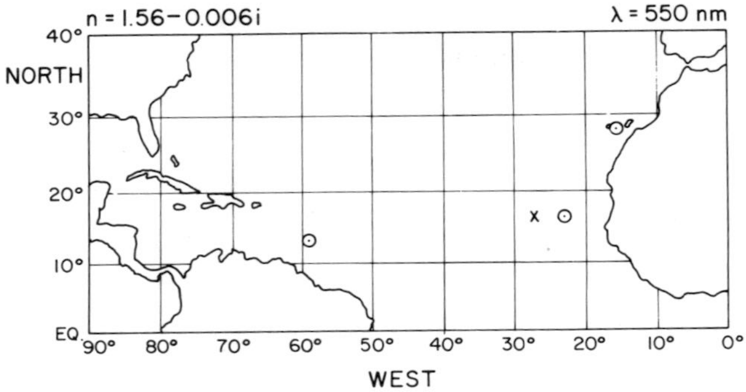

Figure 4.7 Index of refraction for Saharan aerosols over the Atlantic. Crosses and circles indicate locations for particle collections

2. Index of refraction for Saharan aerosols over the Atlantic

The second paper reports values for particle collections obtained at the circled and crossed locations in Figure 4.7 (Patterson, Gillette and Stockton, accepted manuscript). The samples were obtained by stations established by M. L. Jackson and Dale Gillette, by Joseph Prospero, and by Ruprecht Jaenicke and Lothar Schütz on the Research Vessel *Meteor.* The mean value for the index of refraction at all stations was 1.56 – 0.006 *i.*

References to Appendix

Bertrand, J., Baudet, J., and Drochon, A. (1974). Importance des aerosols naturels en Afrique de l'ouest. *J. Rech. Atmos.*, **8**, 845–860.

Charlson, R. J. (1969). Atmospheric visibility related to aerosol mass concentration. *Environ. Sci. Technol.*, **3**, 913–918.

Chepil, W. S., and Woodruff, N. P. (1957). Sedimentary characteristics of dust storms – II. Visibility and dust concentrations. *Am. J. Sci.*, **255**, 104–114.

Pattersson, E. M., and Gillette, D. A. (1977). Measurements of visibility *vs.* mass-concentration for air-borne soil particles. *Atmos. Environ.*, **11**, 193–196.

Pattersson, E. M., Gillette, D. A., and Stockton, B. H., (accepted manuscript). Complex index of refraction between 300 and 700 mm for Saharan aerosols. Accepted by *J. Geophys. Res.*

SECTION 2

(c) Transport

CHAPTER 5

The African Dust Plume: Its Characteristics and Propagation Across West Africa in Winter

A. E. KALU

ABSTRACT

The south-westward advection of Saharan Dust in Winter from the Bilma–Faya Largeau source area across Nigeria into the Gulf of Guinea has been discussed. The African Dust Plume has been defined and described. Three distinct phases of the Saharan dust transport have been identified and discussed. And finally some meteorological factors controlling the emission, transport, persistence and dispersion of the African Dust Plume have been outlined.

5.1 INTRODUCTION

Observation shows that each year between November and March large quantities of dust particles are transported from the Sahara desert towards the Gulf of Guinea across Nigeria. The agency of such dust transport is the harmattan, a cold dry wind which therefore produces one of the severest winter weather conditions in West Africa – the harmattan dust haze.

The harmattan dust particles are very fine opalescent particles which are so minute that they can remain air-borne for a considerable length of time. The dust which subsequently arrives in West Africa as the harmattan dust haze according to measurements by El-Fandy (1953) has a diameter range of 1.3–2.0 μm and concentration of about 300–500 cm^{-3}.

The presence of these harmattan dust particles constitutes a dusty atmosphere and in this paper a dusty atmosphere is defined as an atmosphere in which obstruction to vision is brought about by the presence of extremely dry and minute dust particles in the air with visibility reduced to less than 1 kilometre by natural dust.

It is known that the dust which affects a greater part of West Africa in winter south of latitude 15°N, particularly the Nigerian zone, comes mainly from the northeastern Sahara usually along the alluvian plain of Bilma (18°N, 12°E) (in

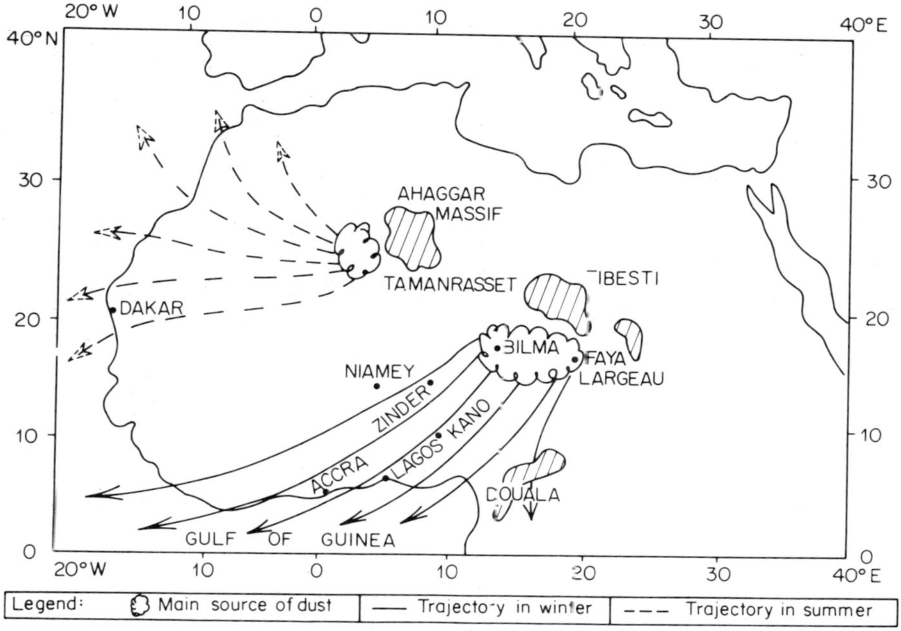

Figure 5.1 Saharan dust trajectories across West Africa

Southern Niger) and Faya Largeau (18°N, 19°E) (Chad) off the western slope of the Tibesti massif (Wilson, 1971; Kalu, 1975) (see Figure 5.1).

According to Wilson (1971) there are several source areas for the Saharan dust, but the one which is responsible for the existence of the dusty atmosphere over Nigeria and adjoining West African countries is the Bilma – Faya Largeau area. This source region is in conformity with the observed south-westerly dust trajectory over West Africa and the north-easterly direction of the mean low-level winds during this season.

The dust particles are then transported down wind from the source region in a 'plume' form towards the Gulf of Guinea by the strong north-easterly winds mainly at the 900 and 850 mb levels particularly north of the Intertropical Discontinuity (ITD), taking a south-westerly trajectory over Nigeria. On the average it takes about 24 hours for the dust to reach the northern border of Nigeria with the dust front moving at 15 knots (Aina, 1972).

In a thick harmattan dust haze which is synonymous with dusty atmosphere as defined above, visibility at the surface can be as low as 200 metres or less as a result of high dust concentration and 3–4 days spell of dusty atmosphere is never unusual particularly in the northern parts of Nigeria. The intensity of atmospheric dustiness defined in terms of dust concentration in West Africa has been found to decrease southwards probably as a result of increasing moisture content of the atmosphere towards the coast.

The occurrence of dusty atmosphere is usually accompanied by an extremely cold weather condition as a result of intrusion of cold air from the middle latitudes into the tropical latitudes across the Mediterranean. This is additive to the characteristic cloudless sky as a result of low humidity values which favours unrestricted loss of terrestial radiation resulting consequently in large diurnal temperature range.

5.2 PREVIOUS WORK

Much has been written about the transport of the Saharan dust in various parts of Africa. Wilson (1971) deals with the deposition and movement of the Saharan dust over Eastern Sahara. El-Fandy (1953) discusses the structure and concentration of the Saharan dust over the Sudan. Carlson and Prospero (1972) have by means of aircraft observations documented the transport of the Saharan dust in summer from North Africa to the Caribbean across the Atlantic.

In West Africa, however, the published materials have concentrated on evolving forecasting techniques for the onset and dispersion of the harmattan dust primarily for civil aviation purposes (Hamilton and Archbold, 1945; Burns, 1961). Their method for forecasting the arrival of the Saharan dust over West Africa depended primarily on first visible evidence of dust emission in the desert source region. This method is unreliable because dust has on many occasions been raised in the desert without being reported because of the scanty network of stations in the desert.

The most recent works on the dust problem are by Adefolalu (1968) and Aina (1972). The principal credit of this approach which is relevant to the Saharan dust transport problem is the introduction of the 'low level jet' technique for the forecasting of the emission and subsequent south-westward advection of the dust. Kalu (1975) also discussed the existence and characteristics of the African Dust Plume and produced a synoptic model for the successful forecasting of the development, persistence and dispersion of dusty atmosphere with or without evidence of air-borne dust in the desert source region.

The present paper discusses the development and transport of the Saharan dust across West Africa in winter. Relevant background information necessary for the understanding of the transport of a dust haze plume has been included.

5.3 THE SOURCE REGION CHARACTERISTICS

There are several source areas for Saharan dust. The dust which eventually affects any area outside the Sahara itself depends on the particular source region and the associated circulation patterns at the lower layers of the atmosphere which are evidenced by the direction of the prevailing winds.

From observation it is believed that the Saharan dust affecting some parts of West Africa in winter comes originally from the alluvian plain of Bilma (Niger) – Faya Largeau (Chad) off the western slope of the Tibesti massif of North Africa. The choice of the plain of Bilma and Faya Largeau is without prejudice to any

other existing source areas for the Saharan dust in other parts of the Sahara desert. Our present interest lies on the source and the meteorological factors favourable for the emission, movement and deposition of the Saharan dust particles which subsequently affect Nigeria as the harmattan dust haze.

It is necessary to point out that the Saharan dust also affects other parts of northern Africa during different periods of the year. During summer, for example, when the dust is essentially absent in West Africa south of the surface position of the ITD, the trajectory of the Saharan dust changes to a westward direction to affect southern Algeria, Morocco, the Spanish Sahara, etc. and further west to the Caribbean Islands across the Atlantic (Martin, 1975; Prospero, 1968).

A lot of studies on the composition of the Saharan dust have been made by many workers, for example, the mineralogical analysis of the Saharan dust samples collected on mesh at Barbados by Prospero and Carlson (1970), a microscopical analysis of the harmattan dust at the Imperial Institute, London (Hamilton and Archbold, 1945), the mechanical sifting analysis by the Agricultural Chemistry of Ibadan (Hamilton and Archbold, 1945).

Wilson (1971) has shown that the subsoil in the alluvian plain of Bilma – Faya Largeau consists mainly of light fine and loose clay particles also of alluvian type which have settled in the area as a result of denudational action from the mountainous Tibesti east of the plain (Figure 5.1 above).

The study also indicates that the neighbouring areas of the Bilma – Faya Largeau has a hilly topography and therefore contains mainly sand or rock debris which by their nature are heavy and not easily raked up by the wind excepting of course some denudational weathering process which collects sand and feeds into the Bilma – Faya Largeau plain.

In order to understand the behaviour of the Saharan dust it is necessary for us to explain the following situations namely (1) the preferential deposition of dust in the Bilma–Faya Largeau source area and (2) the periodic emission of the dust into the atmosphere.

To explain the first case we shall examine the wind pattern over the Bilma–Faya Largeau areas. It has been found that the wind field in the Bilma–Faya Largeau plain on the mean is convergent and bi-directional below 1500 m, with one stream north of the Tibesti mountain and the other from the southern slope of the mountain (Figure 5.2). This pattern is also reflected from the sand flow patterns of the Sahara (Figure 5.3) if we regard sand flow direction as identical to wind direction. A major factor responsible for the observed large-scale deposition of sandy ergs is the resulting velocity convergence as a result of decrease of the katabatically funnelling wind stream down the western slope of the Tibesti massif on arrival at the plain (Samways, 1976). This typical wind field creates a favourable condition for the formation of ergs as a result of the deceleration or convergence of the air streams which diverge from the northern and southern slopes of the Tibesti mountain into the plain of Bilma–Faya Largeau (Figure 5.2).

The fact that dust continues to be raised from the source region every year only

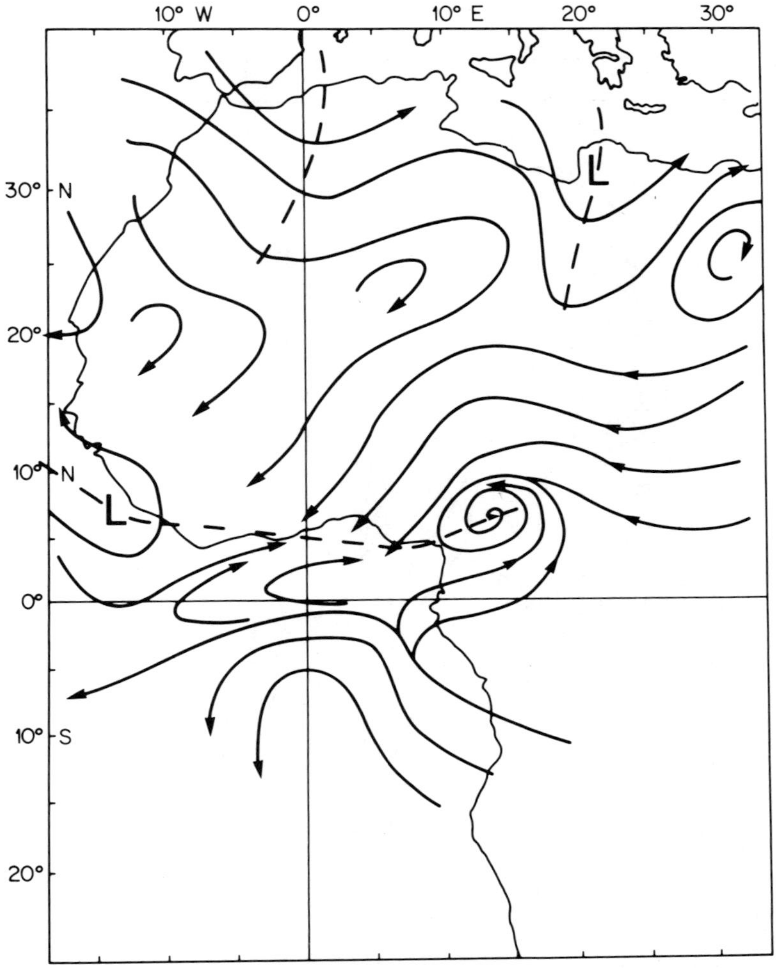

Figure 5.2 Typical midwinter streamlines (850 mb)

implies that there must be an efficient resupply mechanism or feed back system, otherwise, with time as a result of continued deflation process, there would be no longer dust deposits in the source region and so no harmattan dust would occur over West Africa. Since we know that this does not happen, it therefore means that the deflation rate in this region is usually exceeded by a resupply of dust into the plain.

Available dust particles are constantly fed into the wind at the western slope of the Tibesti and, when blowing through the plain the air becomes gradually filled with dust. Deposition occurs under favourable meteorological conditions such as

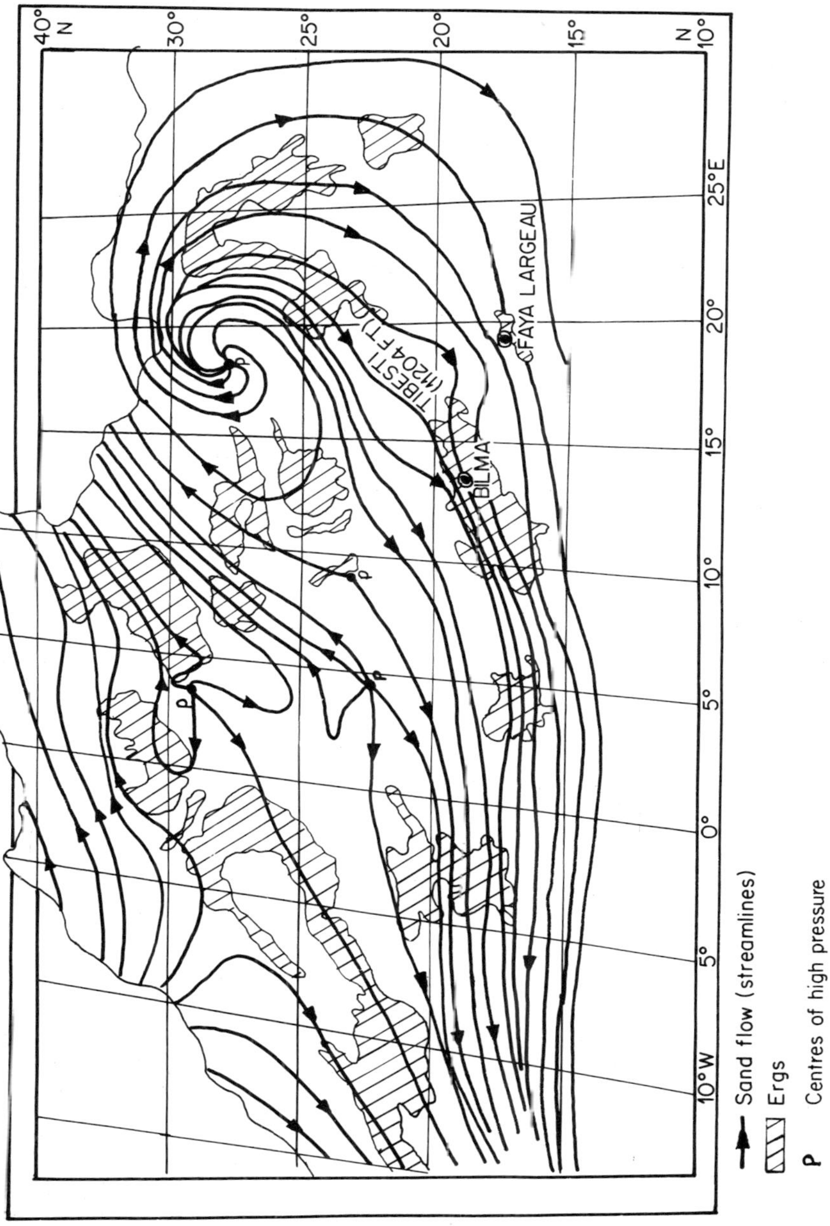

Figure 5.3 Sand-flow in the Sahara 1925–60 (after Wilson, 1971)

convergence or deceleration of the wind stream. The reverse process can also occur, i.e. no ergs in places characterized by accelerating divergent wind system.

Resupply of dust may come from fluvial action from the Tibesti massif east of the alluvian Bilma–Faya Largeau lowland. From Figure 5.3 one thing is clear – that the Bilma–Faya Largeau source area is constantly being resupplied with fresh dust deposit from the east. In fact the dust may have a history further east than we know.

The second explanation we shall consider is the periodic emission of dust from the source region. Dusty atmosphere in West Africa during the winter period results usually from a large-scale discharge into the atmosphere of dust particles. This does not occur every time, only when certain synoptic features must have been established over the source region.

One of such conditions is the development of high wind speed at the surface to create the necessary turbulence and instability as has been discussed by Adefolalu (1968) which will keep the dust air-borne for a considerable length of time. A study by the author (Kalu, 1975) shows that such atmospheric situation is created by the development of an induced meso-scale phenomenon of pressure surge arising from intense low level anticyclogenesis west of the intruding mid-latitude trough. This has been known to be associated with large scale intrusion of 'cold outbreaks' from the middle latitudes into the tropical atmosphere in winter in the form of upper troughs, moving usually eastward along the Mediterranean latitudes (Adefolalu, 1976). This is depicted in Figure 5.4.

Intensification of the subtropical anticyclone at lower levels produces a state of

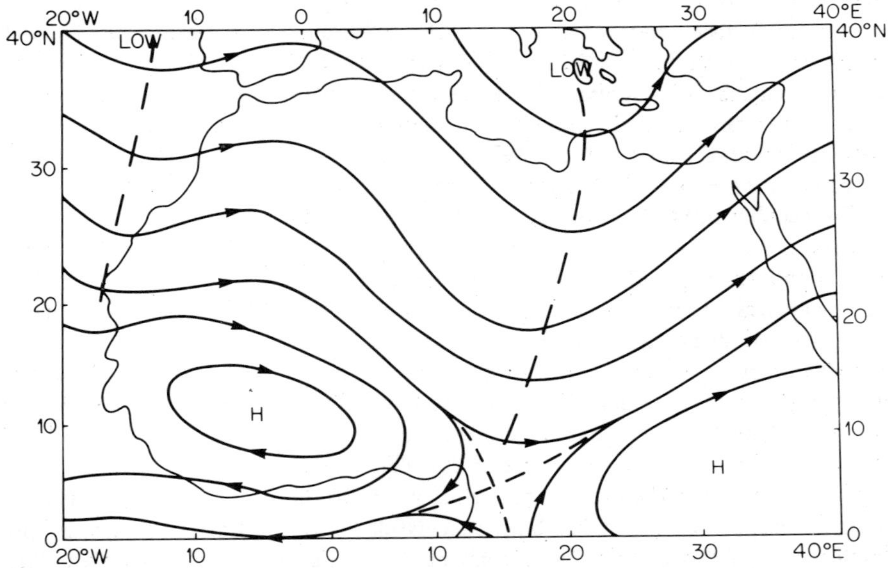

Figure 5.4 An intrusion of an upper level trough (200 mb, 9th February 1974)

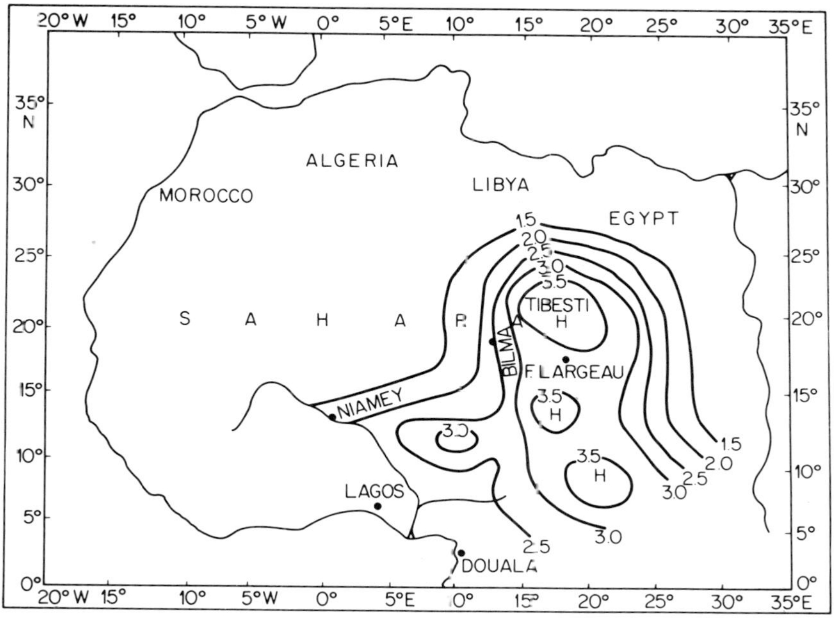

Figure 5.5 Pressure surge indicated by isollobaric high centre (24 hour tendencies)

pressure surge (Figure 5.5) which is associated with strong low level winds particularly at the surface where wind speed of about 30 knots or more can be observed (winds of 40 knots or even 50 knots are not unusual (Figure 5.6). Whenever such synoptic features as described above are evident on weather charts at the appropriate season dust is usually raised at the source region. In Nigeria this has been a useful forecasting tool for the development of the harmattan dust haze in West Africa from the Bilma–Faya Largeau source area.

The peculiar thing about duststorms is that it is not always that such a high wind velocity as about 30 knots or more is needed so as to raise the dust particles from the ground surface and keep them air-borne for some time (Figure 5.7). The meteorological conditions must first be satisfied before there can be a reasonable and sustained discharge of particulate substance into the atmosphere. Thus the periodic development of strong surface wind which arises as a result of low level anticyclogenesis with its associated pressure surge and instability causes the periodic emission of dust into the atmosphere and hence the periodic occurrence of dusty atmosphere down wind of the source of dust (Figure 5.8).

It is to be remarked that the velocity of 30 knots or more is based on observational evidence and has been successfully used in Nigeria to forecast the development of dusty atmosphere from the Bilma–Faya Largeau source area. With such a wind speed enough instability will be generated to keep the dust particles air-borne for a considerable length of time.

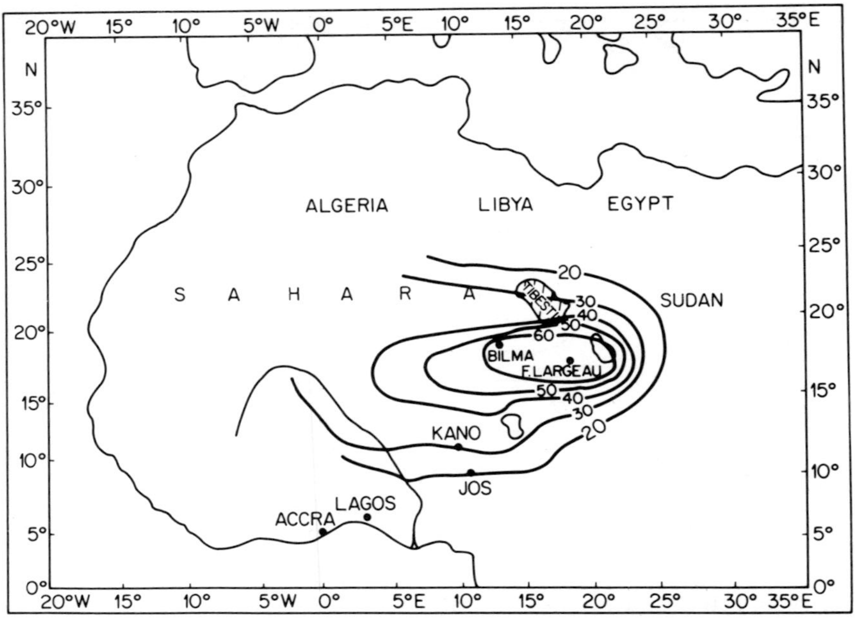

Figure 5.6 Low-level 'jet' 900 m on 31st January 1975, 12^h GMT (isotachs in knots)

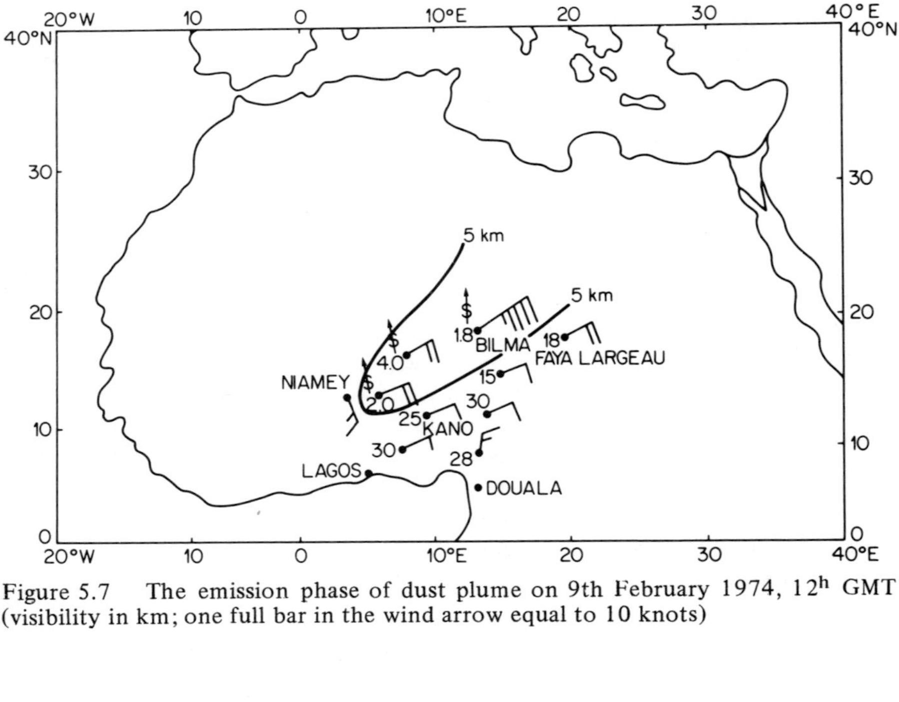

Figure 5.7 The emission phase of dust plume on 9th February 1974, 12^h GMT (visibility in km; one full bar in the wind arrow equal to 10 knots)

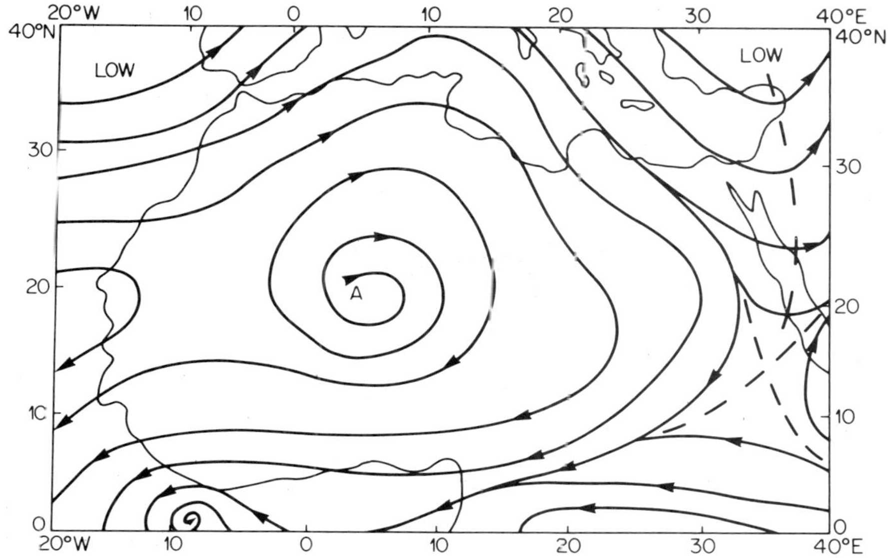

Figure 5.8 Low-level anticyclone with pronounced cyclonic shear at its eastern periphery on 11th February 1974, 12^{h} GMT (850 mb)

5.4 THE AFRICAN DUST PLUME

The harmattan dust haze is a synoptic scale system which has been observed to affect simultaneously a large area of the West African sub-region as a particulate body. The minute Saharan dust particles which in essence constitute a dusty atmosphere have been quantized into an idealized entity which we call 'The African Dust Plume' (ADP) (Carlson and Prospero, 1972) with certain distinct physical structure and properties. The plume concept has been construed and upheld in this paper because it has been observed that the Saharan dust propagates across West Africa in this form. It is not the atomistic conception of dusty atmosphere that is of great interest to us, but in a group because it is in this plume form that the harmattan dust affects us as a significant weather element during the dry winter months in West Africa.

Martin (1975) in his detection of overland and overwater dust in visible and infrared pictures of the Synchronous Meteorological Satellite (SMS) during GATE in 1974 observed and concluded that dust clouds over-land appeared as 'tongues' and as 'plumes' with the forward edge 'bending cyclonically and sometimes forming a hook' (Figure 5.9). In the present discussion of the transport of the Saharan dust across West Africa in winter, a plume model as envisaged by the author's own observations of the harmattan dust over Nigeria at aircraft altitudes below 700 mb, corroborated with resports from several pilots at the Kano International Airport, seems to confirm the 'Plume' concept of dust propagation. A dust plume as

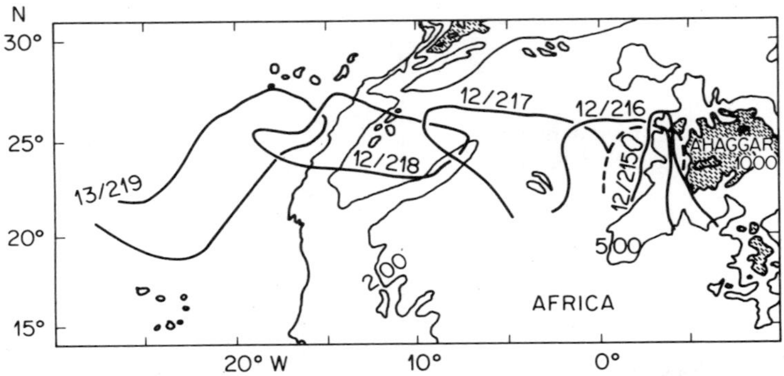

Figure 5.9 Path of a typical outbreak of dust haze in the Western Sahara (after Martin, 1975)
Note 12/215 ...... 3rd August 1974, 12^h GMT
12/216 ...... 4th August 1974, 12^h GMT
12/217 ...... 5th August 1974, 12^h GMT
12/218 ...... 6th August 1974, 12^h GMT
13/219 ...... 7th August 1974, 13^h GMT

construed here is defined as an aerodynamic enclosure in space containing dust particles in addition to the normal constituents of the air in such a mixture that both dust particles and air molecules are indistinguishable by an unaided vision.

For such a model the concentration of dust particles varies both in the horizontal direction as well as in the vertical and has the singular advantage of illustrating the 3-dimensional structure of dust extent in the atmosphere (Figure 5.10). Observations show that in a dusty atmosphere improvement in visibility usually starts from the source region of a particular dust plume and progresses down-wind and never the reverse unless of course if another spell has started before clearance begins upstream. This suggests that the forward edge of the plume may have the highest dust concentration and lowest concentration restricted to the rear of the plume as has been actually observed (Aina, 1972; Kalu, 1975).

Secondly the fact that dust is propagated by the agency of the upper winds implies that concentration would be maximum at levels where the wind is strongest. This level corresponds to 900 mb at Kano, although the dust top may be at higher levels. (See Figure 5.11).

In an effort to define an aerodynamic three-dimensional configuration which would conform to the observed plume characteristics as outlined above, a careful analysis of the visibility distribution at the surface was made and the results showed that the 1 kilometre isopleth or less is approximately ellipsoidal (Figure 5.12). This result is also evident in the work by Aina (1972) and Samways (1976). Also satellite observations confirm a possible ellipsoidal shape of the Saharan dust plume as can be noticed from Figure 5.13 which shows some Application Technology Satellite (ATS) pictures of tracks of a dust cloud from Africa to the Caribbean in July 1969.

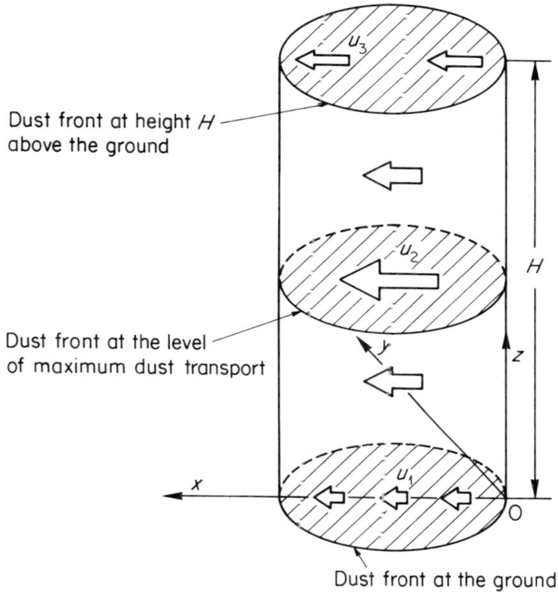

Figure 5.10 A schematic hypothetical model of a dust column showing the ellipsoidal dust fronts at various levels of the atmosphere ($u_2 > u_3 > u_1$)

Since I had no facility to measure dust concentrations at heights above the earth's surface, I depended on aircraft reports all of which seemed to confirm one thing – that the concentration of dust decreased with height particularly within an atmospheric layer which is thermally mixed with a possible discontinuity at the level of maximum wind speed.

If however the atmospheric column is not thermally mixed as usually the case during the night when temperature inversion may exist below 700 mb, the inversion layer will show a greater dust concentration as a result of the dust-trapping characteristic of an inversion layer. Insolation usually destroys such a temperature inversion resulting in a thermal mixing of the atmospheric column which may set-up vigorous convective current whereby the upper part of the layer is rapidly brought down to lower layers and the lower part taken up.

The above accounts would explain why visibility in a dusty atmosphere usually deteriorates in the morning hours around 0900 GMT (Figure 5.14). In most cases it is observed that dust first arrives at a station at higher levels and when mixing occurs within the atmospheric column as described above, dust is then brought down to lower levels through subsidence and turbulent mixing.

The forward edge of a dust plume is anvil-shaped as schematically depicted in Figure 5.15. This is in agreement with the dust 'tongue' as observed by Martin (1975) from satellite dust imagery (Figure 5.16).

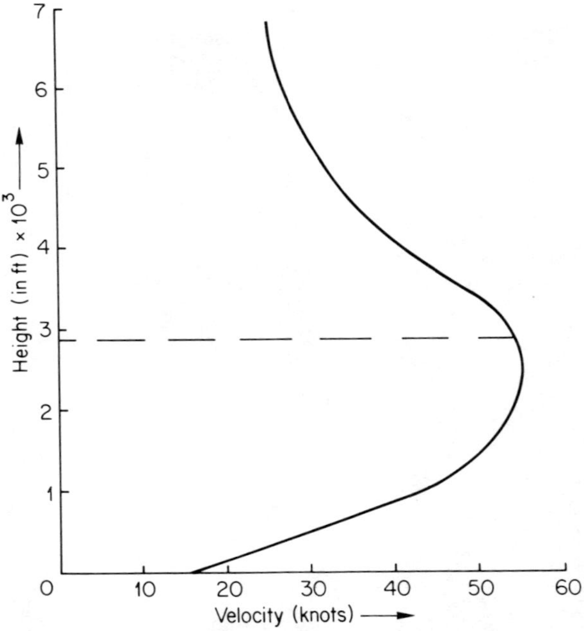

Figure 5.11 The mean velocity profile over Faya Largeau

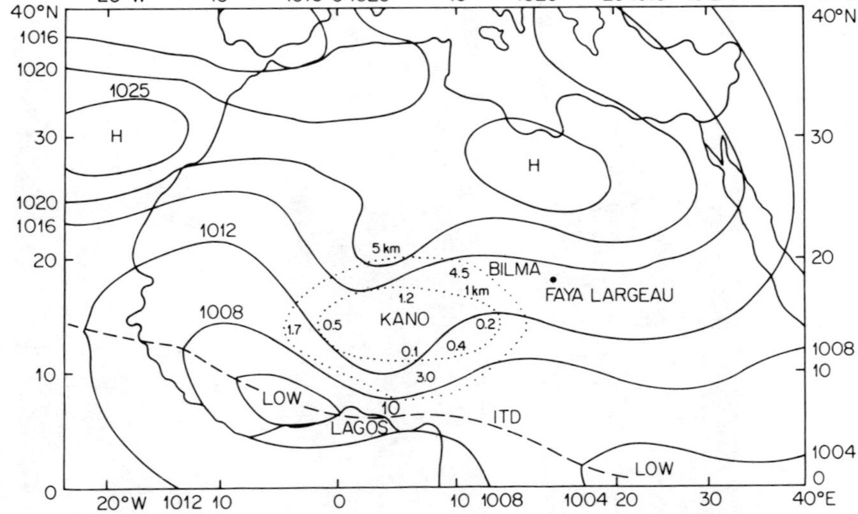

Figure 5.12 Visibility distribution. Surface 0900 GMT, 11th February 1974 (Visibility in km)

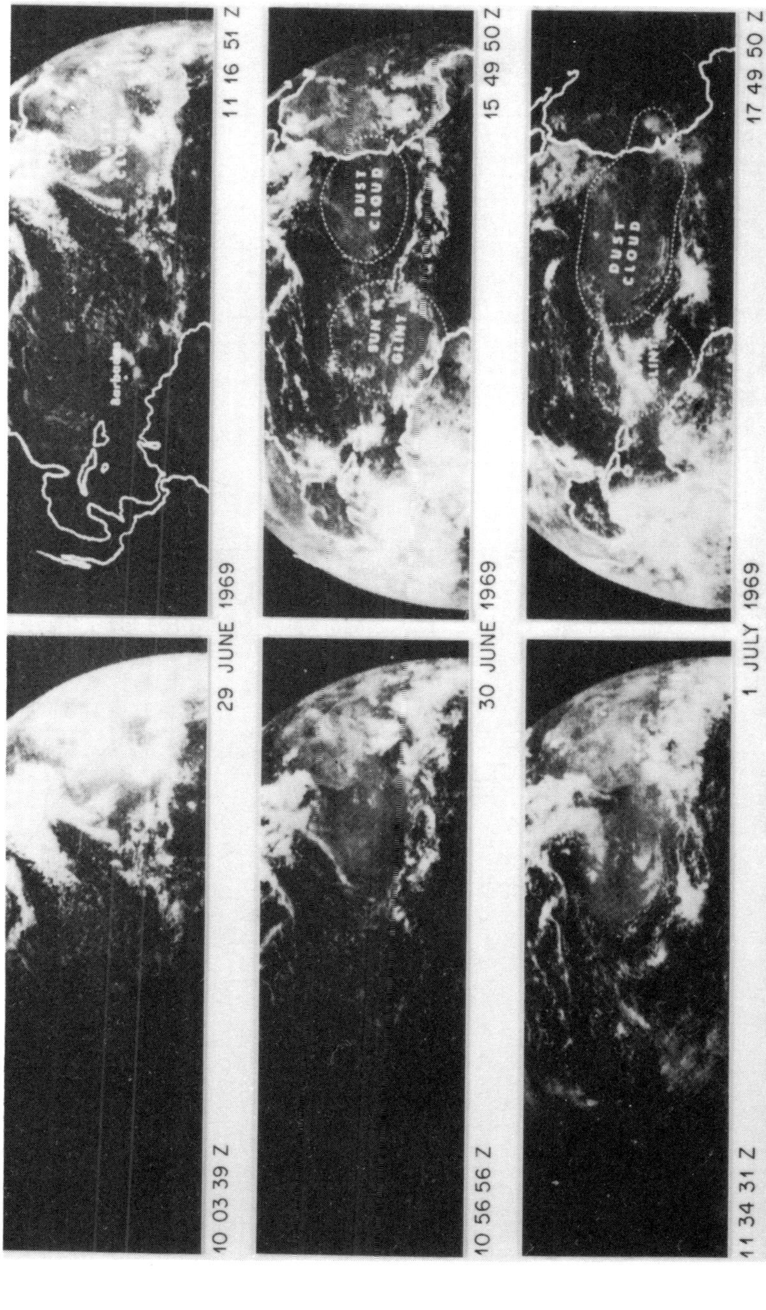
10 03 39 Z
29 JUNE 1969
11 16 51 Z
Barbados
10 56 56 Z
30 JUNE 1969
15 49 50 Z
DUST CLOUD
SUN GLINT
11 34 31 Z
1 JULY 1969
17 49 50 Z
DUST CLOUD

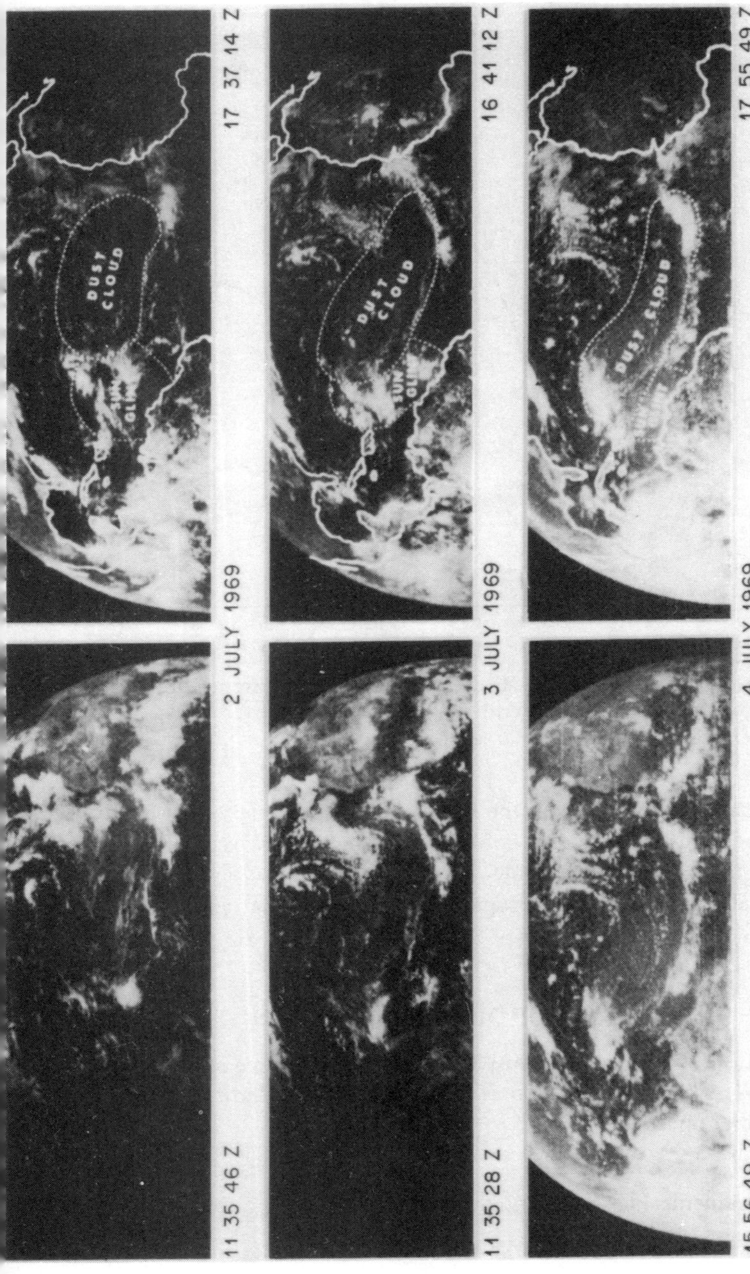

Figure 5.13 ATS III tracks a dust cloud from Africa to the Caribbean. Abnormally high dust counts were recorded on 3, 5, 6, and 11 July 1969 at Barbados in the Lesser Antilles. The dust source is the Sahara, as these Application Technology Satellite (ATS) III pictures indicate. Dust clouds can significantly increase the normal amount of incoming solar radiation absorbed in the atmosphere; consequently less solar energy reaches the surface. All dust cloud effects on weather are not known. The ATS III camera should help clarify dust cloud/weather interrelationships

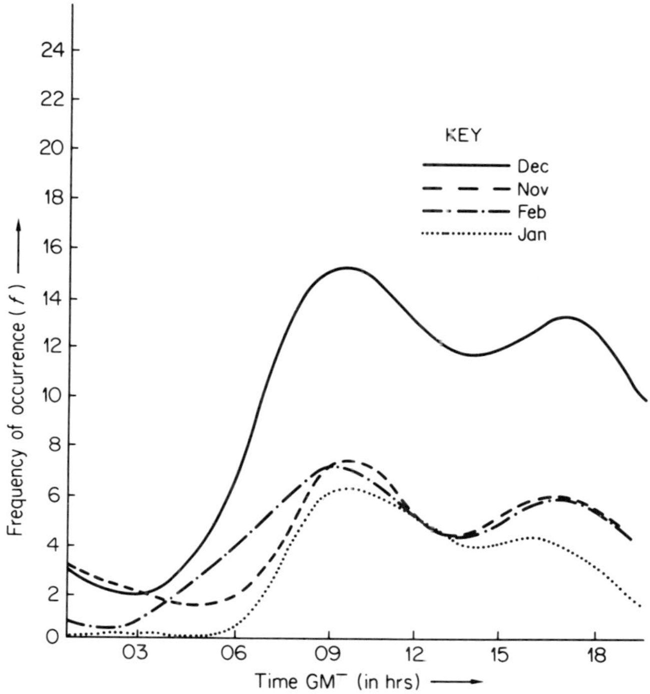

Figure 5.14 The diurnal frequency of occurrence of Harmattan dust storm (visibility 200–1000 m over Kano (1971))

The ellipsoidal configuration of the harmattan dust plume is still highly hypothetical. It needs further research before final conclusions can be made on the actual shape of the African Dust Plume. Meanwhile we shall apply the plume model in our present discussions on the propagation of the harmattan dust across West Africa.

5.5 PHASES OF DUST PROPAGATION

The propagation of the Saharan dust from the desert source regions to the time it arrives in the West African subregion as the harmattan dust haze may be discussed under three distinct phases.

5.5.1 The Instantaneous Phase

This is the emission stage and marks the beginning of what may constitute the history of the African Dust Plume. The phase marks the period when dust particles

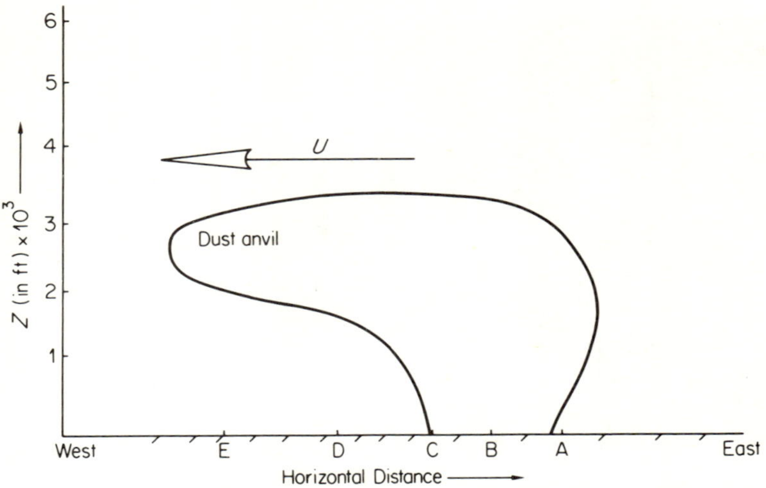

Figure 5.15 Propagation of dusty atmosphere
A: Onset of clearance at station A
B: Station under dusty atmosphere
C: Dust front just reaching station C
D: } Good visibility at night with a chance for
E: } deterioration by mid-morning owing to turbulent mixing

are violently raised from the ground usually in the form of duststorms as a result of very strong surface wind. One important characteristic of the Instantaneous Phase is that it is highly unstable and depending on the degree of instability in the atmosphere. The phase may occur in pulses with very short time intervals and a corresponding fluctuation in visibility. It is more eddy fluctuations, created principally by the low-level pressure surge as discussed in Section 5.3 than normal thermal convection that are responsible for the state of instability and turbulence that has been found to be associated with the emission stage of the dust plume.

The plume of dust during this phase is diffused upward by turbulence and there is pronounced buoyancy and no spreading in the horizontal direction as the mean motion during this phase is directed upwards.

5.5.2 The Spreading Phase

When the dust has been raised from the ground as discussed above and diffused upwards by turbulence to the level where the wind is strong enough, transportation starts in the horizontal direction. This marks the Spreading Phase. During this stage the dust particles within the plume can be assumed homogeneous in terms of grain size as the heavier particles must have sedimented out of the plume body during the first phase as a result of gravity.

It is observed that with time the wind which raised up the dust particles

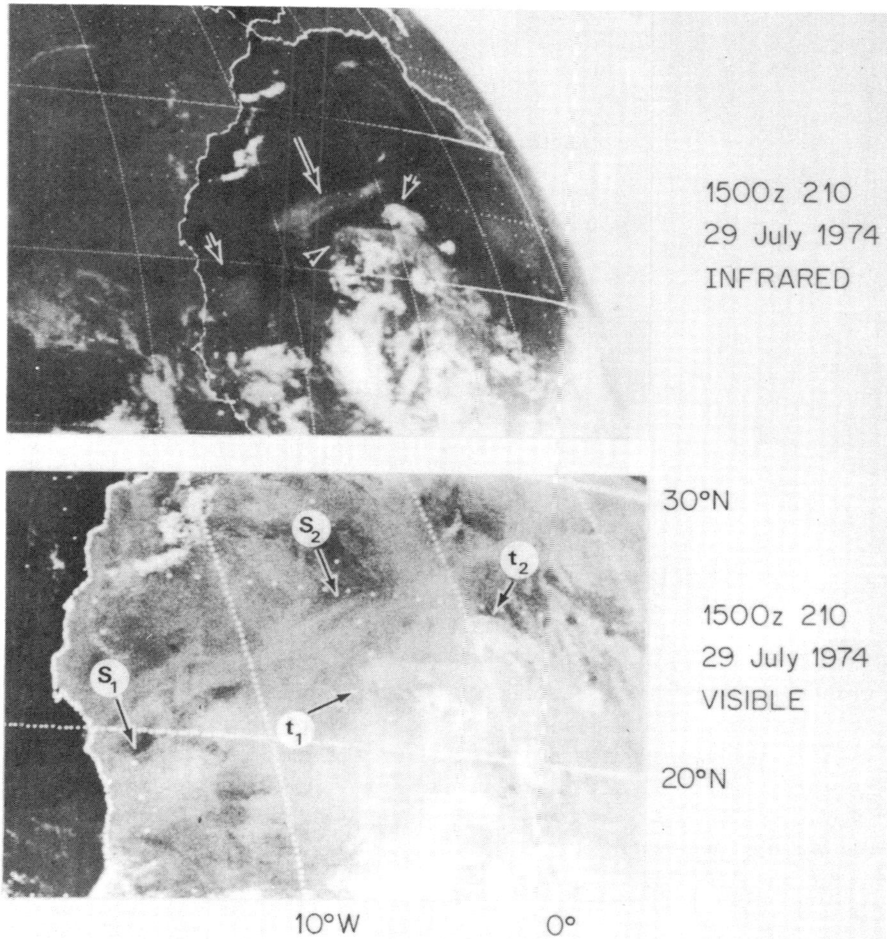

Figure 5.16 Dust tongues (t_1 and t_2) and dust streaks (S_1 and S_2) in simultaneous SMS infrared and visible pictures covering the western Sahara (after Martin, 1975)

gradually weakens and this suggests that the plume also gradually loses its vertical momentum and the eddies developed during the instantaneous phase become less powerful. The overall result of this attenuation in the surface wind strength is a general spreading motion of the dust plume. Whereas the first phase relates to the area of dust emission (i.e. the source region), the second phase usually starts some few kilometres downwind.

5.5.3 The Equilibrium Phase

The Equilibrium Phase is the most stable of the three phases of motion of the dust haze plume. It sets in as soon as the Spreading Phase has been completed. It is assumed that at this stage the plume has finally lost independence and moves entirely under the influence of the prevailing winds.

In West Africa and Nigeria in particular this stage is usually marked by a gradual reduction in visibility from, say 30 km (under normal weather) to about 5 km (in dust haze) within, say 6 hours. Thereafter the visibility deteriorates rapidly becoming poorest when perhaps the core of the dust plume comes over the station. The significant thing about the Equilibrium Phase is that it usually takes place some hundreds of kilometres downwind. For example the harmattan dust haze which affects the entire Northern States of Nigeria arrives there at its Equilibrium Phase.

5.6 SOME METEOROLOGICAL FACTORS CONTROLLING DUST TRANSPORT

5.6.1 The Prevailing Wind

Among the meteorological factors that control the propagation of the Saharan dust, the wind comes readily to mind. The prevailing wind therefore plays the greatest role in the movement of the Saharan dust right from the emission stage to the propagation stage. This is associated with the circulation patterns in the Saharan air layer where dust is usually propagated. Upper jets have sometimes been associated with dust transport (Obasi, 1965).

Dust in general is propagated through the atmosphere by the agency of the upper level winds depending on the vertical extent of the dust layer. It is observed that the Saharan Air Layer (SAL) over West Africa in winter does not usually exceed the 700 mb level particularly north of the ITD, but in the more humid part of West Africa south of the ITD the SAL may be found as high up as at the 600 mb level (Carlson and Prospero, 1972).

The fact that the upper winds are directly responsible for the dust transport implies that the concentration of dust at any place and at any level of the atmosphere depends on the quantitative removal (depletion) of dust by wind. If the wind is strong within the SAL, the concentration of dust over any given location on the earth's surface will decrease as the rate of movement of dust aloft across such a surface is large. The reverse is the case for weak winds resulting in high dust concentration.

Over Kano (12°N, 8°E) in Nigeria the 900 mb level was found to constitute on the mean the level of maximum dust transport. This is because in the light of the above discussions the 900 mb level coincides with the level of maximum vertical wind shear. An examination of the velocity profile of various stations in West

Africa shows that there is always a characteristic zone of max. wind (see Figure 5.15 above).

In West Africa during the northern winter, the dominant synoptic feature is the subtropical anticyclone which generates the strong north-easterly winds which have been associated with the transport of the Saharan dust, (Figure 5.2). The persistence of a dusty atmosphere over any location therefore depends on the magnitude of these north-easterly winds.

Wind speed affects the rate of deposition of air-borne dust particles and so while velocity divergence is favourable for faster dust movement, velocity convergence enhances dust accumulation and deposition on the surface. Characteristically light wind and calm conditions are very conducive for longer persistence periods for a dusty atmosphere than stronger winds.

During the first two weeks of March 1977, West Africa experienced one of the severest harmattan spells that have ever been observed in the territory. The contributory synoptic feature was a southward shift and stagnation of the subtropical anticyclone over West Africa such that a greater portion of the subregion was under the influence of the dry continental north-easterly winds. The blocking effect of the anticyclone over West Africa is so pronounced that the dry season seems to be cutting into the rainy period particularly in the southern part of the region with the result that an unusually dry weather is presently being experienced in West Africa. The northward advance of the ITD has been greatly checked by the dominating influence of the blocking anticyclone.

Characteristically the mean flow at lower levels of the troposphere has continued to be the north-easterlies which therefore advect the dry dusty Saharan air across West Africa into the Equatorial Trough.

The amount of dust transported across any given locality is directly proportional to the mean wind velocity over the area. Thus the stronger the wind the faster and greater the volume of dust transported across a given place. The wind is therefore one of the most important meteorological factors controlling the transport of the Saharan dust.

5.6.2 Vertical motion

Another important characteristic of the atmosphere which affects dust transport is the state of air motion in the atmosphere – whether convective upward motion or subsiding downward motion is taking place. The former results in an upward transport as is usually the case for pressure systems such as the low pressure areas and troughs where an upward and convective motion is usually pronounced. One advantage of an upward motion in a dusty atmosphere is that much of the available dust is pushed upwards into the atmosphere and eventually removed by the stronger upper winds to other areas downwind. This process results in a decrease in the concentration of dust particles at the surface and lower levels.

Subsident motion, on the other hand, is associated with the descent of dust

particles and usually results in higher concentration of dust at lower levels of the atmosphere. Pronounced large-scale subsidence results generally in a greater persistence of a dusty atmosphere. It is usually an effective instrument for forecasting the persistence of the harmattan dust haze in West Africa.

5.6.3 Thermal stratification of the atmosphere

Temperature variation particularly in the vertical direction affects the transport of particulates in various ways and the harmattan dust is not an exception. Thermal stratification of the atmosphere is therefore an important parameter that controls to a large extent the transport of the Saharan dust.

Thermal mixing of the atmosphere is most pronounced when the temperature lapse rate attains or nears the dry adiabatic value and as Pasquill (1971) has observed strong lapse enhances vertical mixing of the atmosphere.

Observations in Nigeria show that a low level temperature inversion of limited depth develops during the night usually somewhere below 700 mb when the atmosphere is stable, and when this occurs in a dusty atmosphere dust is usually trapped within the inversion layer. It is known that an inversion temperature profile of the atmosphere suppresses vertical mixing and this goes to explain the reason for the existence of high concentration of dust within an inversion layer.

However when insolation becomes strong in the morning with the lapse rate becoming steeper the transfer of dust particles from one layer to another increases reaching maximum when the temperature profile attains or nears the dry adiabatic value. This results in a rapid free up-and-down motion within the atmosphere and under such conditions the inversion is destroyed leading to the redistribution of the dust to lower layers of the atmosphere. The overall result of thermal mixing in a dusty atmosphere is usually the occurrence of poor surface visibility. The author's observation in Kano (Kalu, 1975) shows that this happens around 0900 G.M.T. (Figure 5.14).

This temperature stratification of the atmosphere affects the diurnal cycle of dust in the atmosphere and by this we mean the increasing concentration of dust (poor visibility) at the surface and adjoining layers of the atmosphere in the morning when insolation results in the mixing of the atmospheric column and the decreasing concentration (good visibility) when dust is trapped within the inversion layer as a result of suppressed vertical mixing of the atmospheric column.

5.6.4 Moisture Content

The moisture content of the atmosphere or humidity is perhaps one of the most important factors controlling dust transport in the atmosphere. Characteristically the tiny dust particles behave as condensation nuclei which have a great affinity for water and by absorbing water they grow becoming heavier and sediment out of the atmosphere. Thus if the Saharan dust plume comes over to a section of the

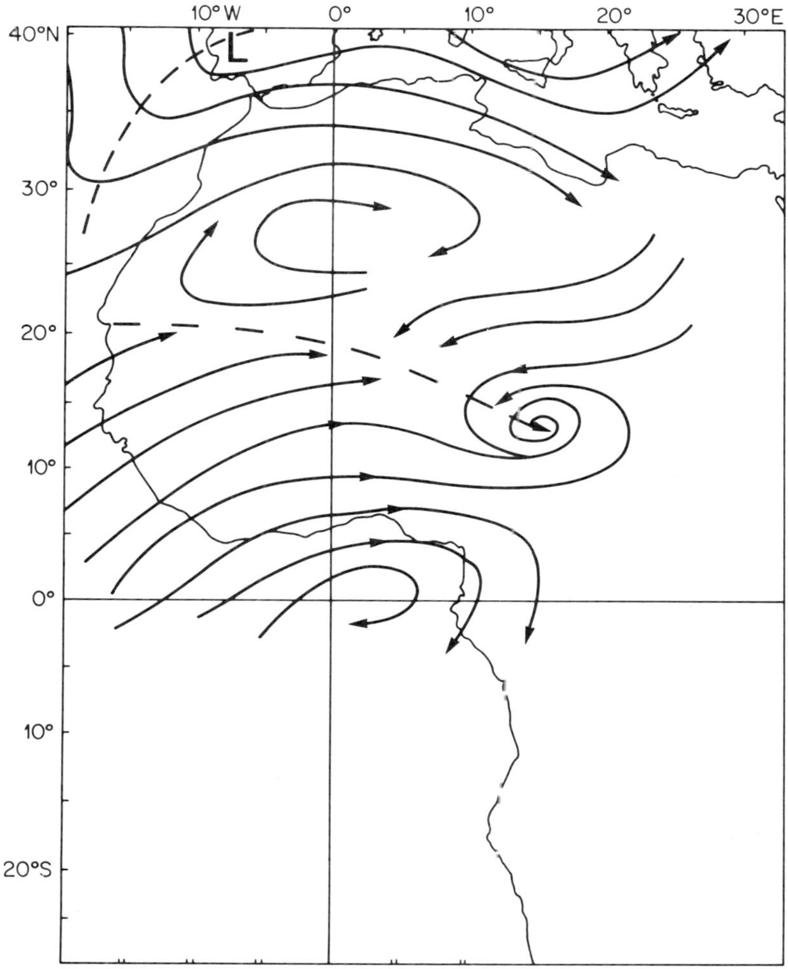

Figure 5.17 Typical midsummer streamlines (850 mb)

atmosphere where the moisture content is high, the immediate result will be that a large amount of it would sediment out giving rise to improved visibility.

In West Africa the intensity of the harmattan dust has been found to decrease south of the surface position of the ITD and this decrease has been associated with increasing moisture content of the atmosphere towards the coast during the winter period when the Saharan dust is observed over the southern part of West Africa.

One significant thing about the Saharan dust transport in West Africa is that in northern summer when the position of the ITD shifts northwards over the southern

part of the Sahara, the trajectory of the Saharan dust across West Africa also shifts to the north-western part of Africa over Mauritania and Morocco. Dust clouds have been frequently observed off the Atlantic coast of North-west Africa and the Caribbean Islands in summer. These have been documented by Carlson and Prospero (1972) and recently by Martin (1975) (Figure 5.9).

During the northern summer a greater part of the West African subregion does not experience any dusty atmosphere because the humid and penetrating south-west – monsoon winds do not allow whatever dust that is available to be transported across West Africa particularly the Nigerian area as the influence of the dry north-easterlies is restricted to Central Sahara during this period (see Figure 5.17).

5.7 CONCLUSION

The Saharan dust is propagated in a plume form across West Africa at about the same speed as the African Easterly Waves and takes on the average about one to two days to advect to the northern Saharan border of West Africa, for example the northern parts of Nigeria, after emission from the Bilma–Faya Largeau source region.

The African Dust Plume (ADP) arises as a result of a state of pronounced atmospheric instability associated with a pressure surge at low levels of the atmosphere resulting from an intense low-level anticyclogenesis and upper convergence.

The dust is transported by the low-level continental north-easterly winds below 700 mb particularly the 900 mb where wind speeds of 30 knots or more are observed. The Saharan air layer is found at lower levels of the atmosphere during winter particularly north of the ITD, but at higher levels in summer (750–600 mb) south of the convergence zone because the Saharan air characteristically overrides the more humid and denser tropical maritime air mass south of the ITD.

While the dust plume maintains a south-westerly trajectory across West Africa in winter with the dust originating from the north-eastern part of the Sahara, in summer, as a result of the penetrating south-westerly monsoon winds, the dust disappears south of latitude 15°N and is then mainly restricted to the north-western part of Africa with possibly a different source history from that which affects Nigeria in winter (Figure 5.1).

Whatever dust that eventually arrives in West Africa from the Sahara depends on the direction and strength of the low-level winds below 700 mb, especially the 900 mb flow as well as the thermodynamic characteristics of the lower tropical troposphere such as thermal stratification and moisture content.

Some aspects of the African dust plume are still hypothetical, for example the three-dimensional aerodynamic structure of the plume. It is however hoped that the proposed West African Monsoon Experiment (WAMEX) will throw more light on the characteristics of the African dust plume.

5.8 ACKNOWLEDGEMENT

The author wishes to thank the staff of the Research and Training Institute, Oshodi, Nigeria, for their immense assistance during the preparation of this paper. My gratitude also goes to Professor G.O.P. Obasi, the Adviser of Research and Training, Nigerian Meteorological Department, and to the Director of Meteorology, Mr. C. A. Abayomi, whose recommendation and approval respectively made my participation in the Workshop possible

REFERENCES

Adefolalu, D. O. (1968). Two case-studies of the vertical distribution of dust during occurrence of Harmattan haze over Nigeria. *Tech. Note*, **21**, Meteorol. Dep., Lagos, Nigeria, 13 pp.

Adefolalu, D. O. (1976). Scale interactions in the tropics. Part I: On the newly discovered 4–6 day Meridional Oscillation of the Monsoon Trough (MONT). To be published in *Mon. Weather Rev.*

Aina, J. O. (1972). A contribution to the forecasting of Harmattan dust haze. *Q. Meteorol. Mag.*, 2, Meteorol. Dep., Lagos, Nigeria, 77–90.

Burns, F. (1961). Dust haze in relation to pressure gradients. *Meteorol. Mag.*, **90**, 223–226.

Carlson, T. N., and Prospero, J. M. (1972). The large-scale movement of Saharan air outbreaks over the northern equatorial Atlantic. *J. Appl. Meteorol.*, **11**, 283–297.

El-Fandy, M. G. (1953). On the physics of dusty atmosphere. *Q. J. R. Meteorol. Soc.*, **79**, 284–287.

Hamilton, R. A. and Archbold, J. W. (1945). Meteorology of Nigeria and adjacent territory. *Q. J. R. Meteorol. Soc.*, **71**, 231–265.

Kalu, A. E. (1975). On the meteorology of dusty atmosphere. *Q. Meteorol. Mag.*, Meteorol. Dep., Lagos, Nigeria. Under review for publication.

Martin, D. W. (1975). Identification, tracking and sources of Saharan dust – an inquiry using Synchronous Meteorological Satellite (SMS). In *WMO GATE Rep.*, 2, 217–219.

Obasi, G. O. P. (1965). Atmospheric Synoptic and Climatological Features of the West African Region. *Tech. Note 28*, Meteorol. Dep., Lagos, Nigeria, 45 pp.

Pasquill, F. (1971). The meteorological aspects of local high-concentration air pollution. *Spec. Environment Rep. 2*, WMO-312, 49–54.

Prospero, J. M. (1968). Atmospheric dust studies on Barbados. *Bull. Am. Meteorol. Soc.*, **49**, 645–652.

Prospero, J. M. and Carlson, T. N. (1970). Radon-222 in the North Atlantic trade winds: Its relationship to dust transport from Africa. *Science*, **167**, 974–977.

Samways, J. (1976). Wind over Africa. *Geogr. Mag.*, **48**, 218–220.

Wilson, I. G. (1971). Desert sandflow basins and a model for the development of ergs, *Geogr. J.*, **137**, 180–198.

CHAPTER 6

The Use of Meteorological Observations for Studies of the Mobilization, Transport, and Deposition of Saharan Soil Dust

C. MORALES

ABSTRACT

SYNOP reports (weather reports from the meteorological network of so-called SYNOP stations) contain much valuable information in relation to studies of the mobilization, air-borne transport and deposition of desert soil dust. Details of this information are presented in the paper. Further two examples are given on how to make use of the SYNOP reports, one showing the migration of a duststorm over the Sudan area on a series of synoptic weather maps, the other for a study on the threshold wind velocity for raising desert soil dust.

6.1 INTRODUCTION

The object of the Saharan Dust Project is to describe and explain as far as possible the three phases of Saharan dust and sand transport within the Sahara and its pre-Saharan margins, where desertification occurs. These three phases are the mobilization phase, the air-borne transport phase, and the deposition phase. In all three phases meteorological factors play an important part. The object of this paper is to demonstrate the usefulness of meteorological synoptic observations for investigations in relation to the Saharan Dust Project.

To which questions can we hope to get an answer by means of meteorological investigations concerning the desertification process, and what resources do we have to help us in answering these questions?

In this relation the following questions are of importance:

1. Which types of weather mechanisms are responsible for raising and transporting desert soil dust? In this respect a distinction should be made between different intensities of dust transport, i.e. from low dust density with good visibility conditions and weak or moderate winds to high dust density and very poor visibility in connexion with strong winds (duststorm conditions).

2. How are these weather mechanisms linked to the actual broad-scale weather situation?

3. What is the frequency distribution over the Saharan area of dust observations as contained in the meteorological SYNOP reports from meteorological stations?

4. Does this frequency distribution have seasonal changes, and, if so, in what manner?

5. Is this frequency distribution related to the topography of the Saharan area as well as to its vegetational and land use pattern and ground conditions?

6. Which is the critical wind speed for initiating the raising of dust into the air? Does this speed change from place to place? Is it depending on other factors like moisture in the air and soil, etc.?

7. Is it possible to visualize duststorms on synoptic weather maps and to follow their migrations and developments?

8. Is it possible to use synoptic weather observations as contained in SYNOP reports as a substitute for direct measurements of air-borne dust?

In this paper efforts have been made to give an answer to at least two of these questions namely No. 6 and 7.

6.2 SYNOP REPORTS AND DESERT DUST STUDIES

The meteorological observations as reported in the SYNOP code [code form WMO FM-11-V-SYNOP in WMO's Manual on Codes (WMO, 1974)] provide much useful material for investigations of air-borne soil dust, in particular

wind direction	(dd in the SYNOP code)
wind speed	(ff in the SYNOP code)
visibility	(VV in the SYNOP code)
present weather	(ww in the SYNOP code)
past weather	(W in the SYNOP code)

The two digits for VV in the SYNOP code permit a very accurate reporting of the visibility, with intervals of 100 m from 0.0 km up to 5.0 km, and with intervals of 1 km from 5 km up to 30 km.

The 'present weather' [code 4677 in WMO's Manual on Codes (WMO, 1974) for ww – Present weather], gives a lot of information in relation to air-borne dust as demonstrated in Table 6.1.

The information which can be obtained from W (past weather) is much less, but anyhow important. Only one digit, i.e. $W = 3$, provides information on dust: sandstorm or duststorm has occurred during the last three hours or six hours prior to the actual observation (depending on the time of the observation).

Some indirect information about the state of the ground (dry or wet) can be drawn from those figures in the ww code, which indicate rain, drizzle, or showers at the time of observation or during the hour prior to the observation. These code figures should be studied together with the code figure giving the amount of

TABLE 6.1 Data on Air-borne Dust

Code figure ww	Symbol	Text en clair	
00–03		The characteristic change of the state of sky is described, but no hydrometeors (precipitation, fog, hoarfrost, etc.), lithometeors (haze, smoke, drifting dust, etc.) or electrometeors (lightning, thunderstorms, etc.) are present	
04		Visibility reduced by smoke, e.g. veldt or forest fires, industrial smoke or volcanic ashes	
05		Haze	
06		Widespread dust in suspension in the air, not raised by wind at or near the station at the time of observation	
07		Dust or sand raised by wind at or near the station at the time of observation, but no well-developed dust whirl(s) and no duststorm or sandstorm seen	
08		Well-developed dust whirl(s) or sand whirl(s) seen at or near the station during the preceding hour or at the time of observation, but no duststorm or sandstorm	
09		Duststorm or sandstorm within sight at the time of observation or at the station during the preceding hour	
30			– has decreased during the preceding hour
31		Slight or moderate dust-storm or sand-storm	– no appreciable change during the preceding hour
32			– has begun or has increased during the preceding hour
33			– has decreased during the preceding hour
34		Severe dust-storm or sand-storm	– no appreciable change during the preceding hour
35			– has begun or has increased during the preceding hour
98	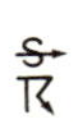	Thunderstorm combined with duststorm or sandstorm at time of observation	

precipitation during the previous 6, 12, or 24 hours (depending on the time of the observation).

The investigations presented in this paper are based on SYNOP reports from the Sudan as contained in Synoptic Bulletins issued by the Sudan Meteorological

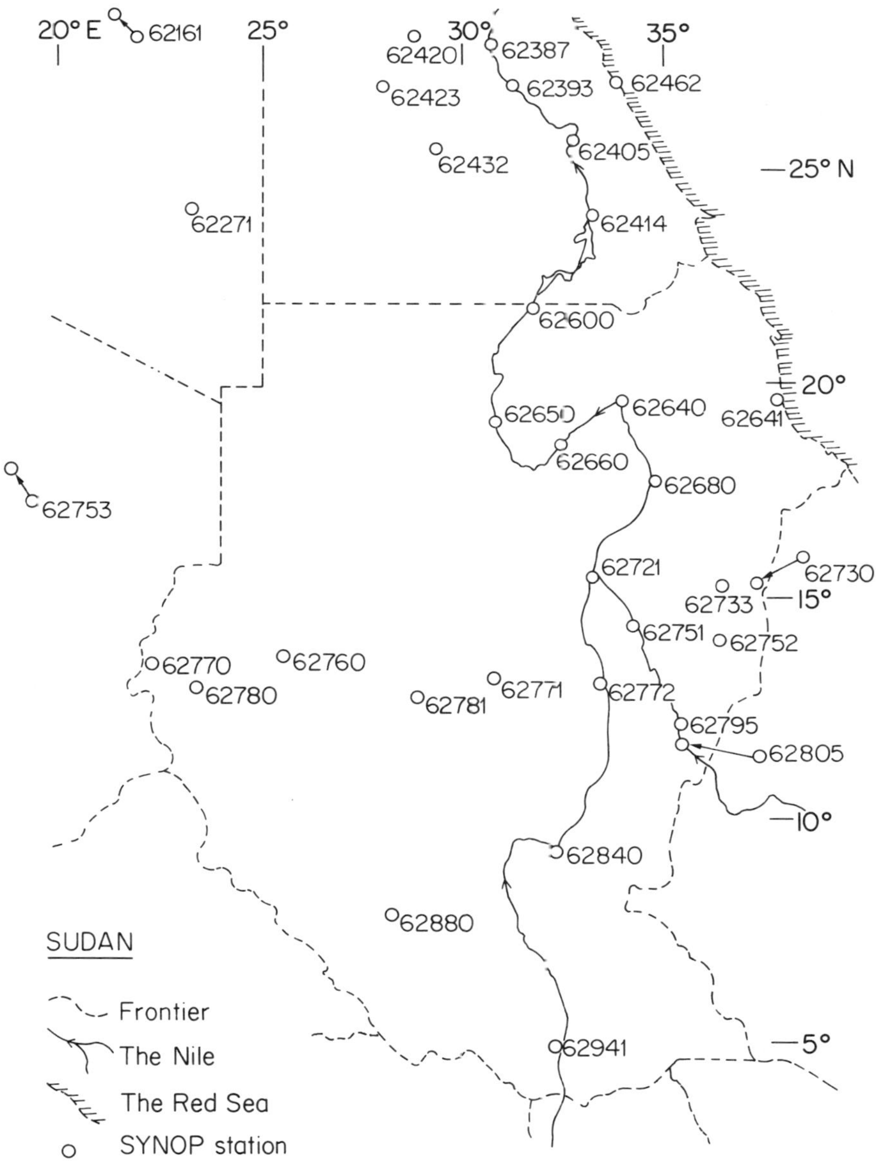

Figure 6.1 The network of meteorological SYNOP stations in the Sudan area. The station numbers are plotted quite close to the station (the small circle). In the weather charts (Figures 6.4 to 6.7) four stations, namely 62161, 62730, 62753, and 62805 have been plotted at a small distance from the true position owing to insufficient plotting space. The true position of these stations is the circle where the arrows end.

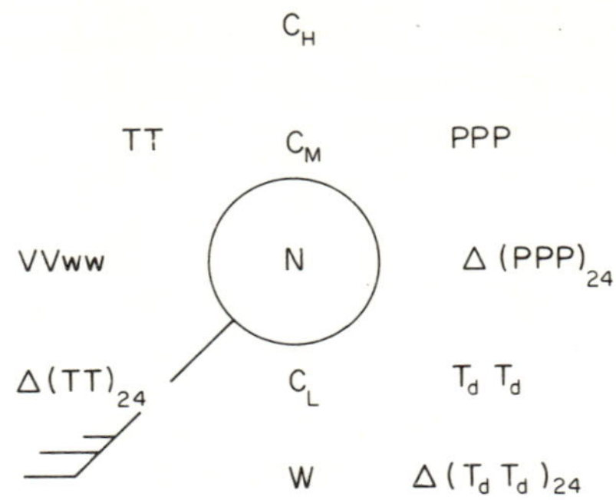

Figure 6.2

Department. SYNOP reports have also been taken from U.S. Northern Hemisphere Data Tabulations issued by the U.S. National Oceanic and Atmospheric Administration (NOAA). In this paper the SYNOP reports have been used for demonstrating their fitness for use in relation to two types of studies, namely

- for a case study of duststorms by means of synoptic weather charts
- for a statistical study in relation to the critical wind speed for initiating the raising of desert soil dust into the air.

Figure 6.1 gives the chart form for the Sudan area in which the SYNOP stations have been plotted together with the relevant station number according to WMO's publication: Weather Reporting, Volume A, observing stations (WMO, 1975).

For the plotting of the SYNOP reports the following station model was used (Figure 6.2).

In the station model below the code letters are explained (Figure 6.3).

The wind direction is plotted as a line from the station circle towards the direction, from which the wind is blowing, and the wind speed as slanting bars on

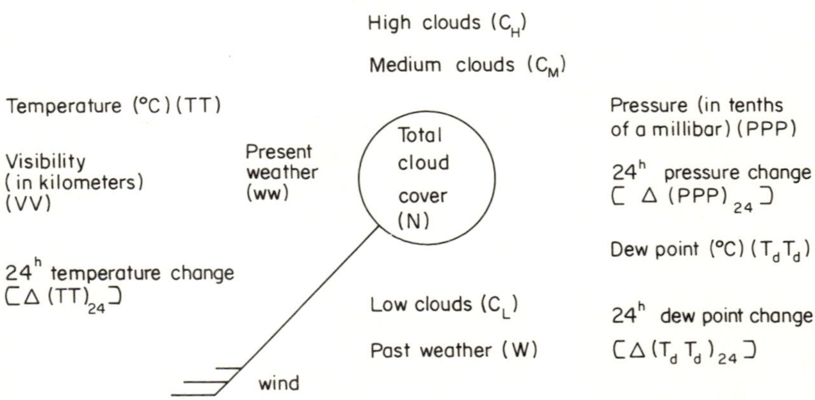

Figure 6.3

that line. One full bar denotes 8 to 12 knots, and a half bar 3 to 7 knots. No bar is plotted for wind speeds below 3 knots. Thus, if the wind is blowing from 230° with a speed of 26 knots this wind is plotted as 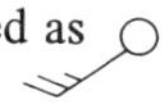

6.3 THE SUDANESE DUSTSTORM FROM 16th TO 19th MARCH, 1974, A CASE STUDY

In Figures 6.4 to 6.7 are reproduced synoptic weather analyses for the Sudan area during a period when a duststorm passed this area. In order to avoid over-burdening of the analyses presented in this paper only isopleths for visibility (in a quasi-logarithmic scale) have been drawn. On account of the sparse SYNOP station network, particularly in the western part of the area, the isopleths for visibility give only an approximate idea of the true visibility pattern.

On the 15th March, 1974, at 00^{h} GMT, the visibility was above 20 km at all Sudan SYNOP stations but for station No 62640 (Abu Hamed) (Figure 6.4). One day later (Figure 6.5) the wind speed started to increase at the northernmost station, No 62650 (Dongola), which reported dust raised by the wind (ww = 07) and wind speed 17 knots, while the neighbouring station No 62660 (Karima) still had good visibility though the wind speed also had increased to 17 knots.

The weather had deteriorated substantially 18 hours later at both stations: Abu Hamed reported severe duststorm with wind speed 30 knots (gale) and visibility between 100 and 200 m, and Karima dust raised by the wind with wind speed 23 knots and visibility 2 km. The duststorm seems to have been connected with a cold front passage (in fact the temperature dropped 8°C in Abu Hamed as compared with 24 hours earlier). On the 17th March at 00^{h} GMT the duststorm had reached Khartoum (station No 62721) (Figure 6.6). The duststorm continued to move south-south-eastwards [see the weather chart for 18th March 12^{h} GMT (Figure 6.7)] and had disappeared from the weather chart 18 hours later.

In the diagram, Figure 6.8, a number of meteorological variables, taken from the SYNOP reports from Khartoum during the actual period (14th March, 12^{h} GMT – 19th March, 00^{h} GMT) are plotted. The variables are wind direction and speed, present weather, visibility, temperature, dew-point and pressure. The diagram supports the assumption that this duststorm was connected with the passage of a cold front. After the arrival of the duststorm (17th March, 00^{h} GMT) both the temperature and the dew-point dropped appreciably while the pressure rose.

6.4 A STATISTICAL STUDY BASED ON SYNOP OBSERVATIONS

The results of a preliminary investigation concerning critical wind speed for initiating the raising of desert soil dust into the air for one SYNOP station in the Sudan [Dongola (62650)] during April 1973 is presented in Figure 6.9. The three

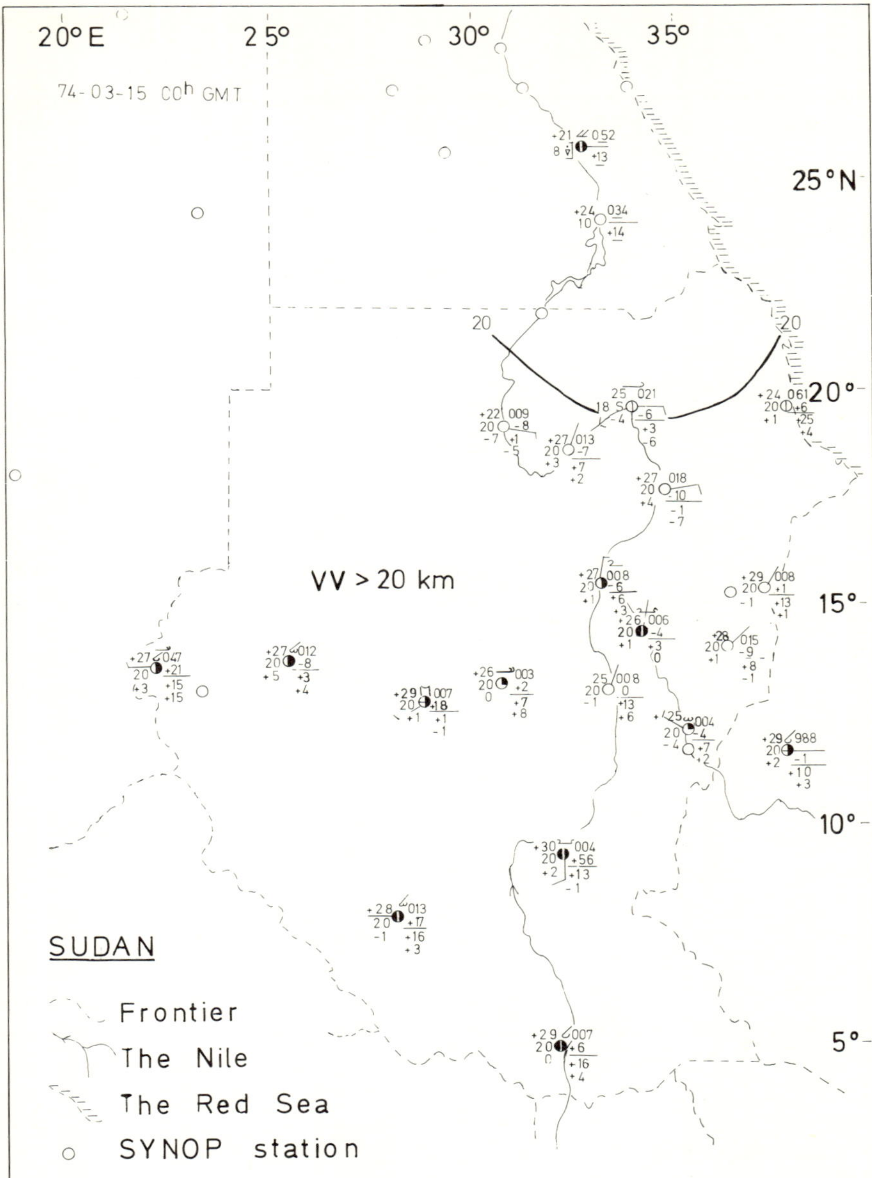

Figure 6.4 Weather charts based on SYNOP observations for the observation time 15th March, 1974, 00^{h} GMT. Only the analysis of the visibility pattern has been introduced in the charts. The isopleths for visibility (VV) have been drawn for the values 20 km, 10 km, 5 km, 2 km, 1 km, 0.5 km, 0.2 km and 0.1 km. The dotted isopleths only give a general indication of the visibility pattern. An area with very low visibility (< 500 m) in connexion with reports of air-borne dust can be followed on the weather charts during its movement sout-south-east.

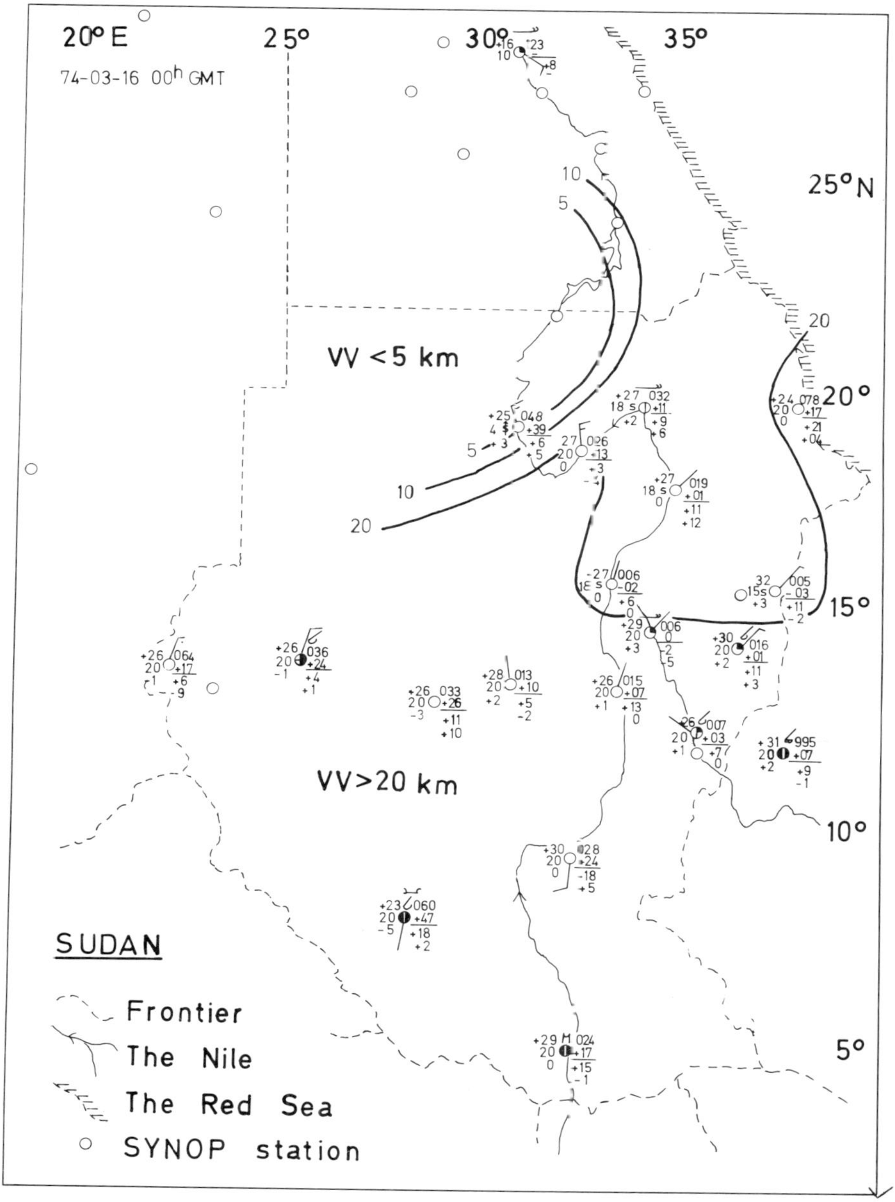

Figure 6.5 As Figure 6.4, for observation time 16th March, 1974, 00^h GMT

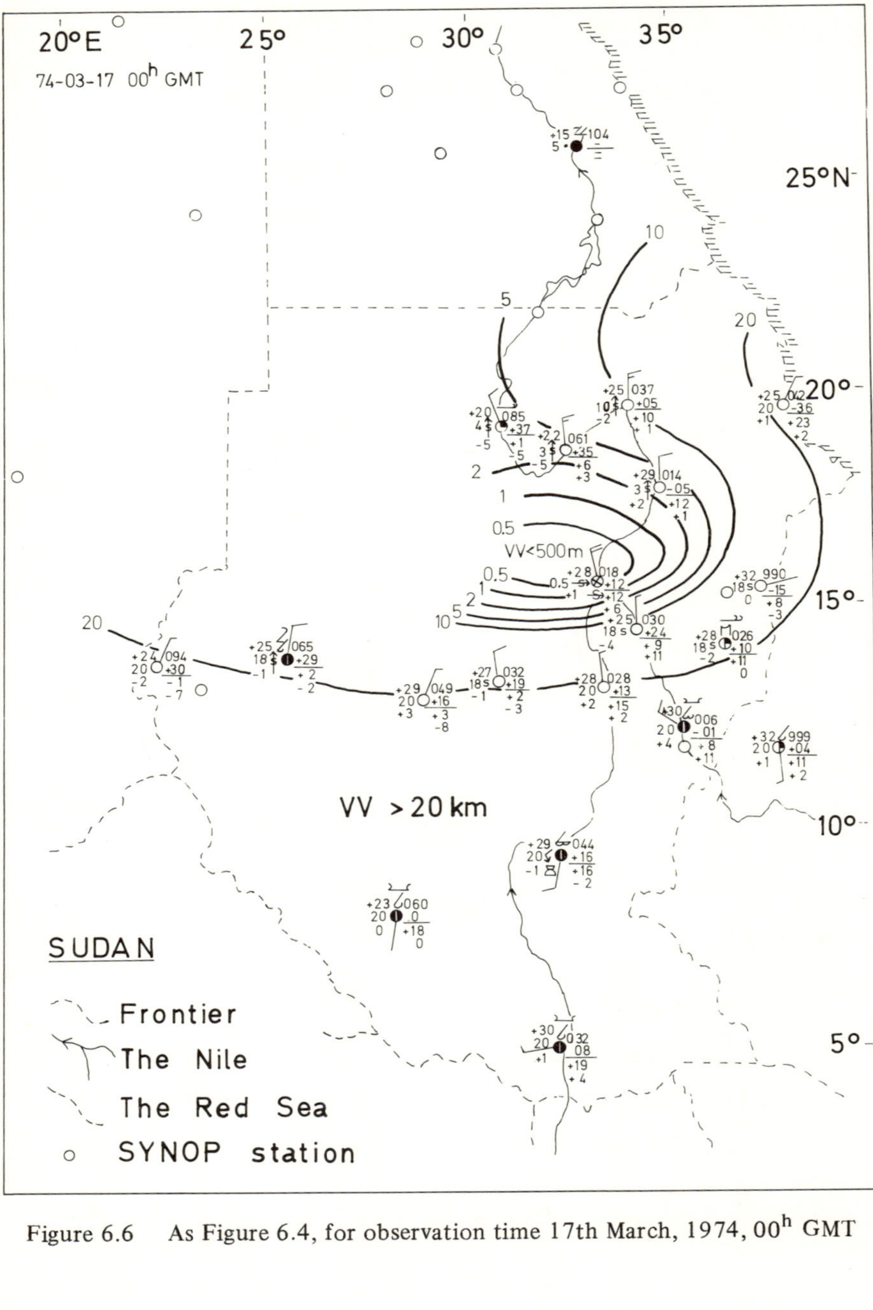

Figure 6.6 As Figure 6.4, for observation time 17th March, 1974, 00^h GMT

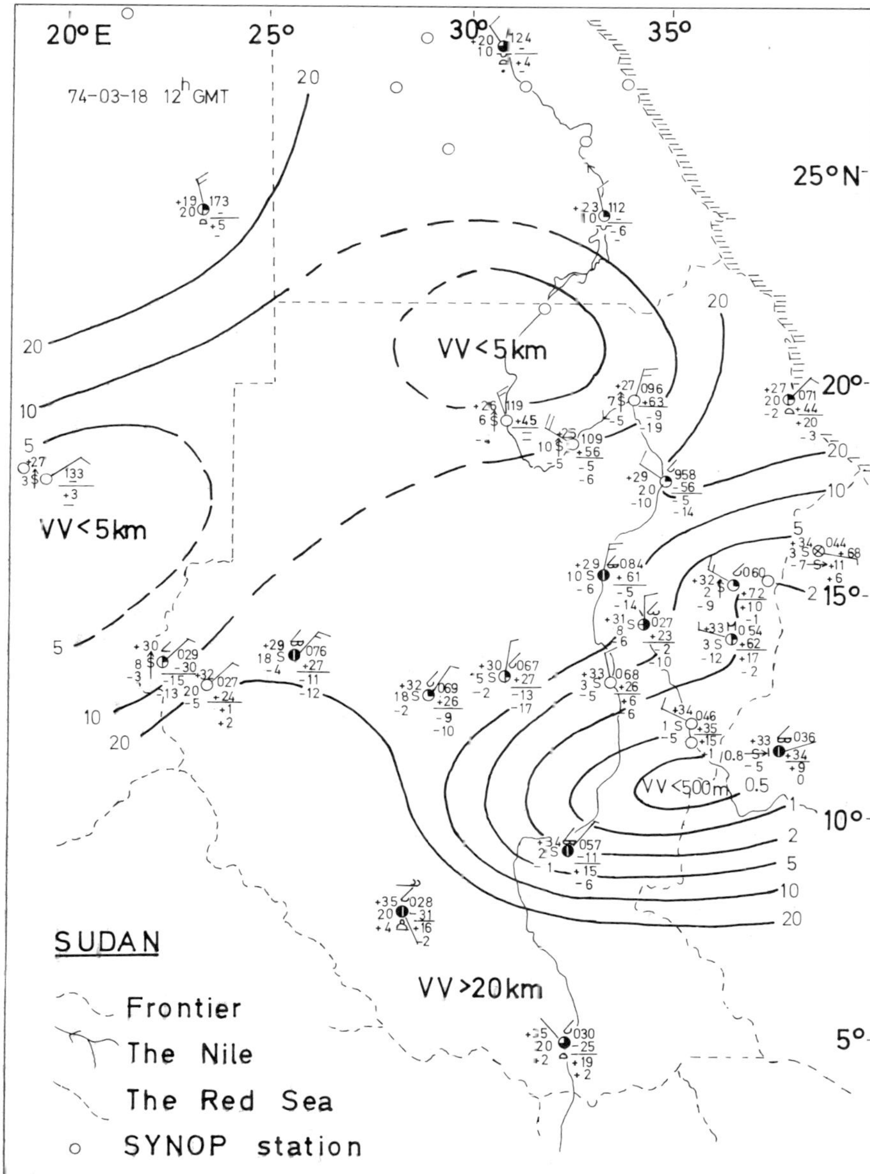

Figure 6.7 As Figure 6.4, for observation time 18th March, 1974, 12h GMT

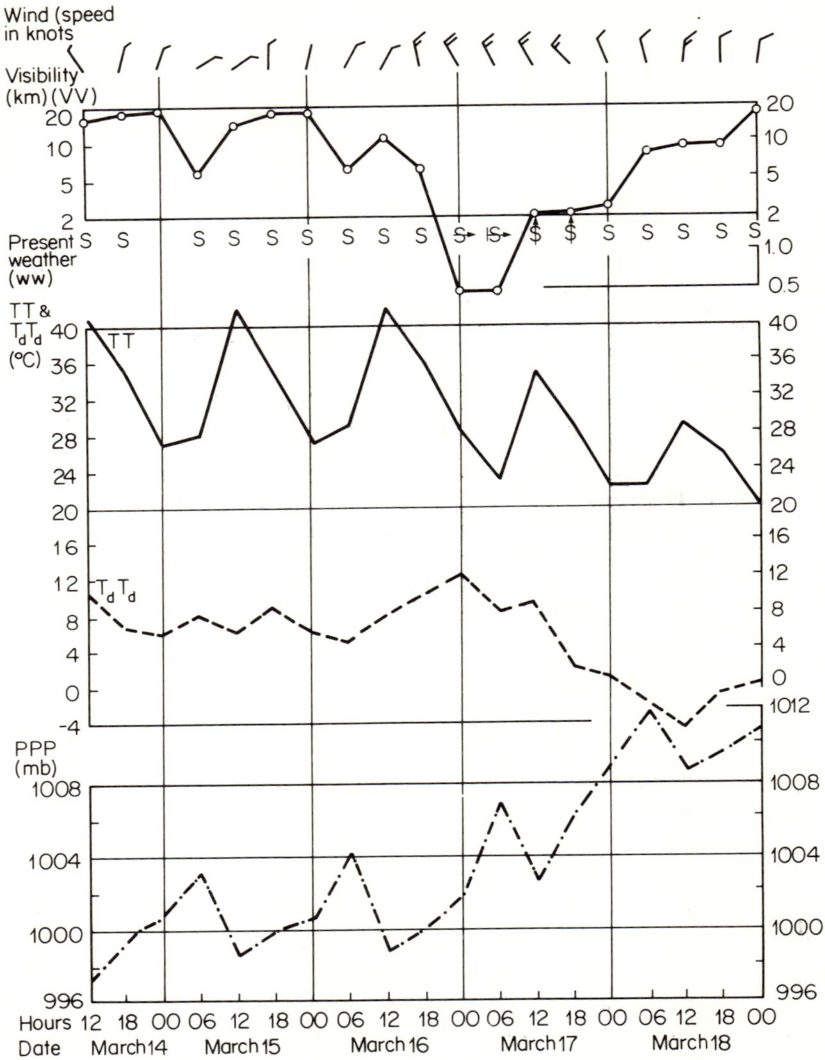

Figure 6.8 Diagram showing the variations in wind, weather, visibility, temperature (TT), dewpoint (T_dT_d), and pressure (PPP) at station 62721 (Khartoum) in Sudan, from 14th to 18th March 1974, when a duststorm moved across the Sudan. The duststorm hit Khartoum at about midnight the 17th March. The storm was connected with the invasion of a colder and drier airmass.

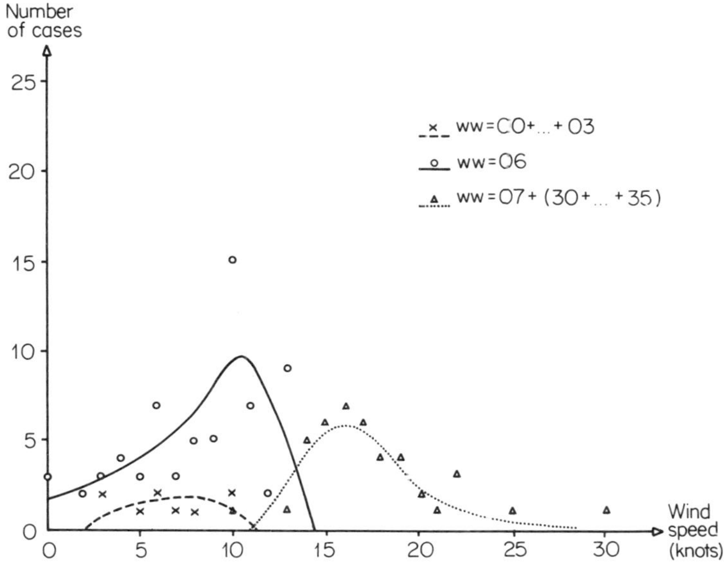

Figure 6.9 The three sets of curves give the frequencies of present weather ww = 00 + . . . + 03 (no air-borne dust, cf. Table 6.1) ww = 06 (dust suspended in the air) and ww = 07 + (30 + . . . + 35) (dust raised by the wind including duststorms) with respect to the wind speed observed at the same observation time. The observations are taken from the SYNOP station Dongola in the Sudan (station number 62650) during April 1973 (observation times 00^h, 06^h, 12^h, and 18^h GMT). The threshold wind speed for raising soil dust into the air is about 12 knots (no reports of ww = 06 above 14 knots and no reports of ww = 07 + (30 + . . . + 35) below 10 knots)

sets of curves give the frequencies of ww = 00 + – – – + 03, ww = 06 and ww = 07 + (30 + – – – + 35) respectively (for explanation of ww *cf.* Table 6.1 above), with respect to the wind speed observed at the same observation time. The mode for ww = 00 + – – – + 03 is about 7 knots, for ww = 06 about 10 knots and for ww = 07 + (30 + – – – + 35) at about 16 knots. The threshold value for dust-raising wind speeds seems to be about 12 knots [no reports of ww = 06 above 14 knots and no reports of ww = 07 + (30 + – – – + 35) below 10 knots].

Since this investigation comprises only one month and one SYNOP station the results should be regarded as preliminary.

6.5 CONCLUSIONS

The mobilization, the air-borne transport and the deposition of desert soil dust are phenomena where meteorological factors are decisive. Since many variables of

importance for studying these phenomena are available in the meteorological observations as reported from SYNOP stations these SYNOP reports can be used for such studies. This is demonstrated with two examples. In the first example a duststorm over the Sudan area is visualized by means of isophleths for visibility on synoptic weather charts. It is possible to follow on the charts the movement of the duststorm in a south-south-east direction over the Sudan area during a four day period.

In the second example the SYNOP reports have been used in a statistical study regarding the threshold wind velocity for raising desert soil dust into the air. According to this study the threshold value for dust raising wind speeds seems to be about 12 knots. This result should, however, be regarded as preliminary since the study only comprises one month and one SYNOP station.

REFERENCES

WMO (1974). International Codes. WMO-306. *Manual on Codes*, Vol. 1.
WMO (1975). Observing Stations. WMO-9. *Weather Reporting*, Vol. A.

CHAPTER 7

The Tropospheric Circulation Over Africa and its Relation to the Global Tropospheric Circulation

R. E. NEWELL and J. W. KIDSON

ABSTRACT

Long-term monthly mean zonal and meridional winds over Africa are presented for levels between 1000 and 100 mb in the form of maps and cross-sections. Desert dust transport depends on these winds in three ways: the winds control the flux and convergence of moisture which in turn governs the rainfall and hence the availability of dust at the surface; the surface wind strength regulates the pick-up of dust; the mean flow pattern controls the path of dust after it leaves the source.

We attempt to explain the classical Arrhenius–Turekian haze maps in terms of these three inter-relationships. Latitude–time sections of the monthly mean 850 mb zonal wind and rainfall along the prime meridian and along 40°E show the most favourable latitudes and months for dust transport. Along the meridian 10–15°N in January, 25–30°N in July and 25°S throughout the year are the most favoured regions for westward transport (see Figure 6a). At 40°E eastward transport occurs in June–August in the 5–30°N latitude band and over much of the year at 30–35°S. Rainfall amounts along the prime meridian of less than 1 cm per month are shown shaded in the Figure. The difference in character of the dust (red-brown in summer, grey to black in winter) emerges clearly from these sections which show a Saharan summer source and a Sahelian winter source.

The interhemispheric monsoon northerly flow examined in detail by Findlater is very strong near 40°E in the April–October period at 850 mb, with a return flow in the upper troposphere.

The mean flow data may be used directly for dust transport calculations, with air concentration data, as it seems likely that eddy flux contributions are small as has been found for water vapour. Such computations need base maps of soil state *vs.* season.

Examples of flow differences between a wet period, 1958–1962, and a dry period, 1970–1973, are given. It has been shown elsewhere (Kidson, 1977) that the 850 mb trough in the Sahelian region is more marked and the 200 mb flow is stronger in wet than in dry years. Here we try to assess the influence of these changes on dust transport patterns. It seems evident that dust in Atlantic deep-sea cores is governed both by the rainfall and the wind patterns: a change in the seasonal excursion of the Hadley cell in ice ages, for example, could influence the observed core concentrations as much as a change in wind strength.

The time variations of the mean winds over the past 25 years are thought to be related to global circulation pattern changes. Some aspects of these possible changes are illustrated using tropospheric temperature data and sea surface temperature data from the Pacific, Atlantic and Indian oceans. Data on a much longer time scale have also been examined and it is concluded that the exact physical cause of changes in flow over Africa has not yet been isolated.

7.1 INTRODUCTION

The tropospheric circulation influences the transport of African desert dust in three ways: the winds control the flux and convergence of moisture which in turn governs the rainfall and hence the availability of dust at the surface together with the washout; the surface wind strength regulates the pick-up of dust; the mean flow pattern controls the path of dust after it leaves the source. In this paper the long-term mean characteristics of the tropospheric circulation and rainfall are discussed. Changes in the tropospheric circulation over Africa are related to changes in the global atmospheric circulation and hypotheses to explain the changes are put forward.

The classical books by Kendrew (1961) and Griffiths (1972) give a detailed summary of the surface climatology of each country and the Atlas of Thompson (1965) covers much of the free-air climatology as well as rainfall by a beautiful set of maps. The climate of the Sahara has been thoroughly discussed by Dubief (1959, 1963). Prior work at M.I.T. (Newell *et al.*, 1972) provided seasonal maps of the wind components at levels between 1000 mb and 100 mb together with maps of vertical motion deduced from the continuity equation (Newell *et al.*, 1974). Another source of flow pattern information, which contains examples of daily patterns, is the Munitalp Foundation Symposium on Tropical Meteorology in Africa (1960). Considerable fundamental work on the flow patterns has also been carried out by Flohn and his colleagues (e.g. Flohn, 1965; Flohn and Korff, 1965; Struning and Flohn, 1969).

The main phenomenon which characterizes the tropospheric circulation is the intertropical convergence zone (ITCZ) and its steady seasonal migration. The cloudiness and rainfall associated with this phenomenon move to the south in northern winter and to the north in northern summer, reaching their extreme northerly position over the Sahel in August. The main features of this zone appear in the surface streamline and pressure patterns of Griffiths and Soliman (1972) and the surface flow is discussed in this volume by Dubief (1979). The ITCZ runs west–east across Africa until at about 25°E it turns southwards in winter and becomes indistinguishable from the surface boundary between the Atlantic Ocean westerlies and the Indian Ocean easterlies. In summer the ITCZ continues eastwards to the Persian Gulf. To the north and south of the ITCZ there are regions of large scale subsidence with the largest region north of the equator in winter being associated with the extreme southern excursion of the rising motion. The zonally averaged mass flux patterns shown in Figures 7.1 and 7.2 illustrate the subsidence

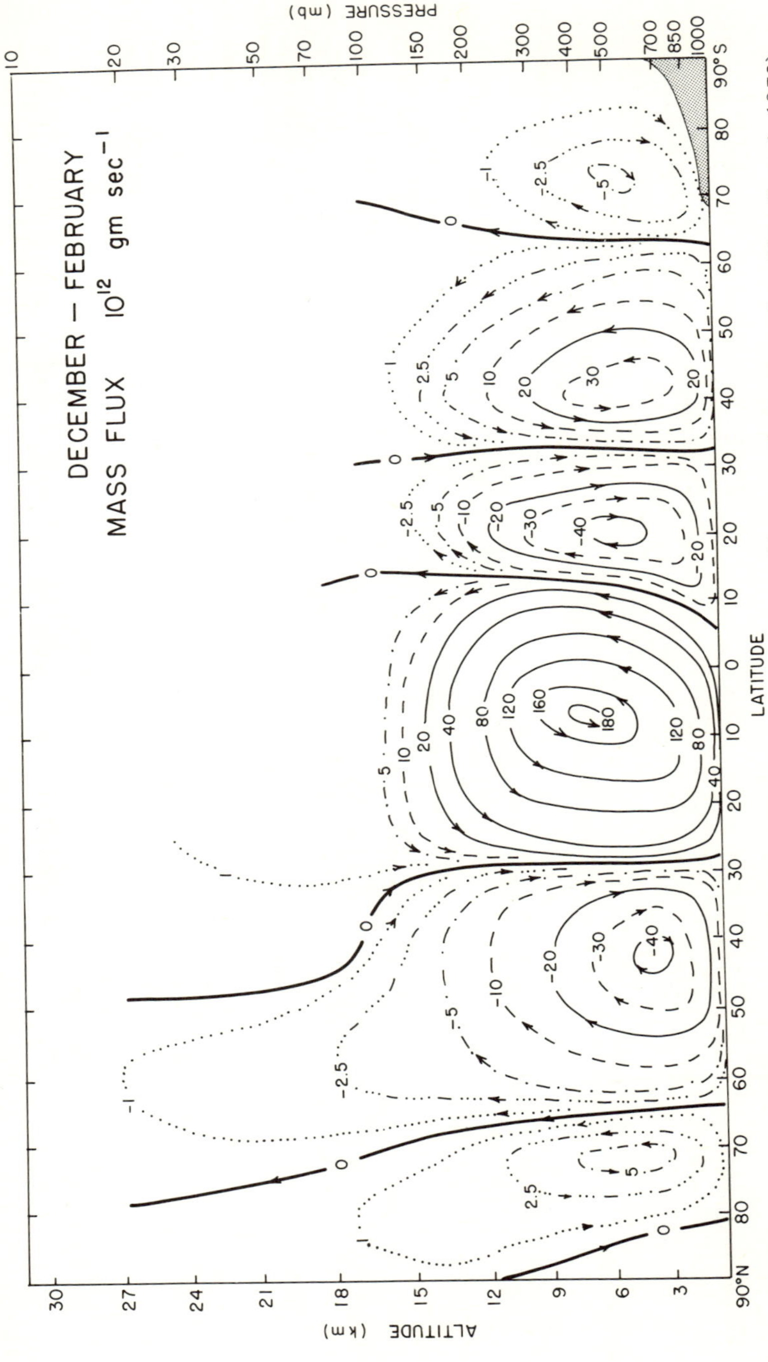

Figure 7.1 Zonally averaged mass flux December–February. Units: 10^{12} gm sec^{-1} (Redrawn from Newell *et al.*, 1972)

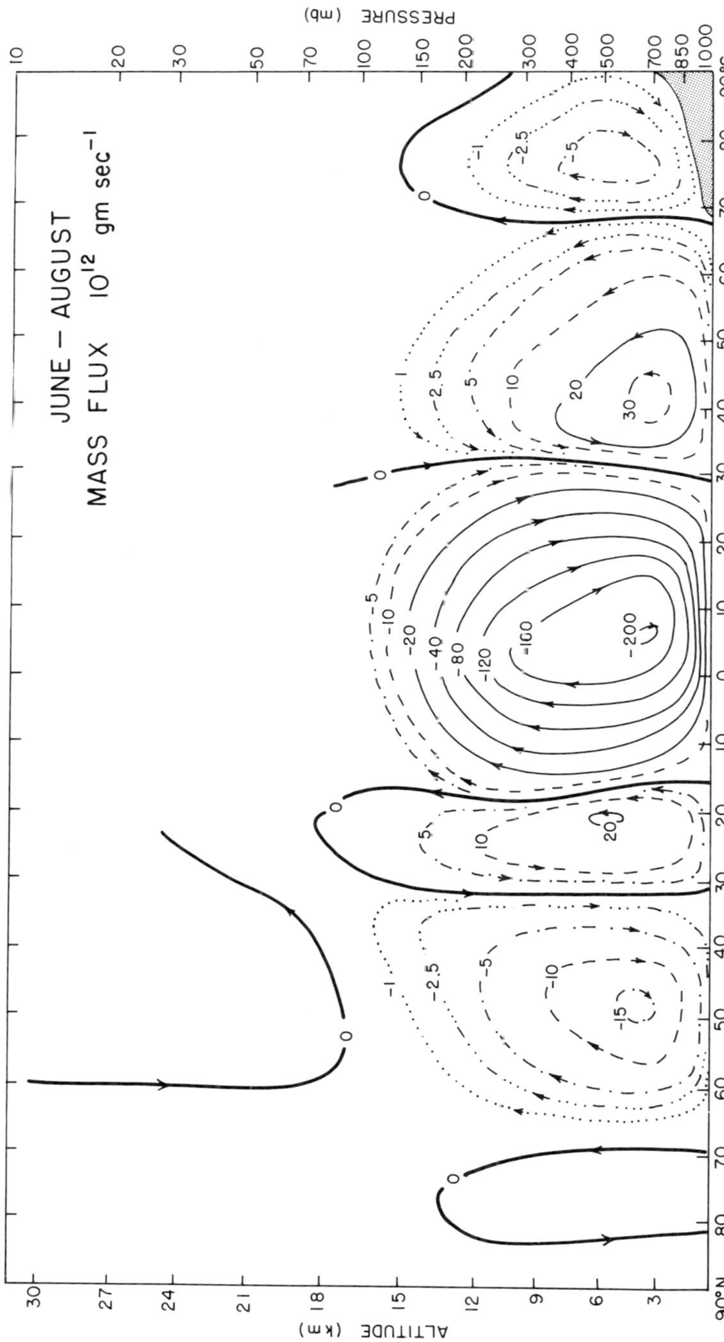

Figure 7.2 Zonally averaged mass flux June–August. Units: 10^{12} gm sec^{-1} (Redrawn from Newell *et al.*, 1972)

regions and it is, of course, this aspect of the global circulation which is responsible for the occurrence of deserts. At the northern and southern extremities of the continent the phenomena which characterize the circulation are the middle latitude baroclinic zones and the associated low-pressure rain-bearing transient systems, these being closest to the equator in the winter of the appropriate season. The description by means of the boundary layer climatology cannot be extrapolated upwards and there are, in fact, strong longitudinal variations in the upper troposphere which are significant in the dust transport problem.

7.2 LONG-TERM MEAN PATTERNS

7.2.1 Wind velocity

As part of a recent study of African rainfall and possible causes of Sahelian drought (Kidson, 1977) mean winds, moisture, and rainfall records have been assembled for all available African stations. Rainfall data cover the period 1951–73 (some of these were collected as part of the previous study of this problem by Tanaka *et al.*, 1975). Wind and moisture data covered the two periods, May 1958 to April 1963 and December 1969 to December 1973, data for the latter being obtained from the monthly Climatic Data for the World upper air tape. Upper air stations used are those shown in Figure 7.3. Rainfall stations are essentially the same as those shown in the map by Tanaka *et al.* (1975). Long-term mean monthly wind components have been plotted by computer on maps at levels between 1000 mb and 100 mb and the Cressman (1959) analysis procedure has been used to obtain an objective analysis on a 5° grid covering the region 33°N to 33°S and 15°W to 45°E. It was found necessary to use an anisotropic scan radius for the zonal wind component above the surface. While the meridional component is representative of a 5° grid square, the zonal component is representative of a 15° longitude, 5° latitude grid region. The components rather than streamlines are selected for illustration here as these values may be used directly in dust transport computations.

Examples of the analysed maps for 850 mb and 200 mb for four mid-season months are shown in Figures 7.4, 7.5, 7.6 and 7.7. The 850 mb zonal component, henceforth denoted by u, exhibits mainly variation with latitude in January, with easterly 'jets' at 10°N and 20°S and westerly flow north of 25°N, south of 30°S and in the 0–15°S zone. Maximum values occur to the north and to the south of the Sahara. In July there are substantial variations with longitude with a reversal from easterly to westerly winds across the Sahara. Maximum values occur at about 25°N near the prime meridian and at about 15°S. Flow from the west dominates the equatorial zone and both regions of easterlies are further north than in January. The 15° longitude scan radius should be borne in mind whenever appreciable longitudinal variations appear in the actual data. In July, for example, the westerlies

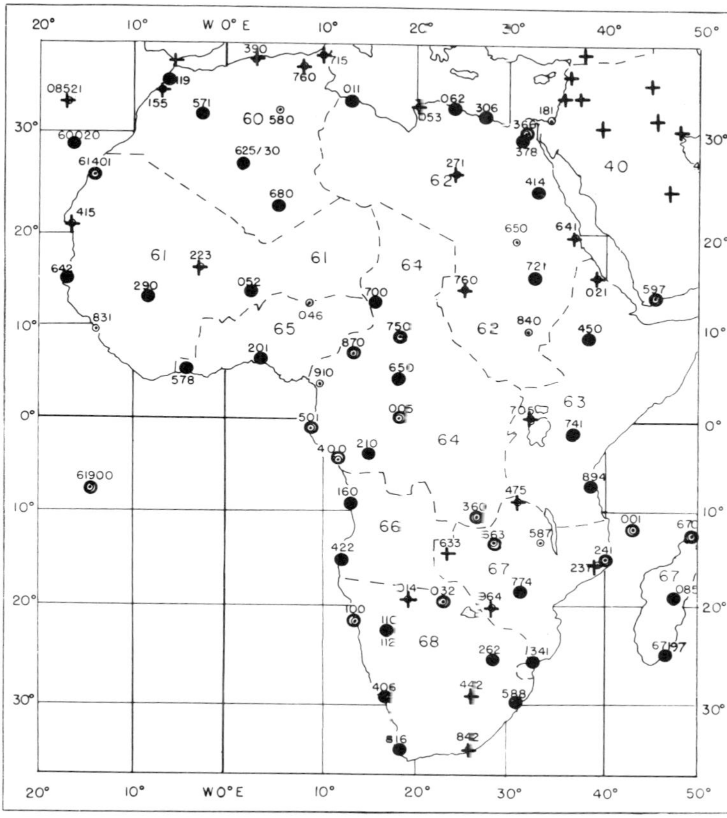

Figure 7.3 Distribution of stations used in the upper air analyses. Open circles show those present during 1958–63, crosses for those present during 1970–73 and solid circles for both periods

over Egypt are carried further westward by this analysis than would be the case if a hand analysis were used.

At 200 mb the zonal wind pattern is simpler than at 850 mb, with a belt of easterlies, centred in the summer equatorial region, embedded in a general westerly flow. The easterly belt is weaker at the equinoxes than at the solstices. While there are variations of speed with longitude there are no major directional changes. Maps of the 850 mb meridional component, denoted by v, show a structure which is much more patchy than that of u. Over the Sahara flow is towards the equator in the east and towards the pole in the west. There is a substantial cross-equatorial drift which is into the Southern Hemisphere in January and into the Northern Hemisphere in July. At 200 mb in January there is fairly strong northward flow from 0° to 20°N and also to the east of the ITCZ in southern Africa. In July the flow is southward from about 5°N to 30°S and generally northward further north.

The patterns at other levels may be viewed from a similar set of maps, but the portrayal of the vertical coherence cross-sections of the components is shown in Figures 7.8 and 7.9. The prime meridian is included, even though data is absent in the Southern Hemisphere, because it is representative of the western Sahara. The figures capture the essence of the basic mechanics namely that the Coriolis torque, acting on the meridional flow, governs the zonal wind field. Polewards motion in both hemispheres gives rise to westerly winds, as is very evident in the upper troposphere; likewise equatorwards flow is associated with easterly winds. The patterns of the zonal wind maps of Figures 7.4 to 7.7 can also be accounted for with this principle. The sections also show that poleward flow in the upper troposphere is accompanied by equatorward flow in the lower troposphere. Figure 7.10 is a cross-section along the equator and beside longitudinal variations this also illustrates the compensation between the upper and lower troposphere.

It seems necessary to consider variations on a month-to-month basis for the dust transport problem and Figures 7.11, 7.12, 7.13, and 7.14 show values of u and v at three longitudes for 850 mb and two longitudes for 200 mb. The main interest here is the Sahara and it is recognized that data is poor south of the equator for 0° longitude. The problem is not so serious for the u-component as for the v-component because of the anisotropic influence radius. In the western Sahara strongest easterly winds at 850 mb occur at 10–15°N in the winter and at 25°N in the late summer. At 40°E there are westerlies from the equator northwards in the summer.

The strong flow at 850 mb from the southern to the northern hemisphere at 40°E, 5°S in summer is worth noting. The low level jet in this region discovered by Findlater (1969a,b; 1972) is a little underestimated in strength by the analysis technique; in fact, Findlater reports a mean of 15 m sec^{-1} (cp. 8 m sec^{-1} here) with some occasions reaching 25 m sec^{-1}.

7.2.2 The mean flow and dust

Dust pick-up was widely debated at the workshop. A desiccated surface favours pick-up and a critical pick-up speed which varies from point to point, is required. Dryness is represented in Figure 7.14 by the shaded regions which correspond to rainfall values of less than one centimetre per month. Over the Sahara highest wind speeds from Figure 7.14 occur at 10–15°N in winter, 20°N in spring and 25°N in late summer. Concomitantly the dry region moves north with season. On these two counts then maximum pick-up is expected to vary seasonally. To show the geographical variation the distribution of mean wind kinetic energy at 850 mb is shown in Figure 7.15. Ideally a better parameter to represent pick-up might be the square of the instantaneous wind speed but this was not available. There is a maximum on each side of the Sahara in January with the largest values south of the Sahara and in West Africa. In July the maximum is in the Northwest Sahara with high values also in the Northeast. The frequency of haze over the ocean is a measure of air-borne

MEAN $\bar{U}$ AT 850mb - JANUARY

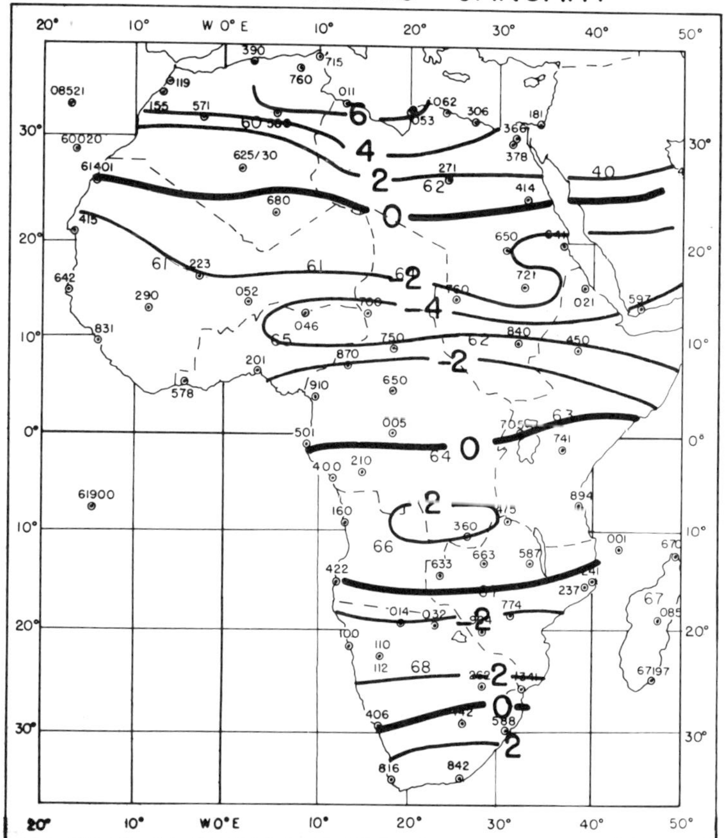

MEAN $\bar{U}$ AT 850mb - APRIL

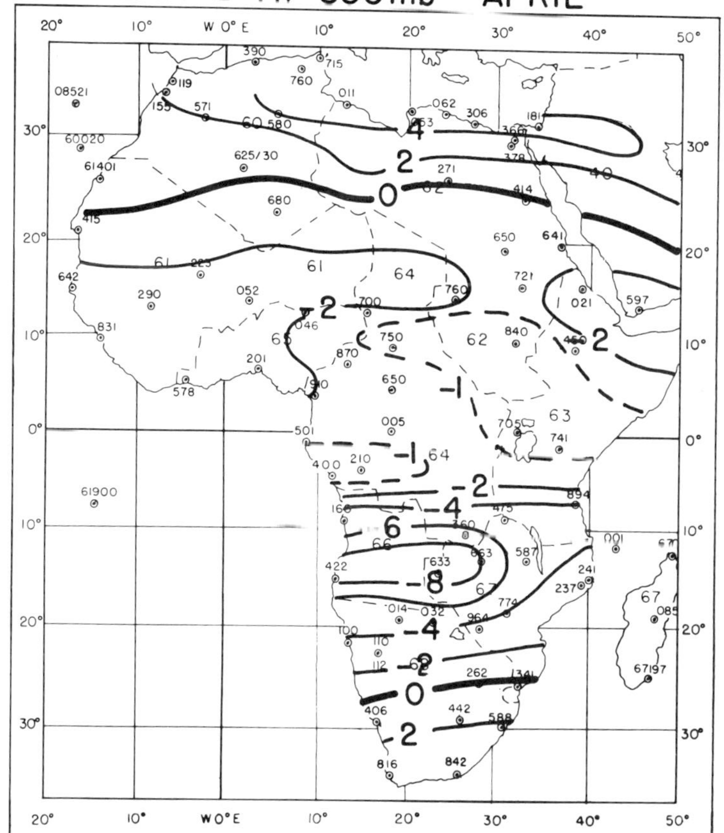

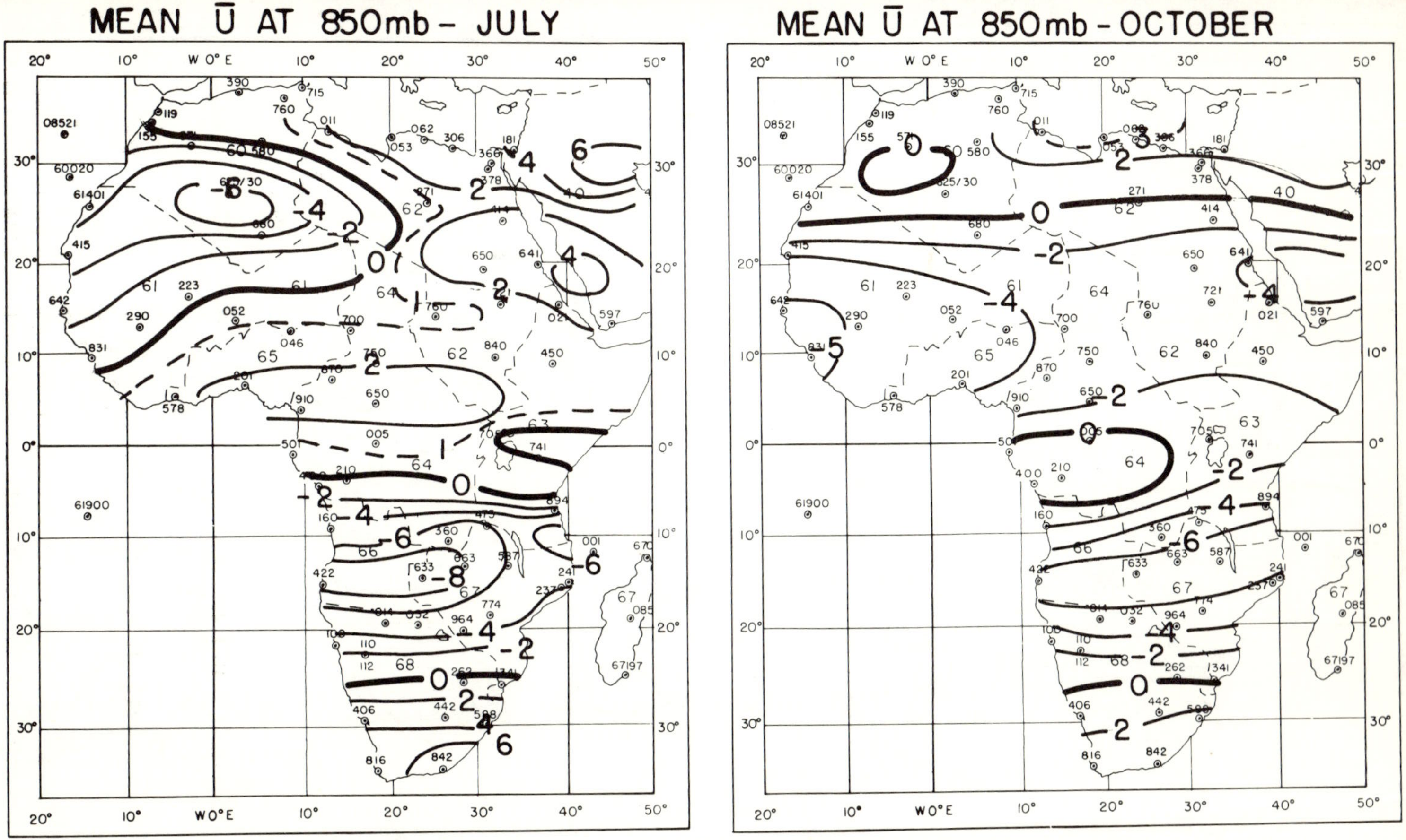

Figure 7.4 Monthly mean wind component $\overline{U}$ at 850 mb for January, April, July, and October

MEAN $\bar{U}$ AT 200mb - JANUARY

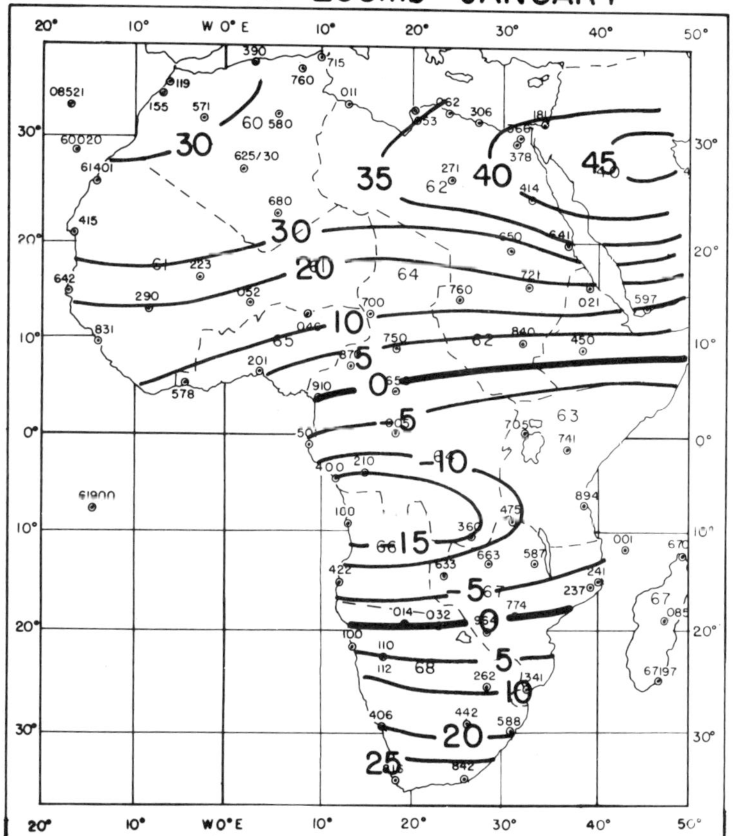

MEAN $\bar{U}$ AT 200mb - APRIL

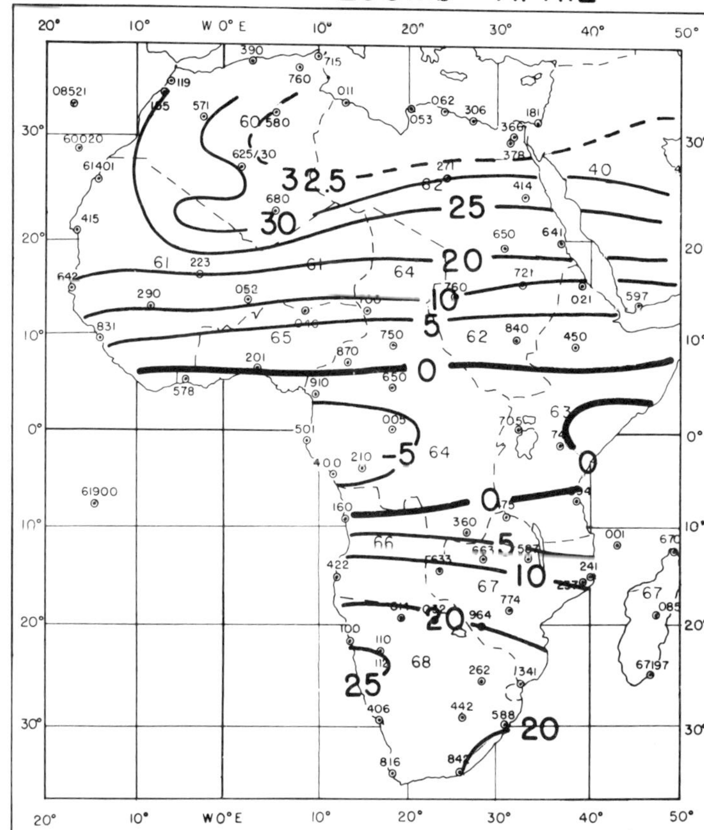

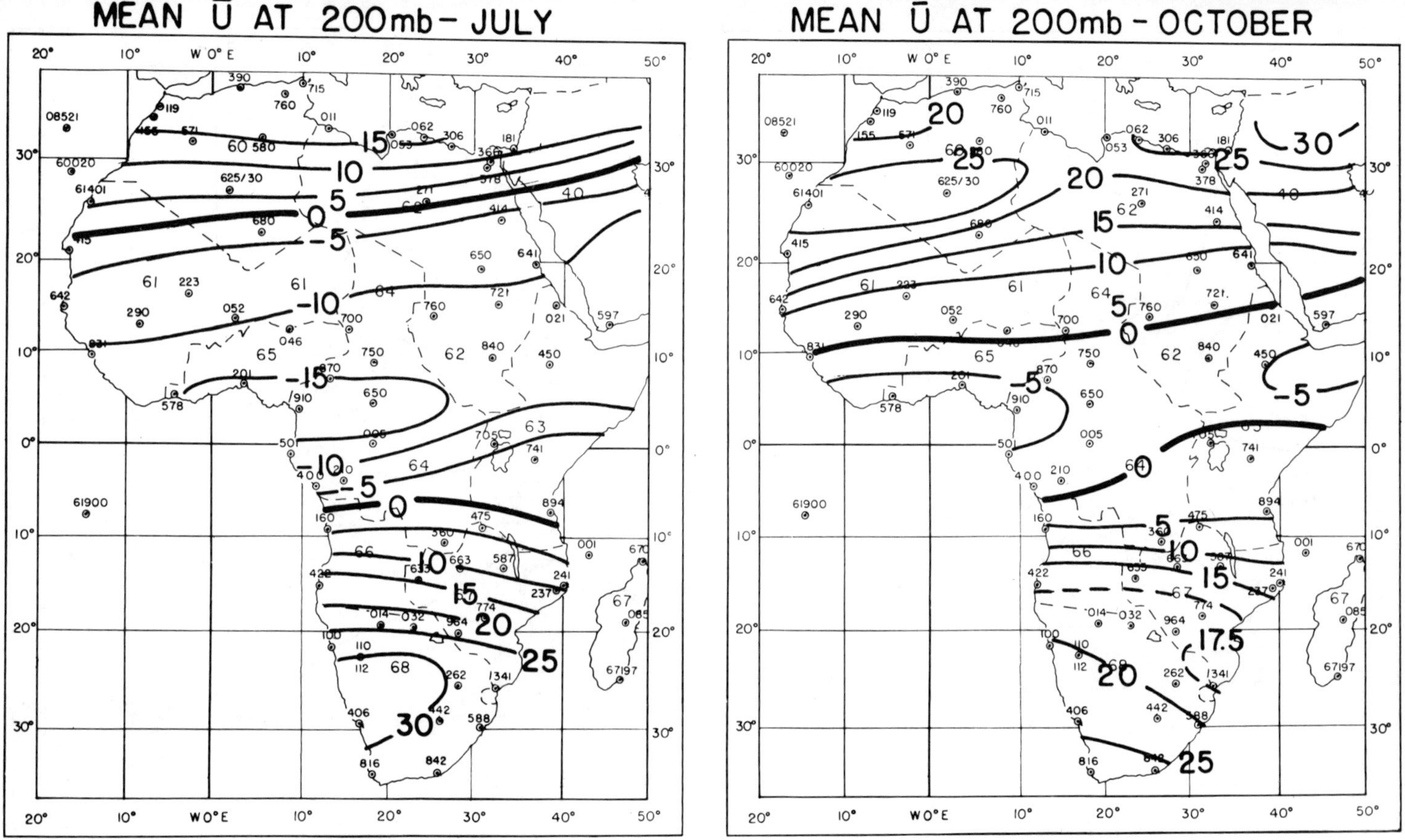

Figure 7.5 Monthly mean wind component $\overline{U}$ at 200 mb for January, April, July, and October

MEAN $\overline{V}$ AT 850mb - JANUARY

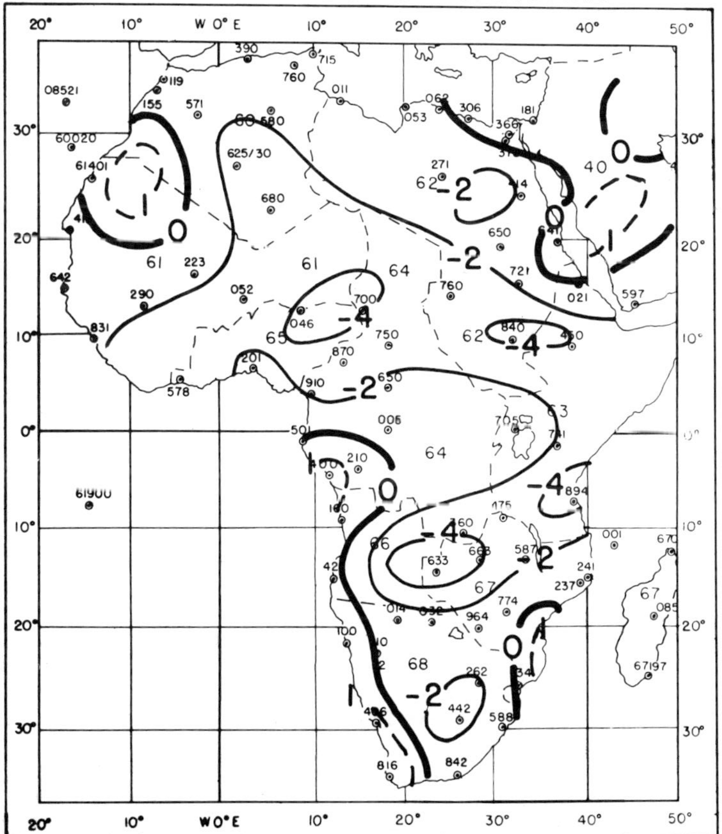

MEAN $\overline{V}$ AT 850mb - APRIL

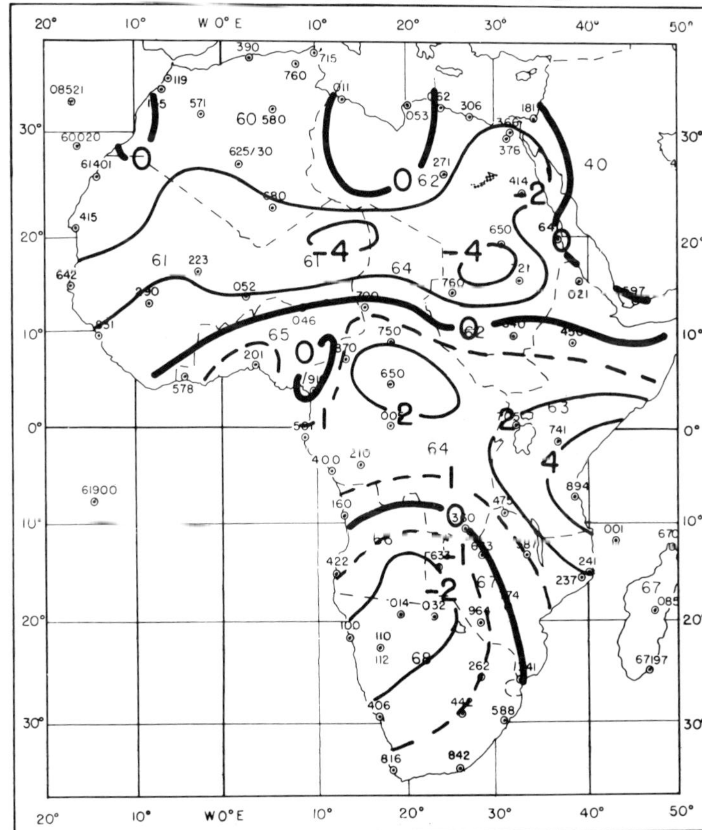

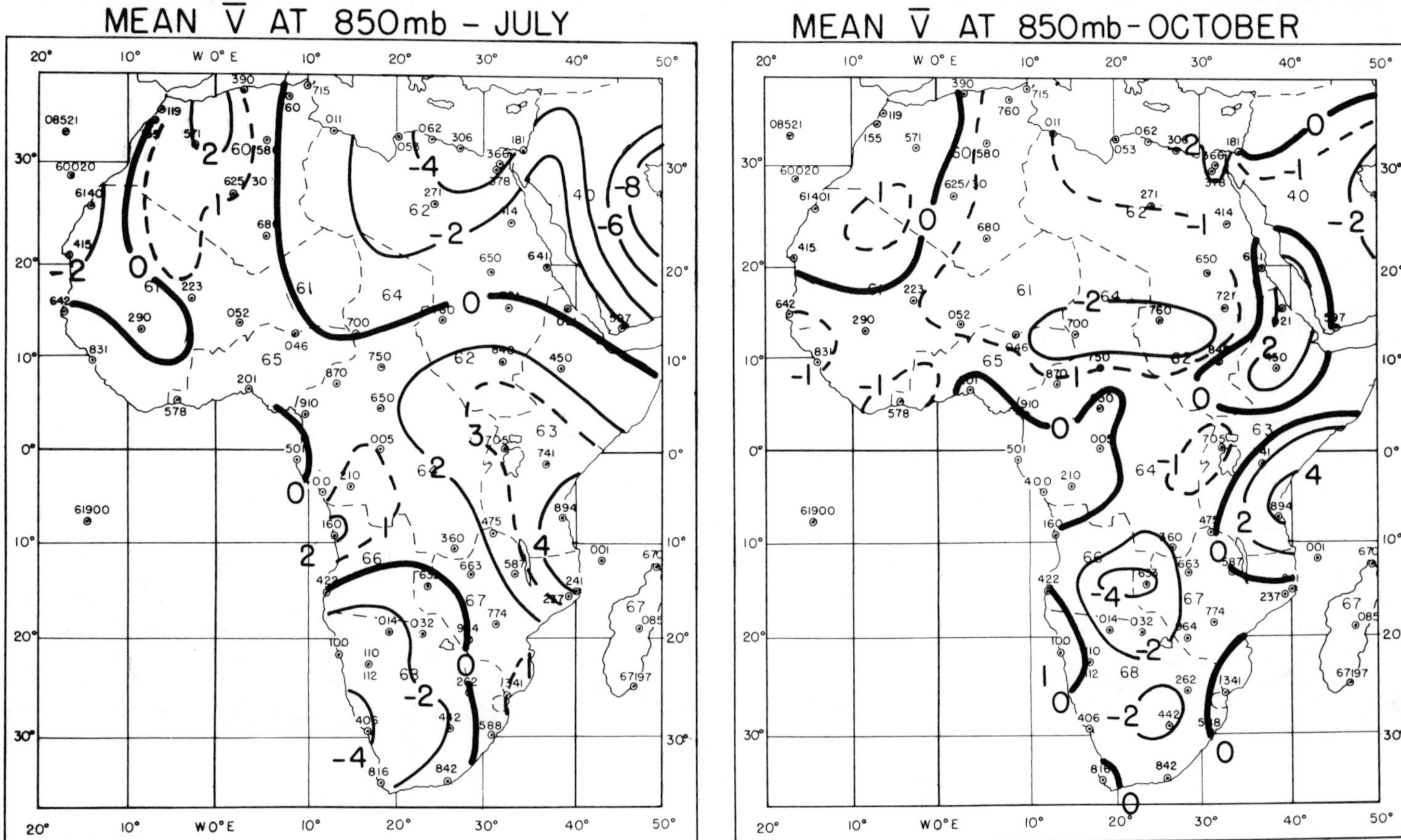

Figure 7.6 Monthly mean wind component $\overline{V}$ at 850 mb for January, April, July, and October

MEAN $\bar{V}$ AT 200mb - JANUARY

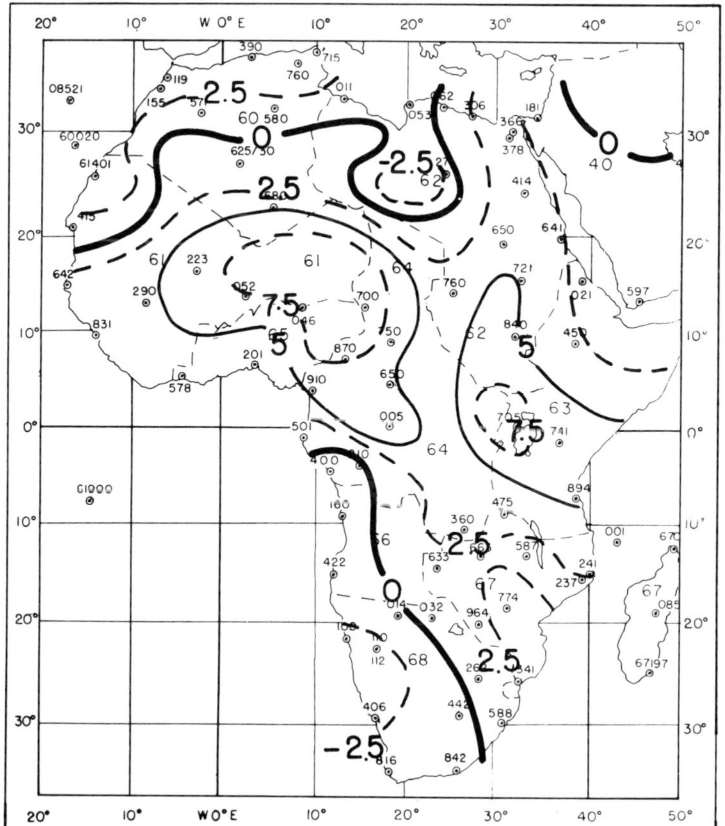

MEAN $\bar{V}$ AT 200mb - APRIL

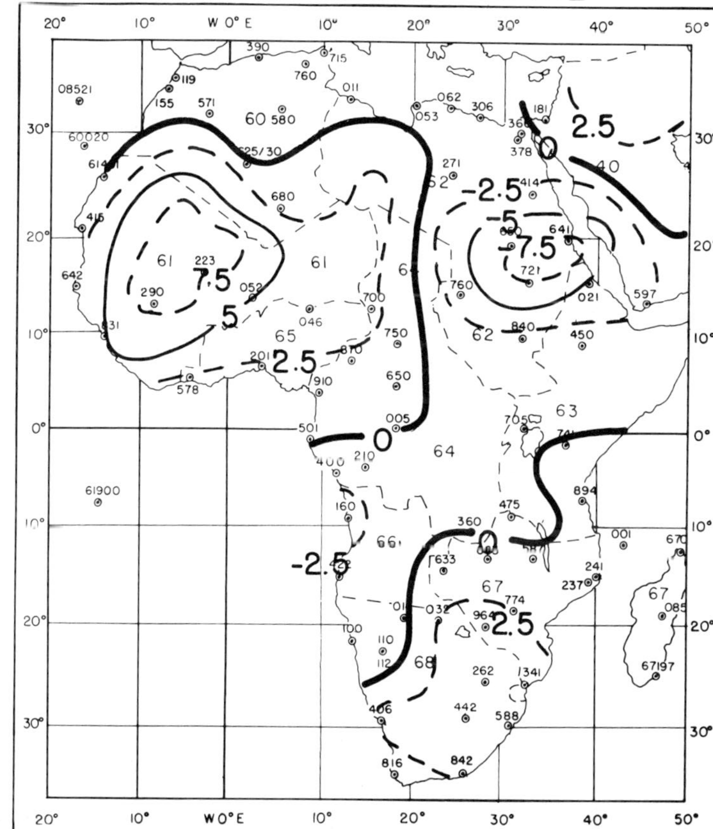

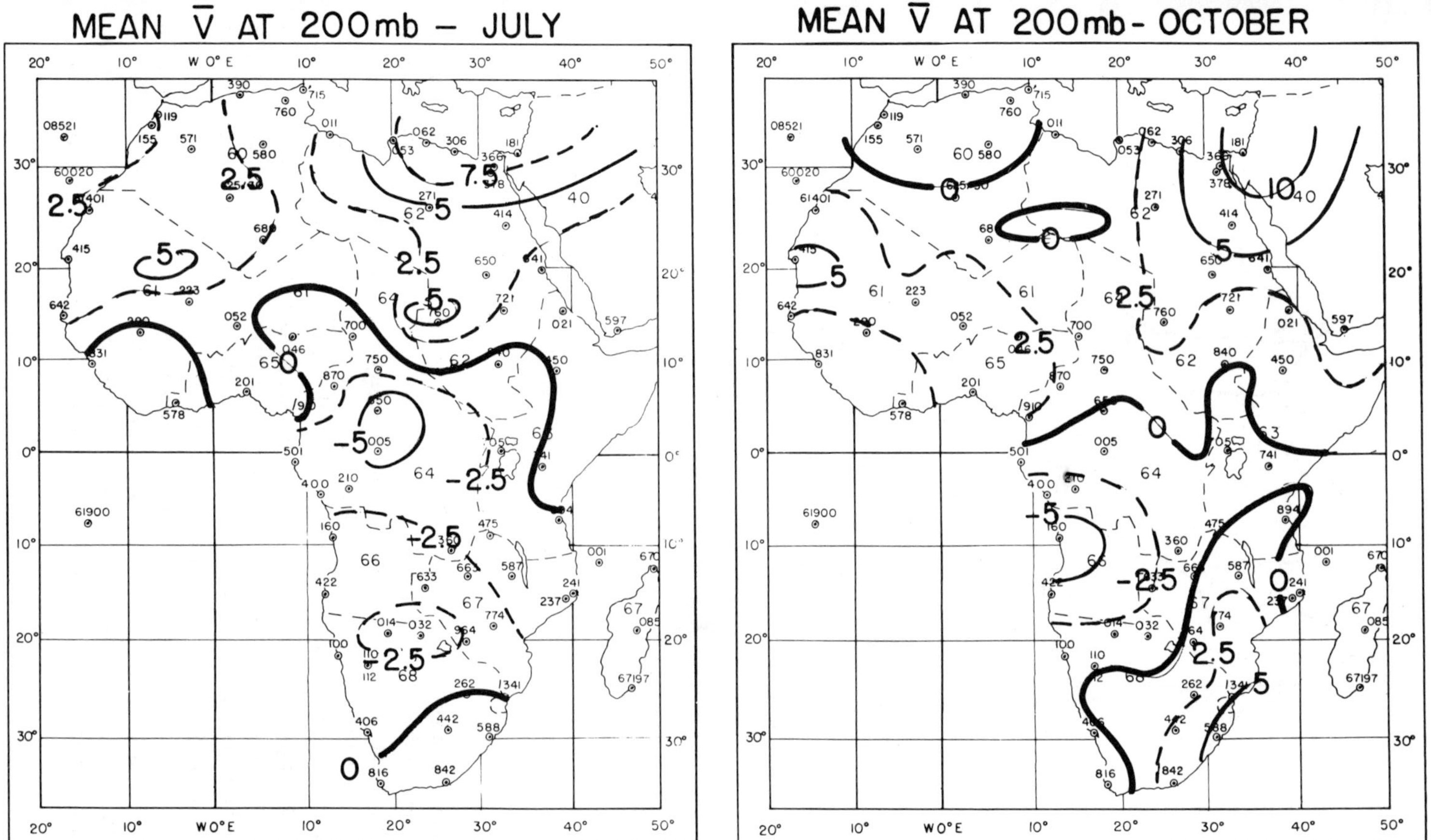

Figure 7.7 Monthly mean wind component $\overline{V}$ at 200 mb for January, April, July, and October

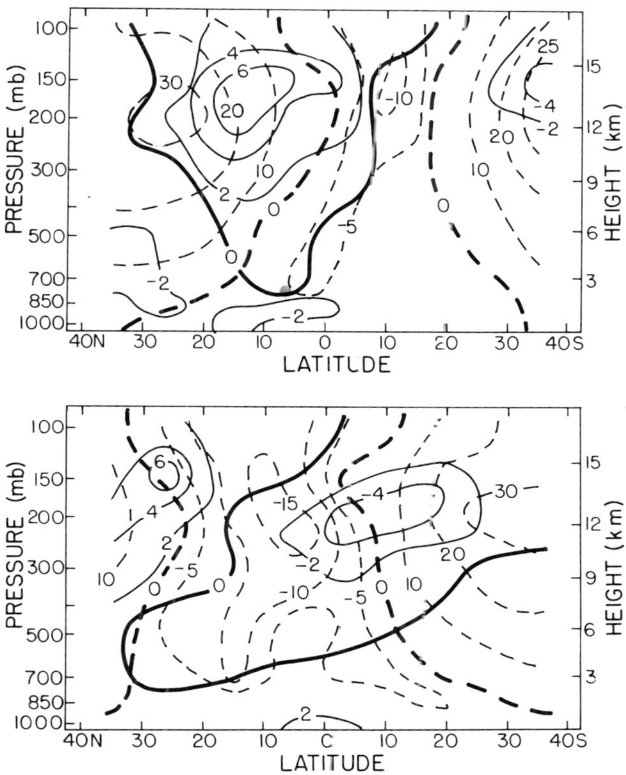

Figure 7.8 Zonal and meridional wind components for 0° longitude (Units: m sec^{-1}) Top: January. Bottom: August Key: ––– Zonal —— Meridional

dust, and data compiled in the 1930's are shown in Figure 7.16. In northern winter considerable dust blows westward from the southern Sahara. Kalu (1979) presented evidence that the alluvian plain of Bilma (Niger)/Faya Largeau (Chad) is a major source in winter. In summer the peak haze frequency lies further north and in addition there is a massive transport towards the east, as might be anticipated from the occurrence of dry westerlies at 40°E longitude in that season (Figure 7.11). In southern Africa there is a correspondence between dry easterlies at 20–25°S throughout the year, and the occurrence of marine haze. The use of 850 mb seems appropriate as much of the dust is confined to the lowest 3 km (Prospero and Carlson, 1972). In general terms then the mean wind kinetic energy pattern, the mean flow pattern and the rainfall can be said to give a first order explanation of the marine visibility maps, but more cannot be said without knowledge of the detailed distribution of the sources and, as Kalu stressed, detailed distribution of

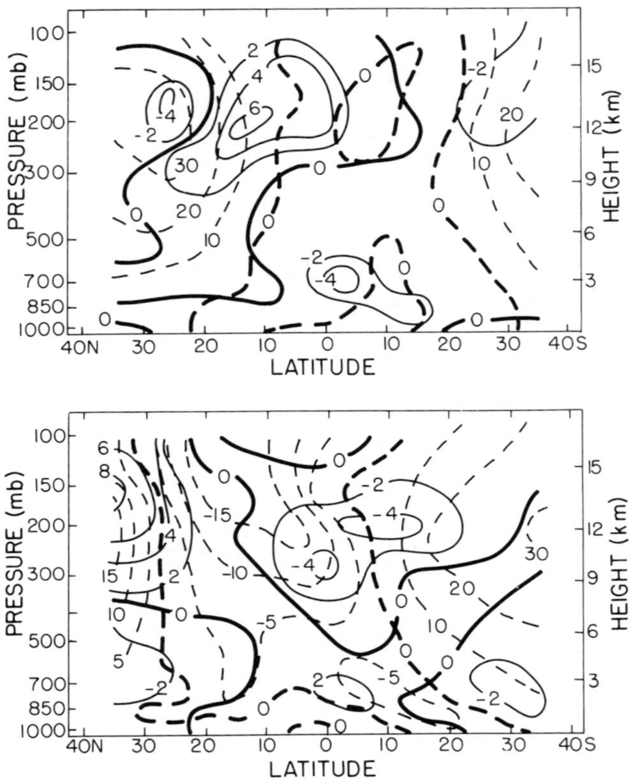

Figure 7.9 Zonal and meridional wind components for 20°E longitude (Units: m sec^{-1}) Top: January. Bottom: August
Key: ———— Zonal ——— Meridional

the surface wind. Insofar as the mean flow pattern represents the sum of the passage of transient eddies then it can be used to represent the dust transport, but for pick-up it was stressed at the meeting that power law values of the instantaneous surface wind speed may be involved (e.g. Gillette, 1979), which would not be adequately represented by mean values.

7.2.3 Moisture

There is a substantial variation with season in moisture over the desert as shown in Figure 7.17. Lowest values occur in association with the largest sinking motion (cf. Figure 1) as might be expected.

In the early studies of water vapour transport problems by the Planetary Circulations Project directed by the late Professor Victor P. Starr, transport over

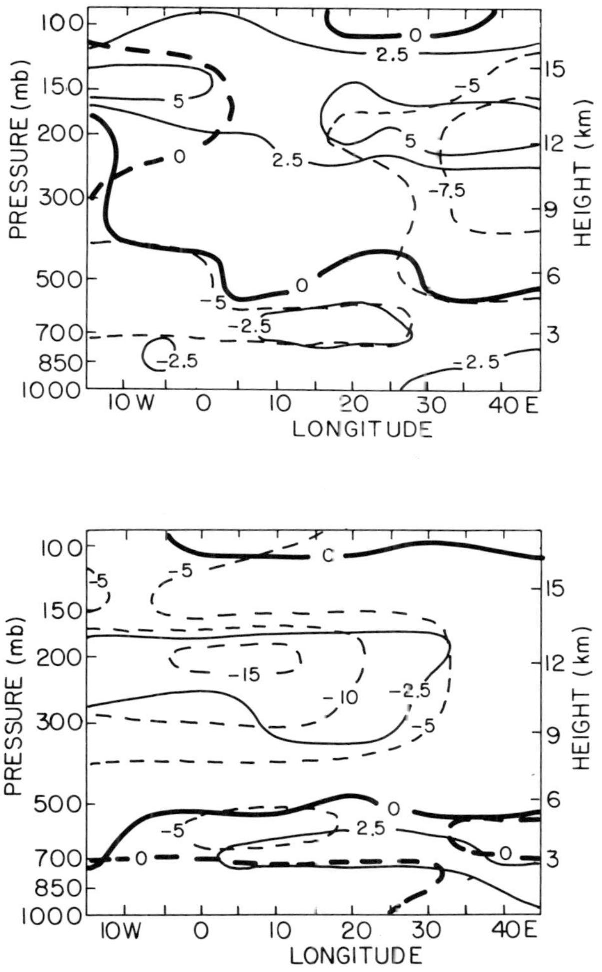

Figure 7.10 Zonal and meridional wind components for equatorial plane (Units: m sec^{-1}) Top: January. Bottom: July.
Key: ---- Zonal ——— Meridional

Africa was treated as a part of the hemispheric or global problem (e.g. see Peixoto, 1973). A few works have focused on Africa, for example, those by Flohn *et al.* (1965) and Peixoto and Obasi (1965). A general finding was that, as for other tropical regions, mean motions contributed most to the total flux, as opposed to day-to-day eddy motions, so that monthly mean winds could be used in association

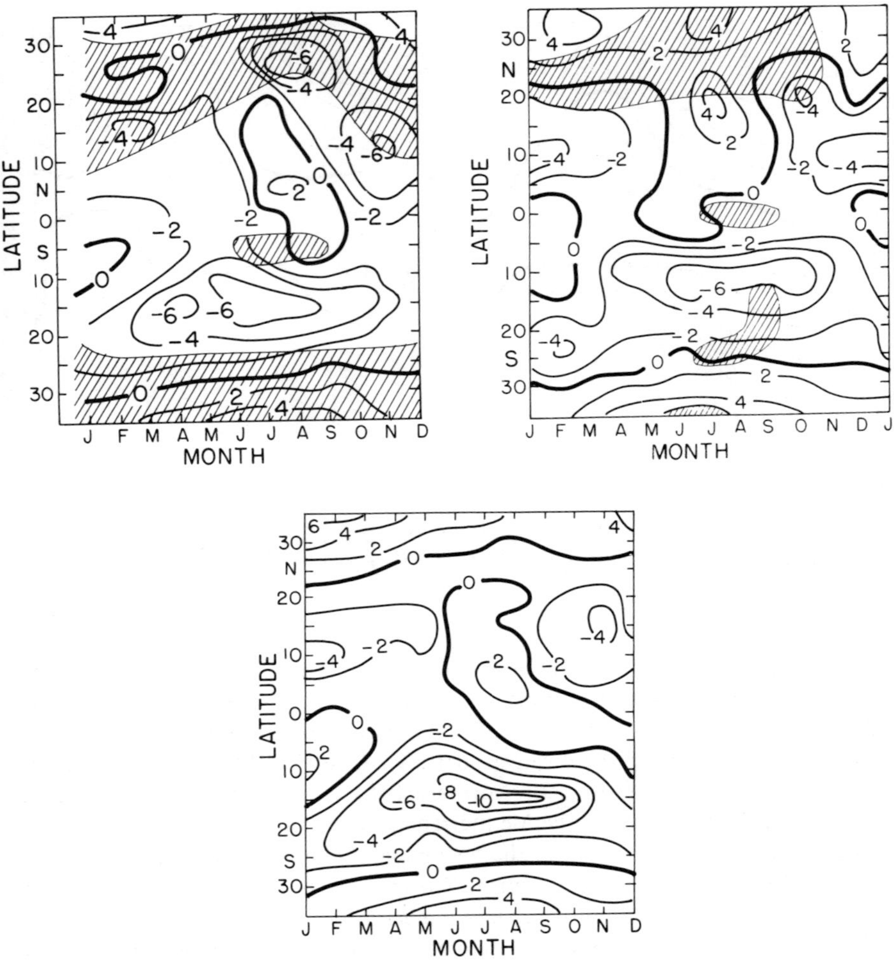

Figure 7.11 Time latitude sections of mean zonal wind at the 850 mb level at 0°, 20°E, and 40°E longitude (Units: m $\sec^{-1}$. Shaded areas: Rainfall <1 cm/month) Left: 0°. Centre: 20°E. Right: 40°E

with monthly mean moisture contents to compute the mean flux. The transport components are:

$$Q_\phi = \frac{1}{g}\int \overline{qv}\, \mathrm{d}p$$

$$Q_\lambda = \frac{1}{g}\int \overline{qu}\, \mathrm{d}p$$

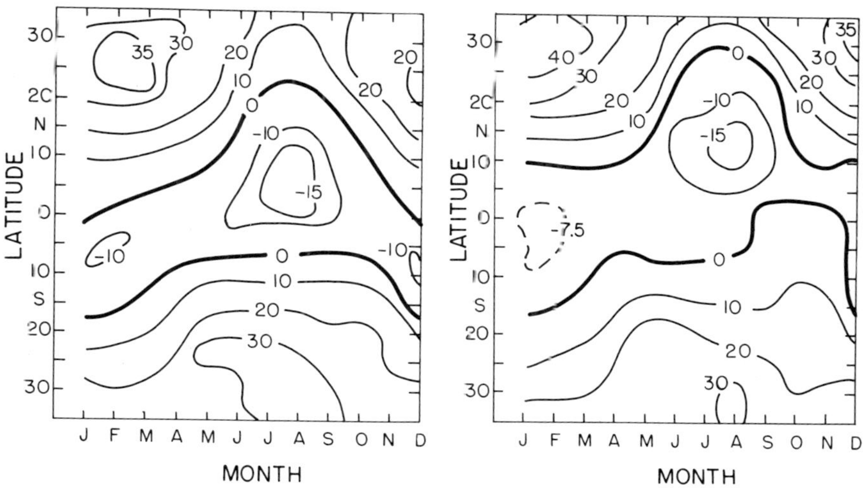

Figure 7.12 Time latitude sections of mean zonal wind at the 200 mb level at 0° and 40°E longitude. (Units: m sec^{-1}) Left: 0°. Right: 40°E

where ϕ and λ are latitude and longitude and $\bar{q}$ is the monthly mean specific humidity (in gm/kg). The streamlines of the vertically integrated flux computed by Kidson (1977) are shown in Figure 7.18. These streamlines represent the non-divergent part of the flow; the divergent part, $\mathbf{\nabla} \cdot \mathbf{Q}$, is also shown. In the early part of the year the moisture over the Northern Sahara originates from the Atlantic while in summer the origin is the Mediterranean area. Moisture over the Sahel also travels from the Mediterranean in the beginning of the rainy season, but later the Indian ocean region becomes a source. The pattern is quite different from that which would be inferred from the surface streamlines.

In general dust, like water vapour, will be concentrated in the lower troposphere and it seems reasonable to assume that the transport is also mainly by the mean winds. The climatological data presented here could therefore be used in the assessment of station sites.

Mean rainfall patterns are shown in a number of climatological texts (e.g. Thompson, 1965; see also Kidson, 1977). These give a guide to regions which are vulnerable to surface pick-up and they also represent a measure of a major sink, washout. In the lower rainfall periods of 1972–73 more dust could be picked up because of dry surfaces and less dust would be removed by rainfall, than in the wetter years in the early 1960's.

Examples of the time variations of rainfall are shown in Figure 7.19 in a form that may be used in wash-out computations.

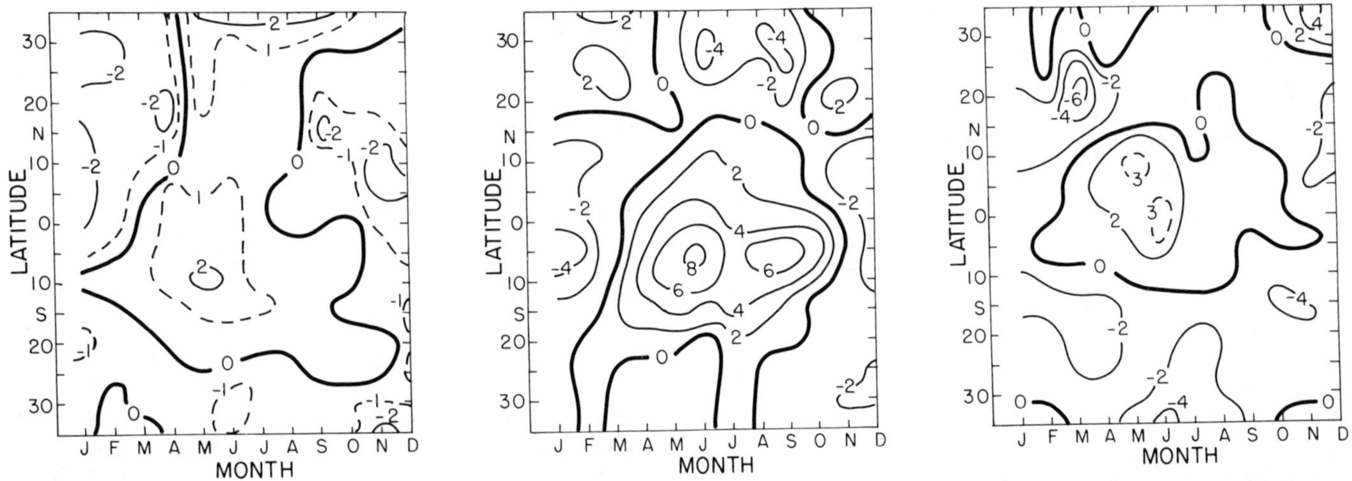

Figure 7.13 Time latitude sections of mean meridional wind at the 850 mb level at 0°, 20°E, and 40°E longitude (Units: m sec^{-1}) Left: 0°. Centre : 20°E. Right: 40°E

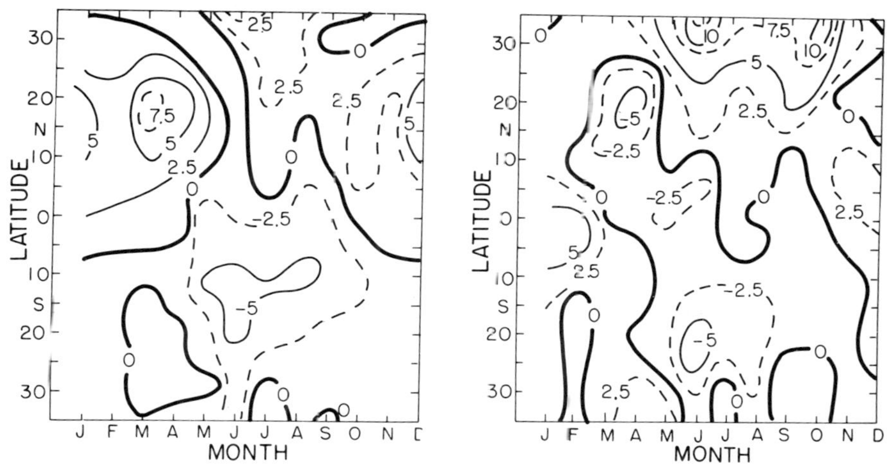

Figure 7.14 Time latitude sections of mean meridional wind at the 200 mb level at 0° and 40°E (Units: m sec^{-1}) Left: 0°. Right: 40°E

7.3 YEAR-TO-YEAR CHANGES

7.3.1 Wind velocity

The details of the monthly mean tropospheric flow patterns vary from year to year. The interest here is between predominantly wet and dry years, the latter with more potential for dust pick-up. Elsewhere Kidson (1977) has presented streamlines for 850 and 200 mb for August 1959 and 1961 (wet years) and August 1972 and 1973 (dry years) (Figure 7.20) and has pointed out that the 850 mb trough normally present at about 8°N in the wet years was absent in the dry years. The easterlies at 200 mb are also a good deal weaker in the dry years as may be seen also from the individual year equatorial cross-sections of Figure 7.21. The meridional flow which produces these easterlies is also a little weaker in 1973.

7.3.2 Moisture

Rainfall variations between wet and dry years of the recent period have been discussed by several authors (e.g. Winstanley, 1973; Sircoulan, 1974; Tanaka *et al.*, 1965; Bunting *et al.*, 1976; Kidson, 1977). The main rain region over West Africa moves north and south in a regular fashion reaching its northernmost position in August. The centre of gravity of this part of the rainbelt (15°W to 25°E) is shown in Figure 7.22. The dashed line shows the long term mean northernmost position in August. The belt was even further north in 1952, 1956, 1958 (for two months) and 1959 and was south of average in 1966 and 1968. But clearly there has not been a

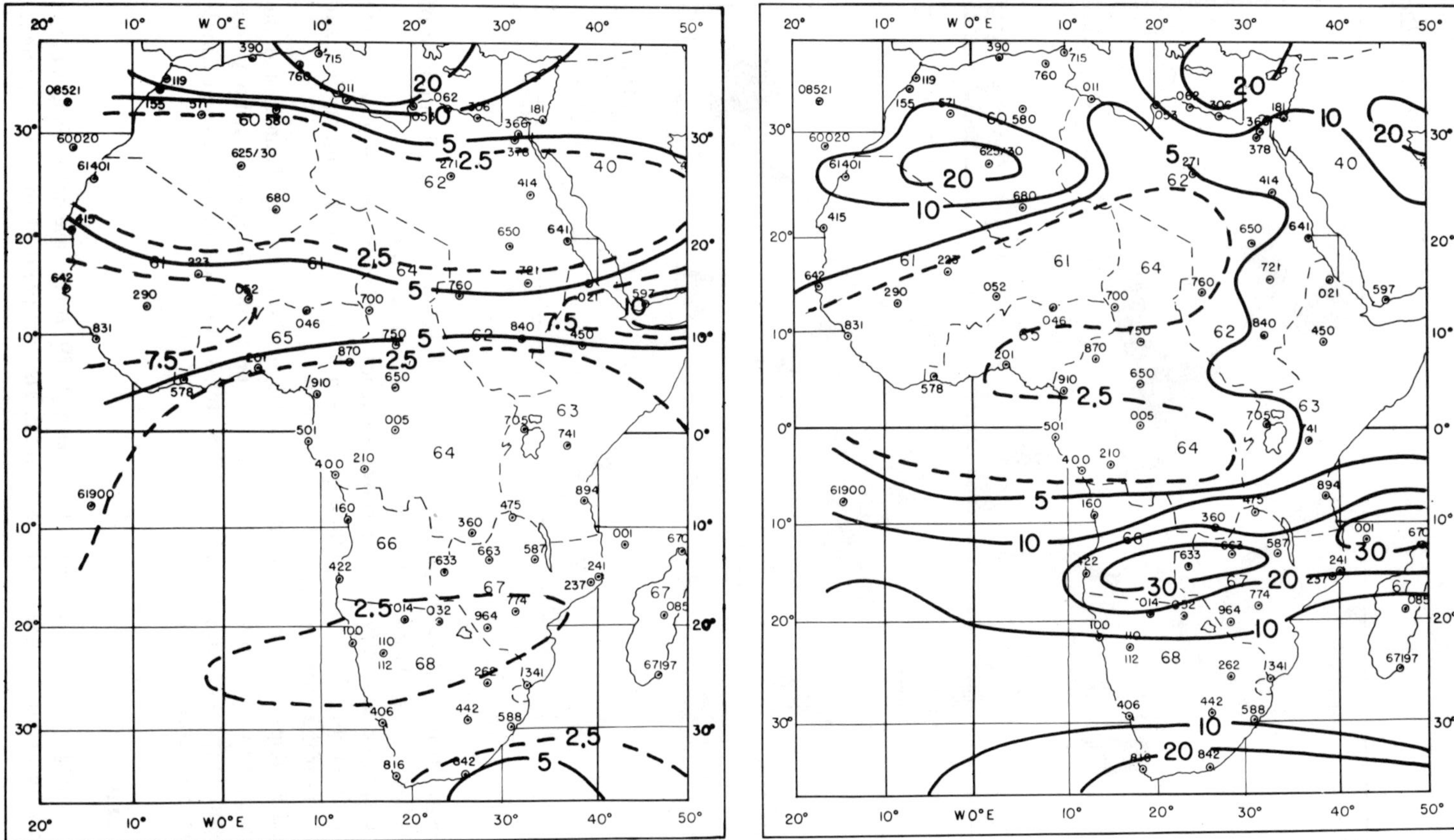

Figure 7.15 Kinetic energy of the mean wind $(\bar{U}^2 + \bar{V}^2)/2$ at 850 mb January and July (Units: $m^2\ sec^{-2}$) Left: January. Right: July

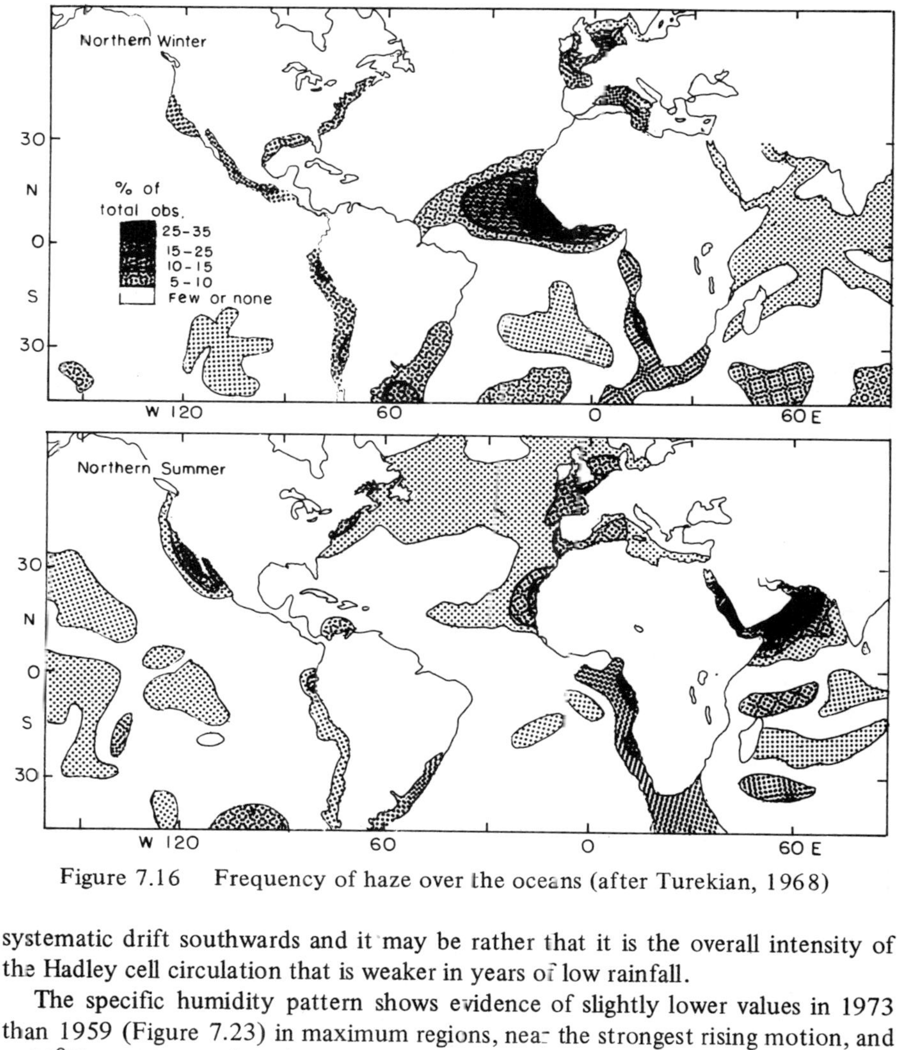

Figure 7.16 Frequency of haze over the oceans (after Turekian, 1968)

systematic drift southwards and it may be rather that it is the overall intensity of the Hadley cell circulation that is weaker in years of low rainfall.

The specific humidity pattern shows evidence of slightly lower values in 1973 than 1959 (Figure 7.23) in maximum regions, near the strongest rising motion, and at 25°N near the northern hemisphere sinking motion.

Relationships with sea surface temperature in the adjoining Atlantic and the Indian Ocean were also sought but there were no strong correlations, although our data sample spanned only the period 1949–69, so we did not have a good sample of sea temperature for the dry years.

7.3.3 Dust

According to the presentation of Prospero (1979) there have been changes in the summer dust content of air over the Atlantic. The summer dust appears to originate

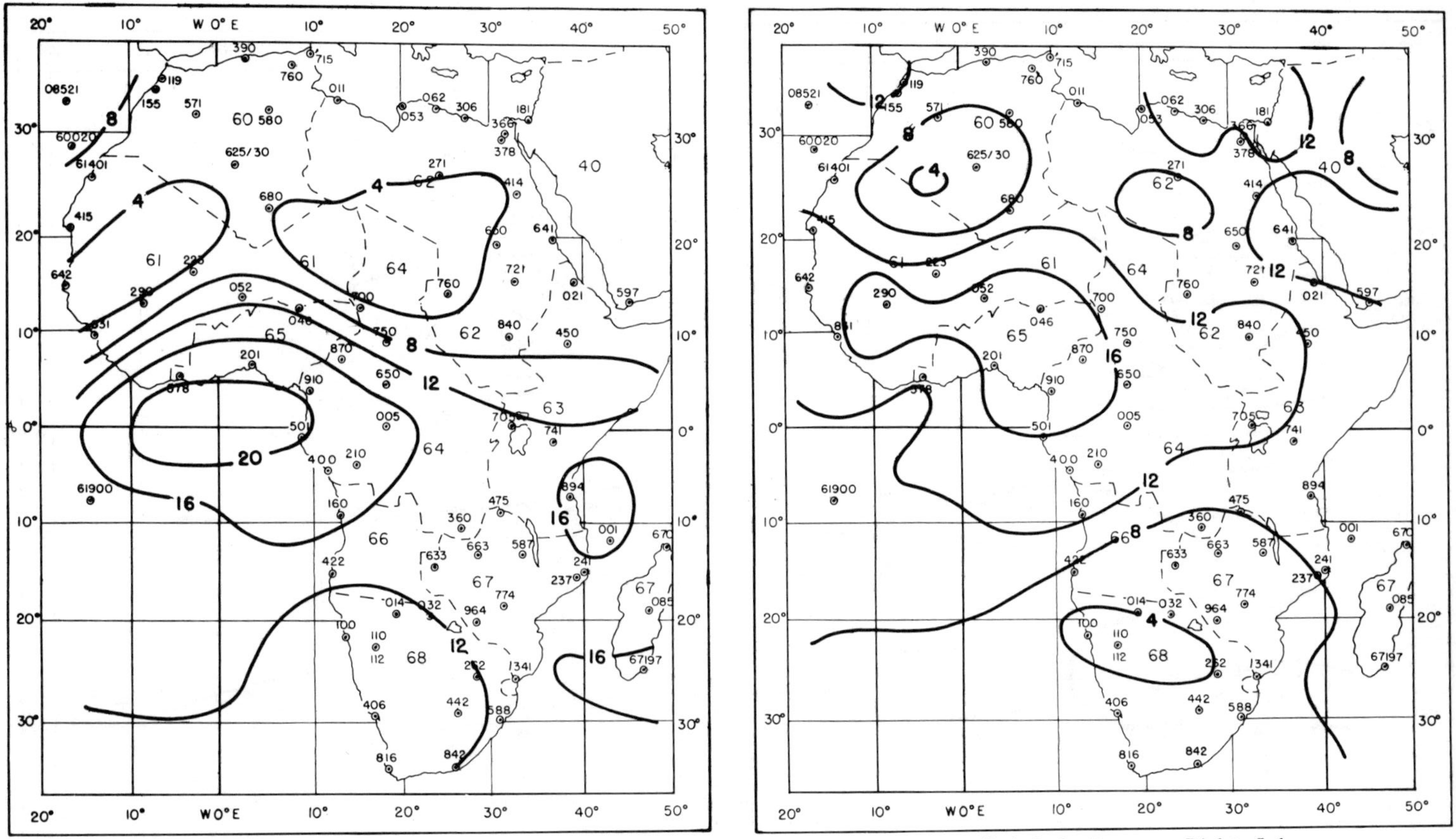

Figure 7.17 Specific humidity at 1000 mb for January and July (Units: gm/kg) Left: January. Right: July

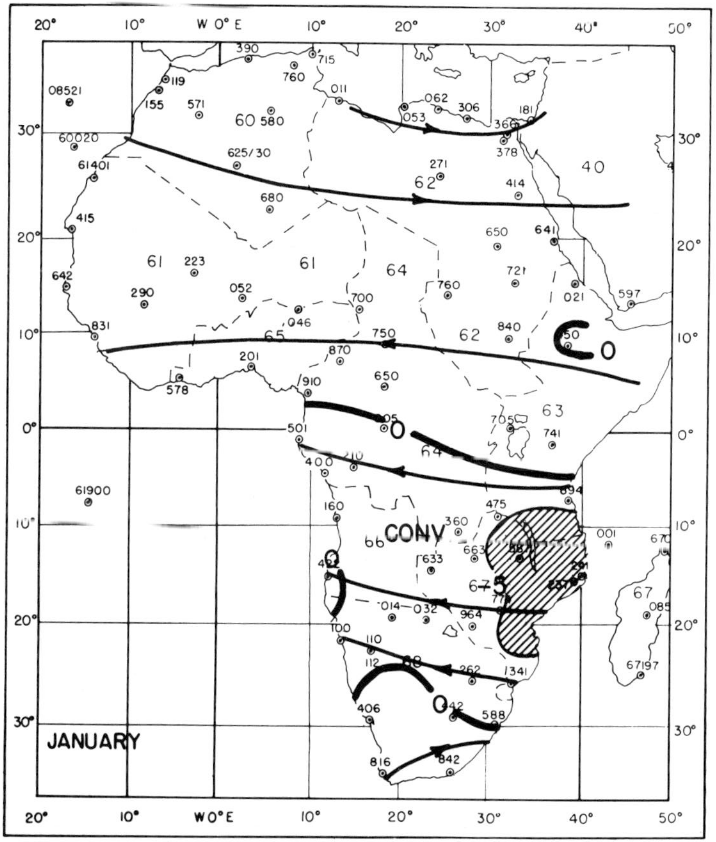
JANUARY
CONV

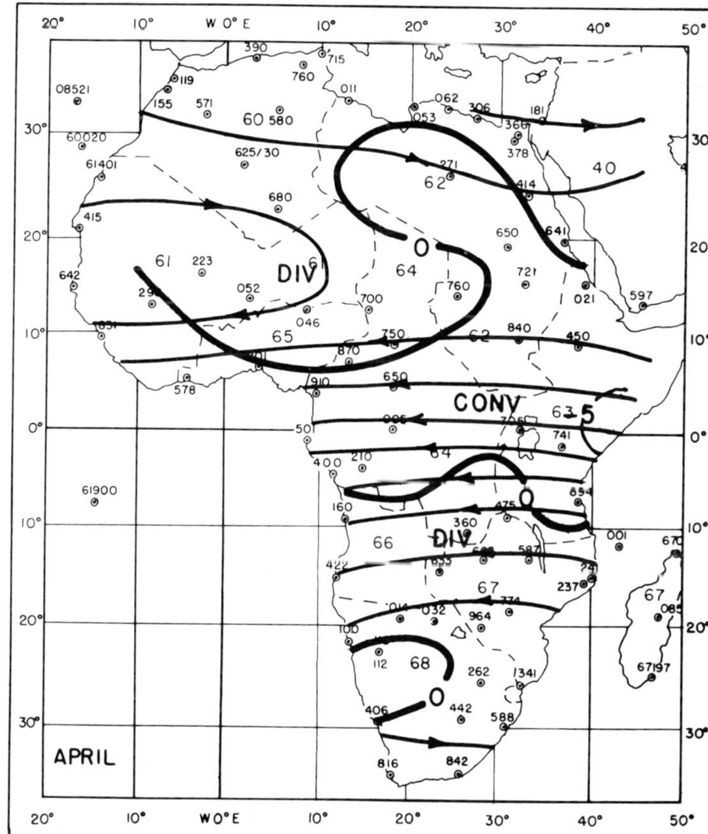
APRIL
CONV
DIV

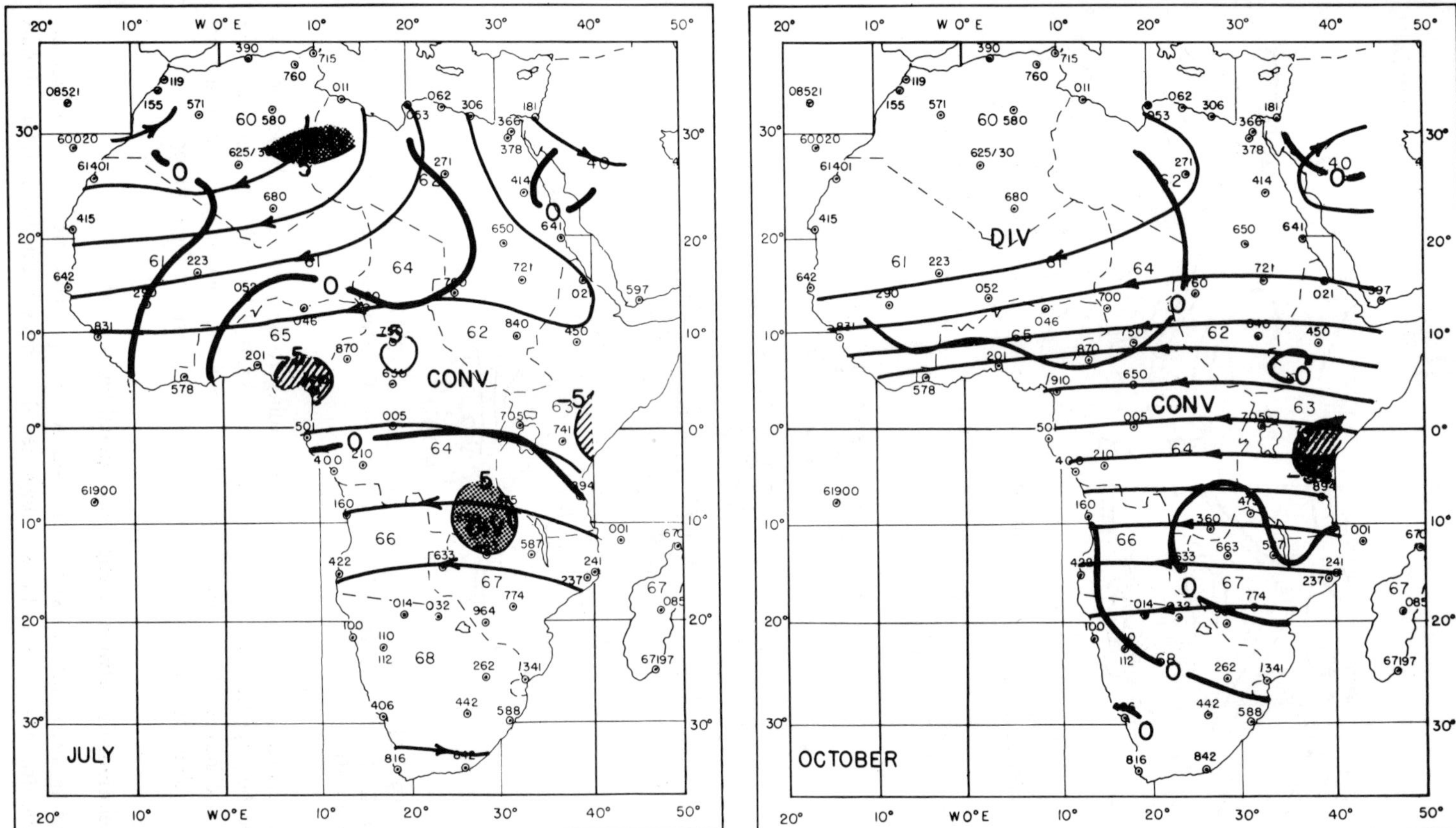

Figure 7.18 Streamlines of the vertically integrated moisture flux (5° spacing corresponds to a flux of 100 kg m^{-1} sec^{-1}). These streamlines represent the non-divergent part of the flow. The thick lines delimit the divergent part of the flow from the convergent part (after Kidson, 1977)

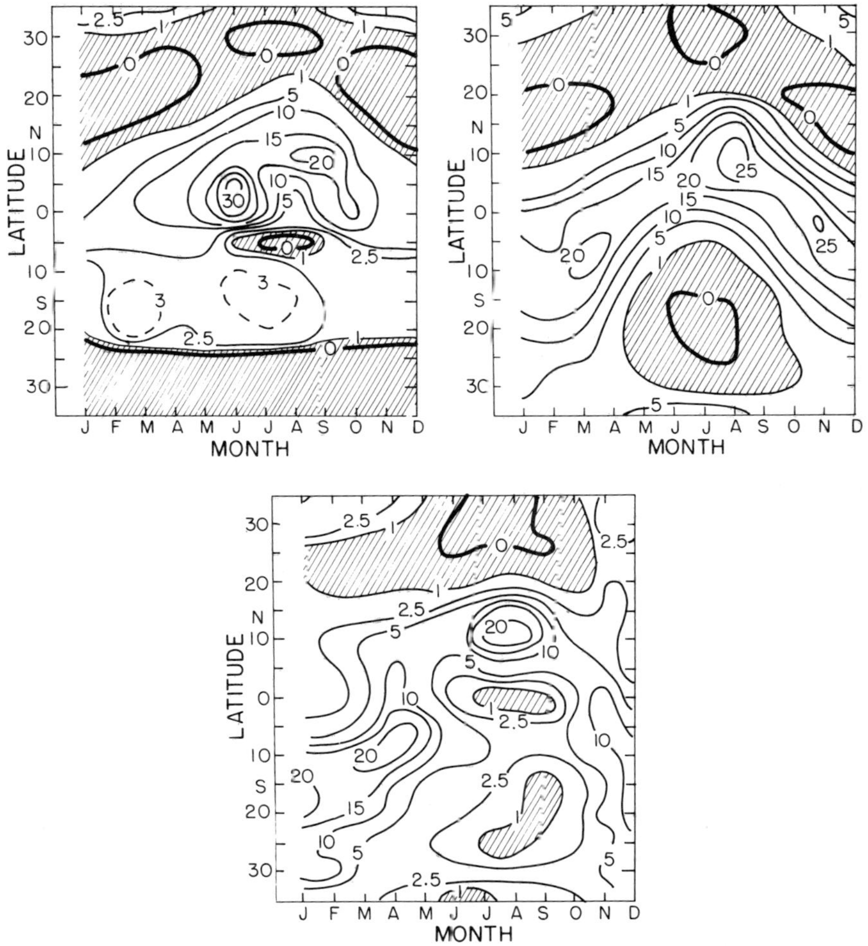

Figure 7.19 Monthly rainfall *vs*. latitude at 0°, 20°E, and 40°E (Units: cm/month) Left: 0°. Centre: 20°E. Right: 40°E. Shaded areas show rainfall < 1 cm/month

from the central and northern Sahara according to Section 7.2.2 above and may therefore be expected to be of the red type (Rapp, 1974). While changes in the summer are not so likely to be related to the Sahel they may be governed by circulation changes which we find over all of Africa north of the equator.

7.3.4 Relationship between drought and global temperature

It was pointed out elsewhere that surface temperature near 15°N increased during the drought years and it was suggested that this was due to planetary albedo

changes (Tanaka *et al.*, 1975). Possible relationships with free air temperature have also been examined. Monthly mean temperature data from 150 upper air stations for the period 1958–74 have been subjected to an empirical orthogonal function analysis to bring out the patterns of the non-seasonal changes. Only stations with at least 80% of the data present were included in the analysis which was performed for levels of 850 mb, 500 mb, and 200 mb. This analysis is to be described elsewhere (Weare, Kidson, Navato, and Newell, *Unpublished Ms.*, 1977). The first ten functions at each level were incorporated in a screening regression procedure to examine their relationship to changes in July to September rainfall in Africa. The patterns most strongly related to the rainfall changes are obtained by combining the EOF patterns with weights given by the screening regression procedure and are shown in Figures 7.24 and 7.25 (after Kidson, 1977).

The 850 mb temperature pattern implies that free air temperature is lower over Africa in the wet years as was found for the surface. In the dry years soil desiccation would be favoured at the higher temperatures. The 200 mb pattern shows evidence of stronger meridional temperature gradients in both hemispheres in wet years. This is consistent with the suggestion of Section 7.3.2 that it is the overall *intensity* of the Hadley cell circulation that is weaker in years of low rainfall.

The source of the enhanced tropospheric temperature gradients is still a mystery. Modulation of sea surface temperature is a possibility, but, as noted, data subsequent to 1969 for the Atlantic and Indian oceans is not yet available.

7.4 CONCLUDING REMARKS

The mean flow patterns over Africa have been presented in horizontal and vertical sections. The broad-scale features of the patterns of dust over the sea can be attributed to the mean flow although the actual pick-up may be more closely related to the transient circulations.

Between the moist regime in the Sahel in the late 1950's and the dry regime of the early 1970's there are significant changes in the flow patterns. There are related changes in the rainfall patterns which seem best described in terms of a reduction in the Hadley cell circulation intensity. Temperature patterns show evidence of high values at the surface and at 850 mb in dry years. The meridional temperature gradients at 200 mb are weaker in both hemispheres in dry years which is consistent with an observed reduction in intensity. We do not know the reason for these gradient changes.

Eolian transport of African dust and its variability in deep sea cores has been used by Parkin and Shackleton (1973) to assess the 'vigour' of the circulation during ice ages. For Atlantic core V23–100 (21°N, 23°W) they deduce stronger trade winds during the cold periods. It has been argued that latitudinal temperature gradients were also greater during cold periods (Newell, 1973). However, from the present findings, it must be borne in mind that trajectory changes could equally well be to blame for the observed dust changes.

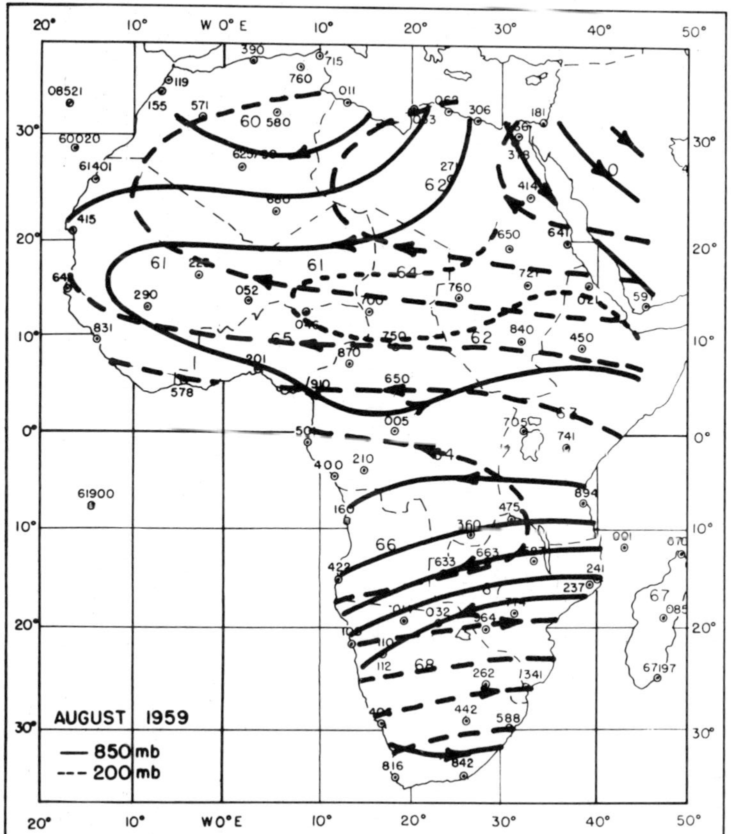
AUGUST 1959
— 850 mb
--- 200 mb

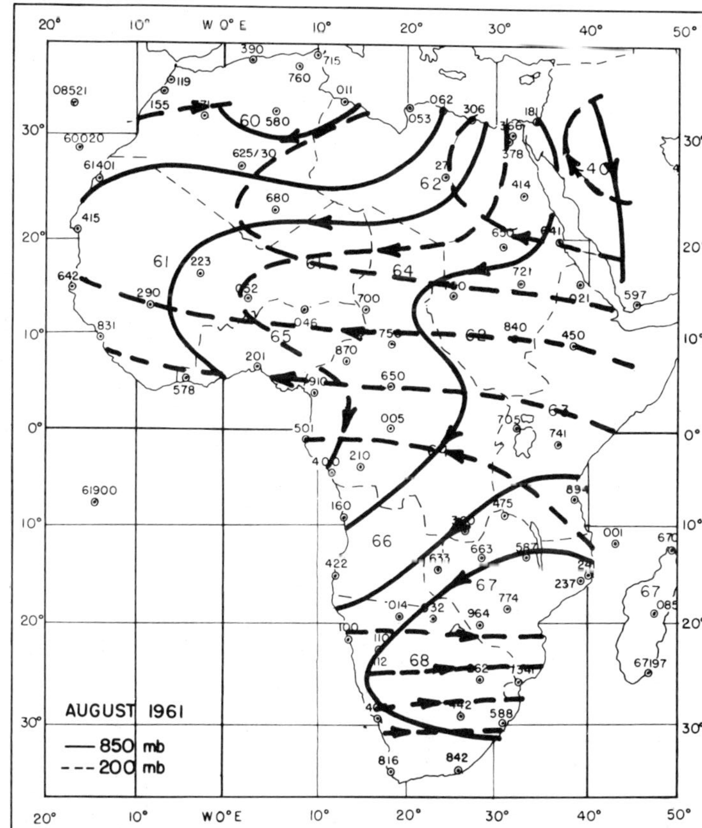
AUGUST 1961
— 850 mb
--- 200 mb

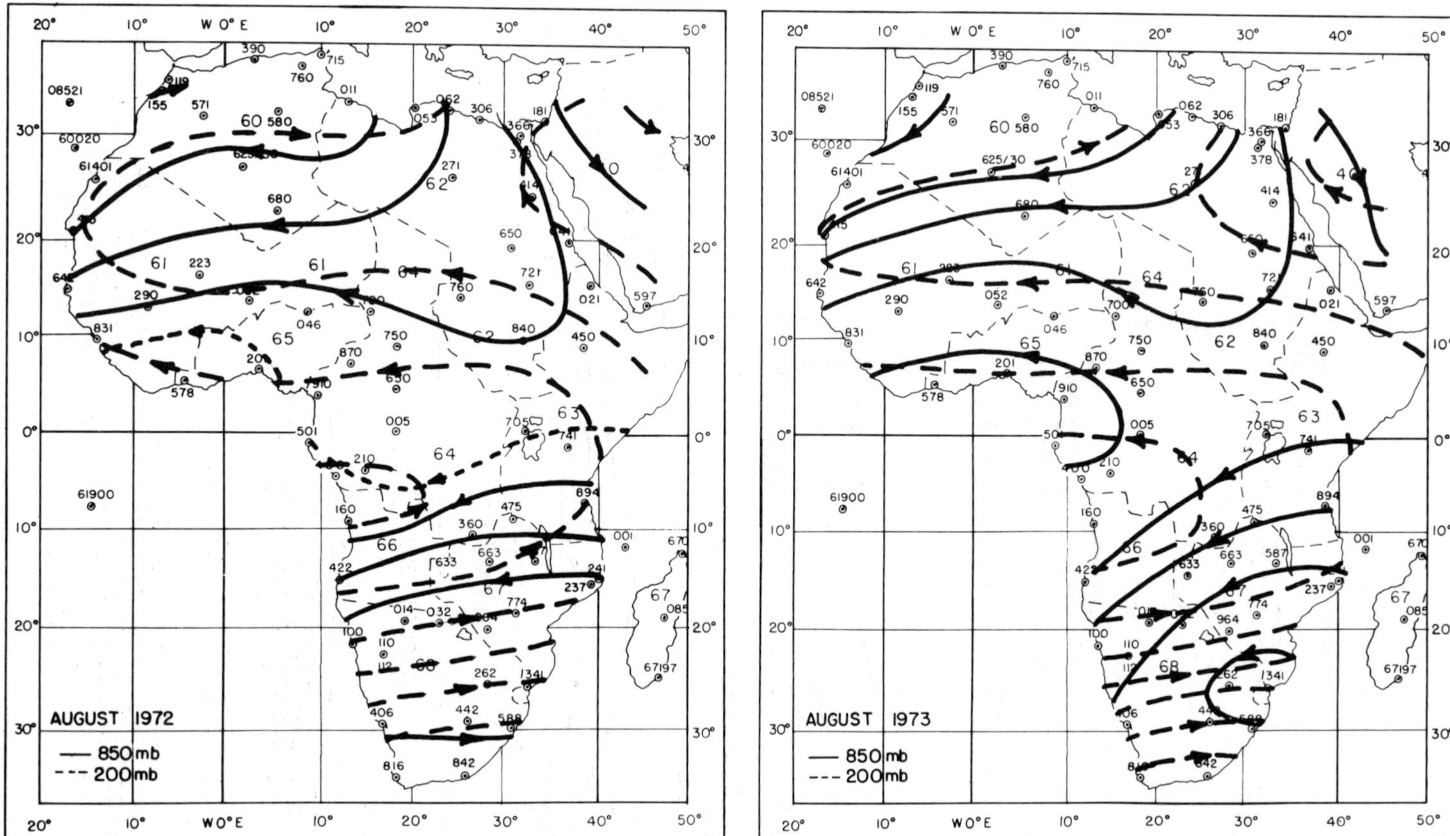

Figure 7.20 Streamlines of the August circulation at 850 mb and 200 mb for wet and dry years (after Kidson, 1977)

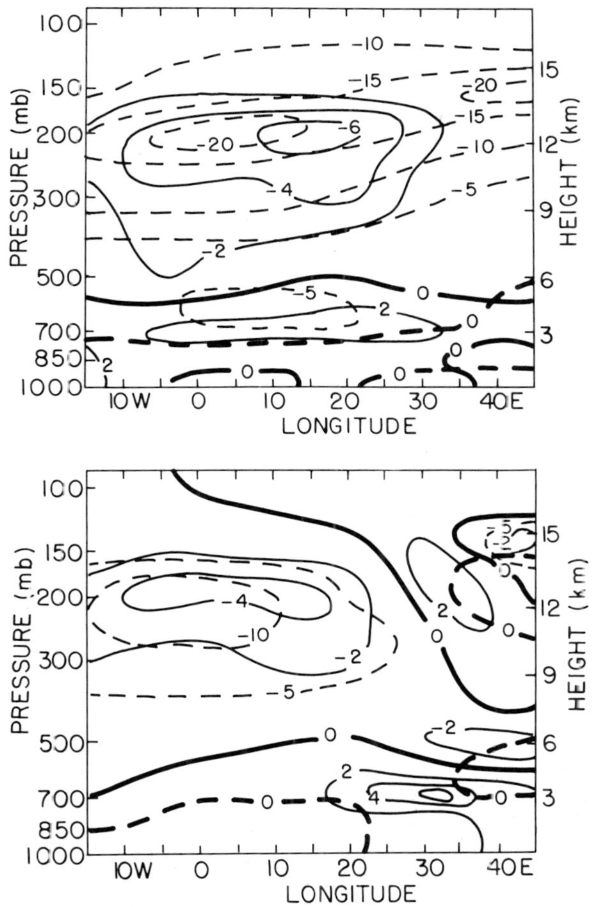

Figure 7.21 Zonal and meridional wind components in the equatorial plane for August 1959 (wet) at top and August 1973 (dry) at bottom (Units: m sec^{-1})
Key: ––– Zonal
—— Meridional

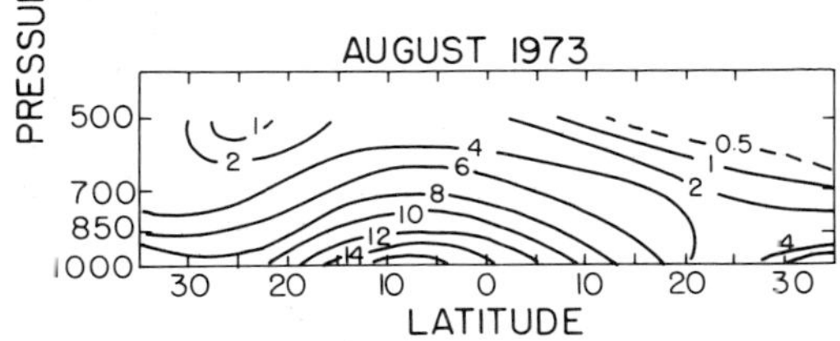

Figure 7.23 Specific humidity for August 1959 (wet) and August 1973 (dry) at 20°E longitude (Units: gm/kg)

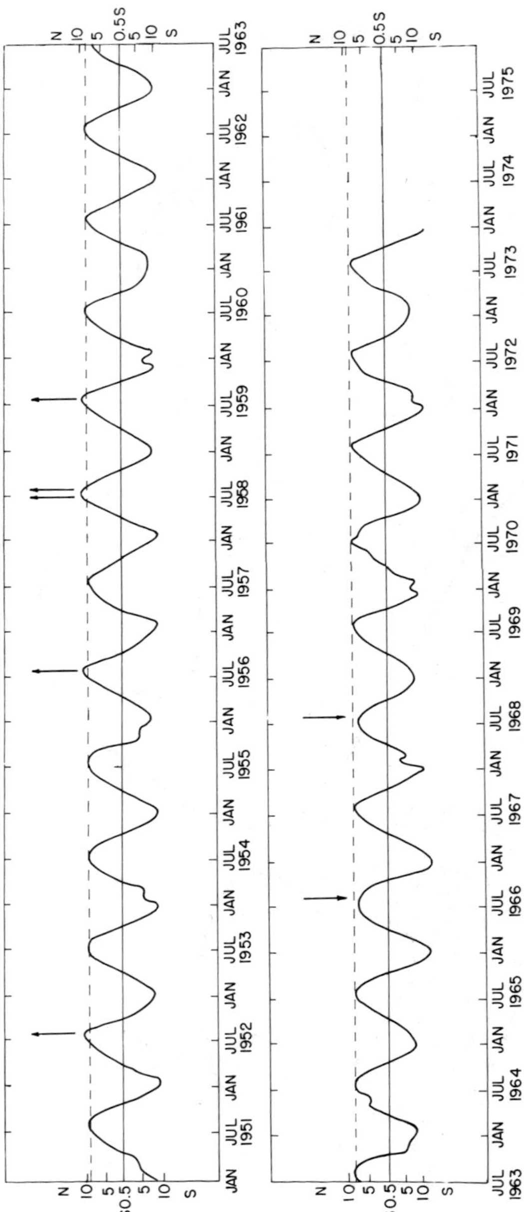

Figure 7.22 Latitude of the centre of gravity of African rainbelt from 1951 to 1973

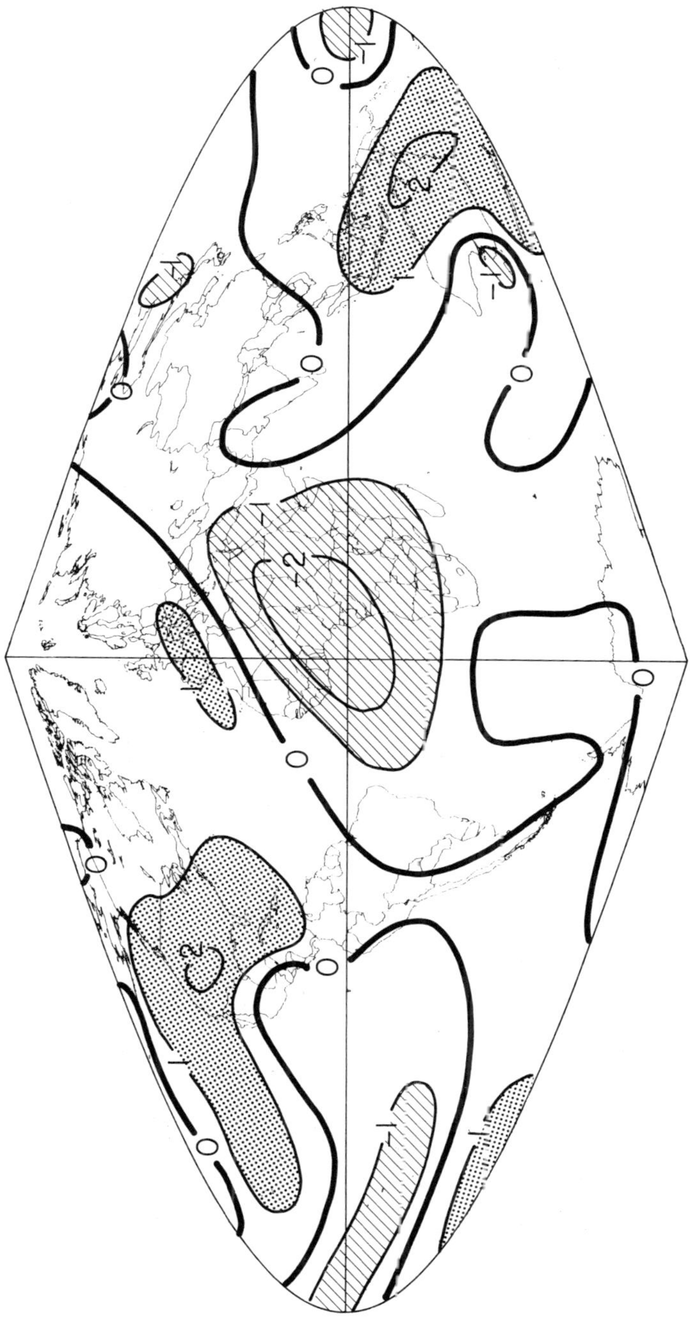

Figure 7.24 Weighted empirical orthogonal function pattern for 850 mb. This pattern results from combinations of the first ten patterns at each level with weights selected by screening regression (after Kidson, 1977)

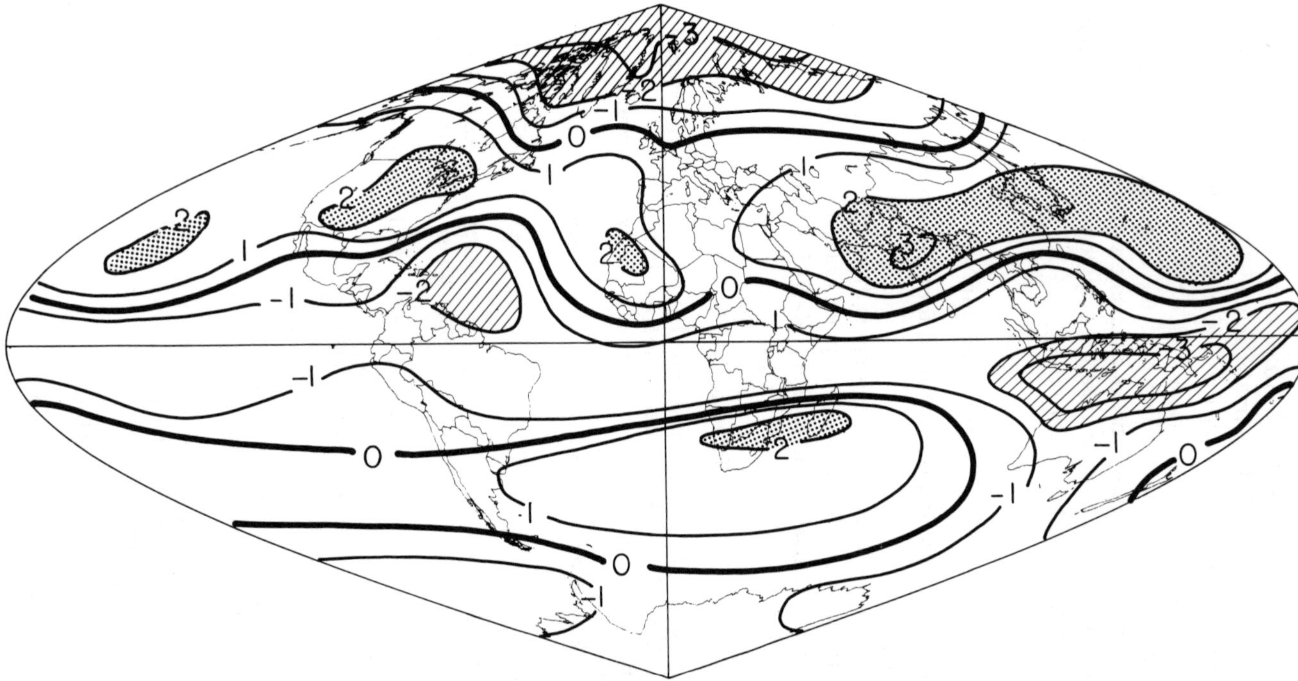

Figure 7.25 Weighted empirical orthogonal function pattern for 200 mb. This pattern results from combinations of the first ten patterns at each level with weights selected by screening regression (after Kidson, 1977)

According to the work of Street and Grove (1975) who have made a careful study of lake-level fluctuations, all of Africa between the equator and 30°N was very wet in the 1000-year period starting at about 9000 B.P. and the dust in ocean cores for that period could be examined as a separate verification. They postulate that sea surface temperature was high at that time. Maley (1976) has also made some pollen analyses which give evidence of the translation of the Hadley cell circulation and even if the intensity change argument is used for the recent period it cannot be extrapolated backwards in time – it is quite possible that the translation argument holds on the 10,000-year time scale.

7.5 ACKNOWLEDGMENTS

We thank Dr. B. C. Weare and Dr. A. Navato for their participation in the free air temperature study.

We gratefully acknowledge research support from the United States Energy Research and Development Administration under Contract E(11-1)-2195 and from the National Science Foundation under Grant 74-09853ATM.

REFERENCES

Bunting, A. H., Dennet, M. D., Elston, J., and Milford, J. R. (1976). Rainfall trends in the West African Sahel. *Q.J.R. Meteorol. Soc.*, **102**, 59–64.

Cressman, G. P. (1959). An operational objective analysis system. *Mon. Weather Rev.*, **87**, 367–374.

Dubief, J. (1959). *Le Climat du Sahara.* Univ. Inst. Rech. Sahariennes, Mem. (hors sér.), Alger, 312 pp.

Dubief, J. (1963). *Le Climat du Sahara.* Univ. Inst. Rech. Sahariennes, Mem. (hors sér.), Alger, 272 pp.

Dubief, J. (1979). *Review of the North African Climate with Particular Emphasis on the Production of Eolian Dust in the Sahel Zone and in the Sahara.* In this report.

Findlater, J. (1969a). A major low-level air current near the Indian Ocean during the northern summer. *Q.J.R. Meteorol. Soc.*, **95**, 362–380.

Findlater, J. (1969b). Interhemispheric transport of air in the lower troposphere over the western Indian Ocean. *Q.J.R. Meteorol. Soc.*, **95**, 400–403.

Findlater, J. (1972). Aerial explorations of the low-level cross-equatorial current over eastern Africa. *Q.J.R. Meteorol. Soc.*, **98**, 274–289.

Flohn, H. (1965). Studies on the Meteorology off Tropical Africa. *Bonner Meteorol. Abh.*, **5**, 57 pp.

Flohn, H., Henning, D., and Korff, H. C. (1965). Studies on the Water-vapour Transport over Northern Africa. *Bonner Meteorol. Abh.*, **6**, 36 pp.

Gillette, D. A. (1979). *Environmental Factors Affecting Dust Emission by Wind Erosion.* In this report.

Griffiths, J. F. (1972). Climates of Africa. *World Survey of Climatol.*, **10**, Elsevier, Amsterdam, 604 pp.

Griffiths, J. F., and Soliman, K. H. (1972). Climates of Africa, Chapter 3. *World Survey of Climatol.*, **10**, 75–131, Elsevier, Amsterdam.

Kalu, A. E. (1979). *The African Dust Plume: Its Characteristics and Propagation Across West Africa in Winter.* In this report.

Kendrew, W. G. (1961). *The Climates of the Continents.* Oxford, 5th ed., 608 pp.

Kidson, J. W. (1977). African rainfall and its relation to the upper air circulation. *Q.J.R. Meteorol. Soc.*, **103**, 441–456.

Maley, J. (1976). Essais sur le rôle de la zone tropicale dans les changements climatiques; l'exemple africain. *C.R. Acad. Sci. Paris*, **283**, 337–340.

Munitalp Foundation. (1960). *Tropical Meteorology in Africa.* Nairobi.

Newell, R. E., Kidson, J. W., Vincent, D. G., and Boer, G. J. (1972). *The General Circulation of the Tropical Atmosphere*, Vol. 1, MIT-Press, Cambridge, Mass., 258 pp.

Newell, R. E., Kidson, J. W., Vincent, D. G., and Boer, G. J. (1974). *The General Circulation of the Tropical Atmosphere*, Vol. 2. MIT-Press, Cambridge, Mass., 371 pp.

Newell, R. E. (1973). Climate and the Galapagos Islands. *Nature*, **245**, 91–92.

Parkin, D. W., and Shackleton, N. J. (1973). Trade wind and temperature correlations down a deep-sea core off the Saharan coast. *Nature*, **245**, 455–457.

Peixoto, J. P. (1973). Atmospheric Vapour Flux Computations for Hydrological Purposes. *WMO/IHD Rep.*, **20**, WMO-357, 100 pp.

Peixoto, J. P., and Obasi, G. O. P. (1965). Humidity Conditions over Africa during the IGY. *Sci. Rep.*, **7**, Planetary Circulations Project, MIT-Press, Cambridge, Mass., 143 pp.

Prospero, J. M. (1979). *Monitoring Saharan Aerosol Transport by Means of Atmosphere Turbidity Measurements.* In this report.

Prospero, J. M., and Carlson, T. N. (1972). Vertical and areal distribution of Saharan dust over the western equatorial North Atlantic Ocean. *J. Geophys. Res.*, **77**, 5255–5265.

Rapp, A. (1974). A Review of Desertization in Africa – Water, Vegetation, and Man. *Secr. Int. Ecology*, Rep. 1, Stockholm, Sweden, 77 pp.

Sircoulon, J. (1974). Les Données Climatiques et Hydrologiques de la Sècheresse en Afrique de l'ouest Sahelienne. *Secr. Int. Ecology*, Rep., **2**, Stockholm, Sweden, 43 pp.

Street, F. A., and Grove, A. T. (1976). Environmental and climatic implications of late quaternary lake-level fluctuations in Africa. *Nature*, **261**, 385–389.

Struning, J. O., and Flohn, H. (1969). Investigations on the Atmospheric Circulation above Africa. *Bonner Meteorol. Abh.*, **10**, 55 pp.

Tanaka, M., Weare, B. W., Navato, A. R., and Newell, R. E. (1975). Recent African rainfall patterns. *Nature*, **255**, 201–203.

Thompson, B. W. (1965). *Climate of Africa.* Oxford Univ. Press, Nairobi, London, New York, 132 pp.

Turekian, K. K. (1968). *Oceans.* Prentice-Hall, New York, 120 pp.

Winstanley, D. (1973). Recent rainfall trends in Africa, the Middle East and India. *Nature*, **243**, 464–465.

CHAPTER 8

Monitoring Saharan Aerosol Transport by Means of Atmospheric Turbidity Measurements

J. M. PROSPERO, D. L. SAVOIE, T. N. CARLSON, and R. T. NEES

ABSTRACT

Atmospheric turbidity measurements were made in two spectral bands (500 nm and 880 nm) in a network of five land stations and ten ships distributed across the northern Equatorial Atlantic during the GARP Atlantic Tropical Experiment (GATE) during the summer of 1974. The turbidities were high off the coast of Africa and decreased toward the west. The mean Volz turbidities (B_{500}) at Sal Island, Barbados, and Miami were: 0.297, 0.130, and 0.100. The values at 880 nm were: 0.250, 0.114, and 0.062. The central and eastern Atlantic turbidities are equal to those of large, heavily industrialized cities. The mean Angstrom wavelength exponent α at these three stations was: 0.348, 0.285, and 0.899. Low values of α indicate that neutral extinction obtains which is consistent with the appearance of the sky in this region. High values of α (i.e. values greater than one) were consistently observed only at Miami and only when Saharan aerosols were not present in the atmosphere. The consistently low α values at all stations in the presence of Saharan air parcels suggests that the optically effective portion of the aerosol size distribution does not change substantially during transit across the Atlantic. The daily mean station turbidities do not seem to be correlated in any simple way with the daily average mineral aerosol concentrations measured in surface level air at the stations.

8.1 INTRODUCTION

In 1974, a large scale aerosol program was carried out as a part of the Global Atmospheric Research Program (GARP) Atlantic Tropical Experiment (GATE). One of the principal objectives was to investigate the geographical distribution of aerosols with respect to meteorological parameters over the entire equatorial North Atlantic Ocean. To this end, we established an aerosol and atmospheric turbidity network which consisted of five land stations and ten ship stations. The design and operation of this network was closely coordinated with a solar radiation program

carried out by T. N. Carlson at Sal Island, one of the network stations situated off the coast of Africa. At this time, we present the turbidity data and some supporting aerosol data from this network.

8.2 SITE DESCRIPTIONS

The land stations in this network were: Dakar, Senegal (14°44′N, 17°33′W); Sal Island, Cape Verde Islands (16°45′N, 22°57′W); Barbados, West Indies (13°10′N, 59°25′W); Miami, Florida, U.S.A. (25°45′N, 80°15′W); Bermuda (32°19′N, 64°46′W).

The Dakar aerosol site was situated on the coast at Pointe des Almadies, 12 km northwest of the city proper. Photometer measurements were usually made at the airport at Yoff, 4 km east of the coastal site.

Sal Island is the northeasternmost island of the Cape Verde Archipelago and is situated 600 km west–northwest of Dakar. The aerosol sampler was positioned on the crest of a hill 26 m above sea level and 500 m from the eastern shoreline.

Barbados, the easternmost island of the West Indies arc, is 3900 km west of Sal. The sampling site is located on a 20 m coral rock bluff on the east coast. Photometer measurements were made at a site 4 km west of the sampling site.

The Miami station is situated on the eastern side of Virginia Key, a small island about 4 km east of the city of Miami. The sampler was placed on the roof of a building 20 m above ground level and 25 m above sea level. Photometer measurements were made within a few hundred yards of this station.

The Bermuda measurements were made at the aerosol station operated by R. Duce (U. of Rhode Island) on the western tip of the island.

The ten ships in the network were: R/V Gilliss (U.S.), R/V Endurer (U.K.), R/V Charterer (U.K.), R/V Jean Charcot (France), R/V Sirius (Brazil), R/V Saldanha (Brazil), R/V Matamoros (Mexico), R/V Dallas (U.S.), R/V Researcher (U.S.), and R/V Oceanographer (U.S.). Ships were assigned fixed stations for each of the three experimental phases of GATE.

8.3 MEASUREMENT OF ATMOSPHERIC TURBIDITY

Atmospheric turbidity is defined as the extinction of solar radiation by suspended particles that are large with respect to the wavelength of light; that is, particles with radii from about 0.1 to 10 μm. The measurement of turbidity is important because it integrates the total loading of aerosol in the atmosphere on the line between the sun and the detector. There are several turbidity indices in existence; in this work we use the Volz turbidity, B, which is defined as:

$$I_\lambda \cdot S = I_{0\lambda}\ 10 \exp - \frac{P}{P_0} (\tau_{R\lambda} + \tau_{0\lambda} + B_\lambda) M \qquad (1)$$

where:

I_λ = irradiance at wavelength λ at the observing point,
$I_{0\lambda}$ = extraterrestrial irradiance at λ at the mean sun-earth distance,
S = correction factor for the mean sun-earth distance,
$\tau_{R\lambda}$ = Raleigh scattering coefficient for air molecules at λ,
$\tau_{0\lambda}$ = absorption coefficient for ozone at λ,
B_λ = extinction coefficient for aerosols at λ,
M = optical air mass,
P = station pressure,
P_0 = standard pressure at sea level.

The values of S for each day of the year can be calculated or obtained from a table. $\tau_{R\lambda}$ and $\tau_{0\lambda}$ are also known and M can be calculated or measured directly by the instrument. Thus, with the known calibration constant for the instrument, $I_{0\lambda}$, the aerosol optical thickness (Volz turbidity) can be calculated:

$$B_\lambda = \frac{\log (I_{0\lambda}/I_\lambda \cdot S)}{M} - (\tau_{R\lambda} + \tau_{0\lambda}) \quad (2)$$

The turbidity is often expressed for base e: $\tau_d = 2.303\ B_\lambda$.

Another useful turbidity parameter is Angstrom's wavelength exponent, α, for total haze scattering where scattering, σ, is given by:

$$\sigma \sim \lambda^{-\alpha} \quad (3)$$

Alpha can be calculated from turbidity measurements at any two wavelengths:

$$\alpha = \ln (B_{\lambda_1}/B_{\lambda_2})/\ln (\lambda_2/\lambda_1) \quad (4)$$

The value of α is determined by the size distribution of the aerosol. Typical continental (non arid-region) aerosols have α's of one to two; arid region aerosols have α's close to zero.

The sun photometers which we used (except as noted below) were manufactured by F. E. Volz. These have two channels, green and red. The green channel has an absorption type filter with a λ_{eff} of 500 nm and a 60 nm half width. The red channel has an interference filter with a λ_{eff} of 880 nm and 35 nm half width. The field of view is 3°. A silicon photovoltaic detector is employed.

8.4 PHOTOMETER CALIBRATION

The evaluation of sunphotometer data from a specific instrument is dependent on the knowledge of the extraterrestrial irradiance value ($I_{0\lambda}$) for that instrument. In practice, $I_{0\lambda}$ is obtained by Langley plots on days with constant low turbidity. Readings are taken throughout the day and log I is plotted against air mass

TABLE 8.1 Mean Atmospheric Turbidity Parameters for the North Atlantic during GATE (All Stations: Summer, 1974)

Site (Photometer No.) Parameters	$B(500)$		$B(880)$		Alpha		Lat (°N)	Long (°W)	n
LAND:									
BARBADOS (No. 281)									
Average (SD)	0.130	(0.081)	0.114	(0.076)	0.285	(0.244)	13.17	59.42	84
Geometric Mean (GSD)	0.106	(1.955)	0.090	(2.087)					
BERMUDA (No. 32)									
Average (SD)	0.097	(0.053)	0.041	(0.025)	1.618	(0.473)	32.32	64.77	59
Geometric Mean (GSD)	0.090	(1.407)	0.036	(1.552)					
DAKAR (No. 278)									
Average (SD)	0.350	(0.125)	0.328	(0.133)	0.152	(0.210)	14.73	17.55	12
Geometric Mean (GSD)	0.330	(1.434)	0.303	(1.547)					
MIAMI (No. 101)									
Average (SD)	0.100	(0.050)	0.062	(0.038)	0.899	(0.403)	25.75	80.25	81
Geometric Mean (GSD)	0.090	(1.545)	0.054	(1.661)					
SAL (No. 273)									
Average (SD)	0.305	(0.117)	0.253	(0.113)	0.368	(0.180)	16.75	22.95	33
Geometric Mean (GSD)	0.281	(1.566)	0.227	(1.679)					
SAL (No. 276)									
Average (SD)	0.266	(0.103)	0.221	(0.093)	0.353	(0.148)	16.75	22.95	45
Geometric Mean (GSD)	0.244	(1.562)	0.200	(1.653)					
SAL (EPPLEY)									
Average (SD)	0.273	(0.109)					16.75	22.95	30
Geometric Mean (GSD)	0.252	(1.518)							

DALLAS (EPPLEY No. 55)									
Average (SD)	0.062	(0.036)					8.53	24.41	33
Geometric Mean (GSD)	0.052	(1.874)							
CHARCOT (EPPLEY)									
Average (SD)	0.299	(0.109)					15.00	35.00	14
Geometric Mean (GSD)	0.273	(1.635)							
CHARTERER (No. 289)									
Average (SD)	0.245	(0.126)	0.214	(0.120)	0.273	(0.162)	14.48	30.55	43
Geometric Mean (GSD)	0.216	(1.661)	0.186	(1.718)					
ENDURER (No. 280)									
Average (SD)	0.263	(0.169)	0.213	(0.168)	0.545	(0.362)	15.33	23.27	39
Geometric Mean (GSD)	0.223	(1.769)	0.166	(2.043)					
GILLISS (No. 282)									
Average (SD)	0.139	(0.119)	0.121	(0.122)	0.492	(0.613)	9.31	24.62	30
Geometric Mean (GSD)	0.111	(1.887)	0.086	(2.272)					
MATAMORAS (No. 271)									
Average (SD)	0.098	(0.057)	0.055	(0.042)	1.168	(0.557)	6.80	47.58	31
Geometric Mean (GSD)	0.085	(1.706)	0.044	(1.907)					
OCEANOGRAPHER (EPPLEY No. 56)									
Average (SD)	0.069	(0.034)					9.21	23.57	11
Geometric Mean (GSD)	0.057	(2.225)							
RESEARCHER (EPPLEY No. 52)									
Average (SD)	0.077	(0.037)					7.47	23.69	41
Geometric Mean (GSD)	0.066	(1.973)							
SALDANHA (No. 275)									
Average (SD)	0.106	(0.036)	0.067	(0.022)	0.821	(0.356)	1.40	36.24	52
Geometric Mean (GSD)	0.100	(1.421)	0.064	(1.418)					
SIRIUS (No. 274)									
Average (SD)	0.196	(0.072)	0.144	(0.072)	0.759	(0.437)	7.86	38.81	56
Geometric Mean (GSD)	0.184	(1.441)	0.127	(1.690)					

thickness; extrapolation of the meter reading to zero air mass yields $I_{0\lambda}$. However, conditions of relatively constant low turbidity of sustained duration are only obtainable at a limited number of locations. Thus, most instruments, including ours, are calibrated against a limited number of these reference instruments which have been properly calibrated by the Langley method.

An extensive series of intercalibrations were made between our Volz instruments and two EPA reference instruments belonging to E. Flowers (NOAA); the EPA instruments are periodically calibrated by the Langley method at suitable locations in the western United States. Unfortunately the EPA and Volz instruments had only the 500 nm channel in common. In order to check the calibration of the 880 nm channel, we followed the Langley procedure.

Instruments were recalibrated in the Fall after the completion of GATE. Stability was found to be good for most instruments. Of the 13 Volz instruments tested and used in GATE, in the green channel, five drifted less than 2% and five drifted between 2 and 5%. In the red channel, six drifted less than 2% and three between 2 and 5%.

Also included in this work is data from two other types of instruments. On the Charcot, a two-channel (380 nm, 500 nm) Eppley sun photometer was used which was the same type employed in the United States EPA network. Also, five specially constructed (GATE) Eppley sun photometers were employed on four ships (Gilliss, Dallas, Researcher, and Oceanographer) and on Sal Island as a part of Carlson's radiation study there. The GATE Eppleys had two bands (380 and 500 nm), were solar tracking and the output was continuously recorded.

8.5 AEROSOL MEASUREMENTS

A detailed discussion of the aerosol sampling and analysis procedure is presented in Savoie and Prospero (1977). In summary, daily 24-hour duration bulk aerosol samples were collected on 6 cm effective diameter IPC-1478 filters at flow rates of 1.1–1.3 standard cubic metres per minute. In order to reduce the filter blank, the filters were subjected to a series of extractions to remove acid- and water-soluble species prior to being used.

In the analysis procedure, a quarter section of each filter was taken and the water-soluble material removed in a series of three extractions. The extractions and filtration of the resulting solutions were performed in a centrifugal filtration device fitted with a 25 mm membrane filter.

The mineral aerosol component was measured by ashing (500°C) the extracted IPC filter and the membrane filter used to filter the extraction aliquots.

8.6 RESULTS AND DISCUSSION

The mean atmospheric turbidity parameters for all stations are presented in Table 8.1. The data for the land stations is broken down into phases in Table 8.2. The dates of the phases were: I – 26 June to 18 July; II – 19 July to 19 August;

TABLE 8.2 Atmospheric Turbidity Parameters for the North Atlantic during Each Phase of GATE (Land Stations: Summer, 1974)

Site (Photometer No.)	Phase I				Phase II				Phase III			
Parameters	$B(500)$	$B(880)$	Alpha	n	$B(500)$	$B(880)$	Alpha	n	$B(500)$	$B(880)$	Alpha	n
BARBADOS (No. 281)												
Average (SD)	0.173 (0.090)	0.154 (0.086)	0.254 (0.178)	16	0.144 (0.087)	0.127 (0.079)	0.262 (0.215)	32	0.093 (0.057)	0.080 (0.056)	0.364 (0.282)	31
Geometric Mean (GSD)	0.149 (1.882)	0.129 (1.948)			0.118 (1.959)	0.102 (2.044)			0.079 (1.807)	0.064 (1.974)		
BERMUDA (No. 32)												
Average (SD)	0.116 (0.012)	0.065 (0.018)	1.067 (0.465)	4	0.088 (0.080)	0.039 (0.033)	1.488 (0.454)	24	0.098 (0.020)	0.038 (0.016)	1.748 (0.415)	25
Geometric Mean (GSD)	0.115 (1.111)	0.063 (1.330)			0.076 (1.535)	0.033 (1.621)			0.096 (1.217)	0.036 (1.464)		
DAKAR (No. 278)												
Average (SD)	0.350 (0.125)	0.328 (0.133)	0.152 (0.210)	12								
Geometric Mean (GSD)	0.330 (1.434)	0.303 (1.547)										
MIAMI (No. 101)												
Average (SD)	0.122 (0.052)	0.081 (0.032)	0.735 (0.412)	17	0.106 (0.055)	0.070 (0.051)	0.871 (0.298)	26	0.074 (0.037)	0.044 (0.024)	0.946 (0.472)	29
Geometric Mean (GSD)	0.112 (1.511)	0.075 (1.525)			0.097 (1.499)	0.059 (1.716)			0.069 (1.431)	0.040 (1.511)		
SAL (No. 273)												
Average (SD)	0.364 (0.136)	0.316 (0.134)	0.295 (0.200)	13	0.300 (0.075)	0.253 (0.076)	0.302 (0.073)	2	0.262 (0.087)	0.208 (0.077)	0.429 (0.154)	18
Geometric Mean (GSD)	0.341 (1.479)	0.288 (1.600)			0.296 (1.286)	0.248 (1.355)			0.242 (1.590)	0.189 (1.673)		
SAL (No. 276)												
Average (SD)	0.331 (0.151)	0.295 (0.144)	0.220 (0.071)	2	0.266 (0.113)	0.224 (0.102)	0.343 (0.158)	25	0.257 (0.087)	0.210 (0.076)	0.382 (0.136	18
Geometric Mean (GSD)	0.314 (1.603)	0.277 (1.667)			0.245 (1.546)	0.202 (1.628)			0.237 (1.607)	0.190 (1.709)		
SAL (EPPLEY)												
Average (SD)	0.336 (0.102)			10	0.241 (0.100)			20				
Geometric Mean (GSD)	0.323 (1.345)				0.223 (1.521)							

Phase I: 26 June–18 July; Phase II: 19 July–19 Aug.; Phase III: 20 Aug.–23 Sept.

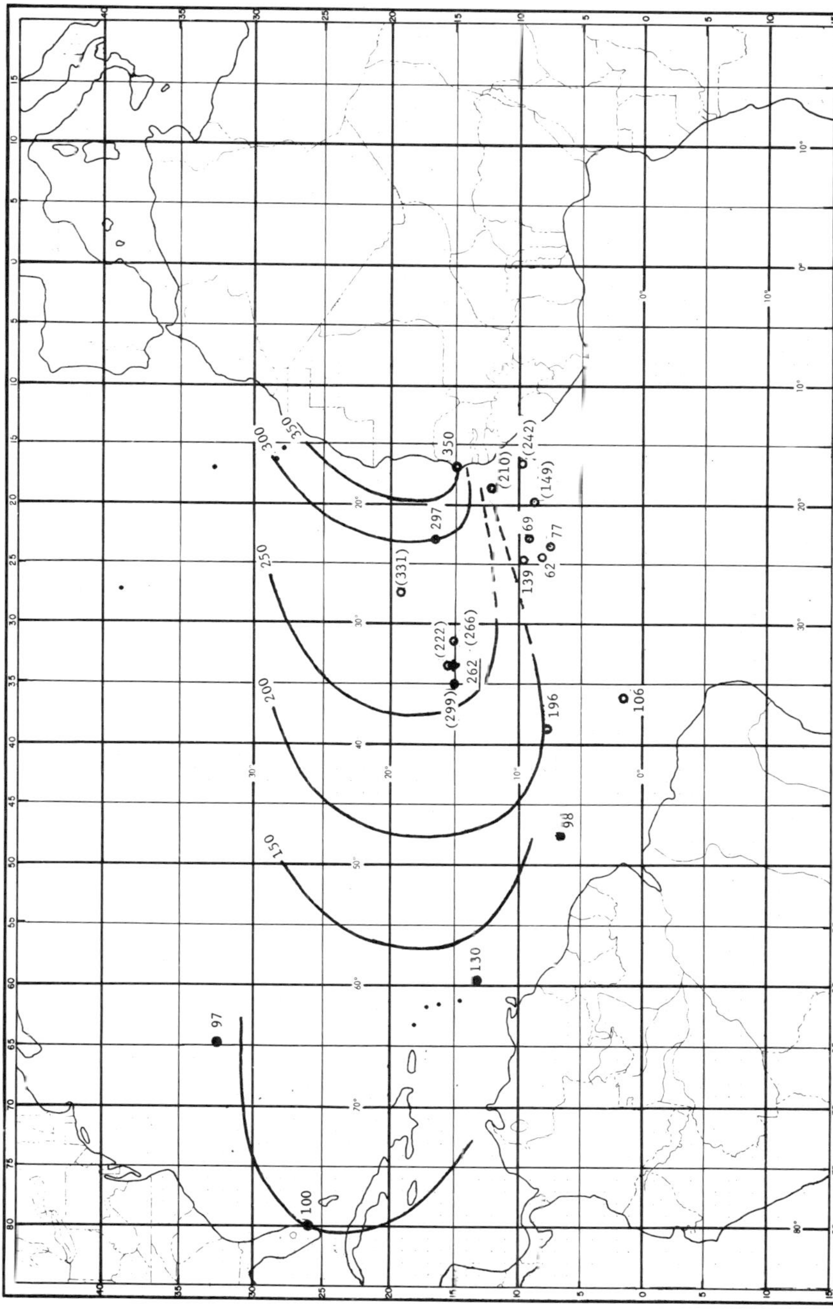

( ) Single phase data — Mean of 3 diff. ships

Figure 8.1 Volz turbidity, arithmetic mean values (x 10^3) for GATE

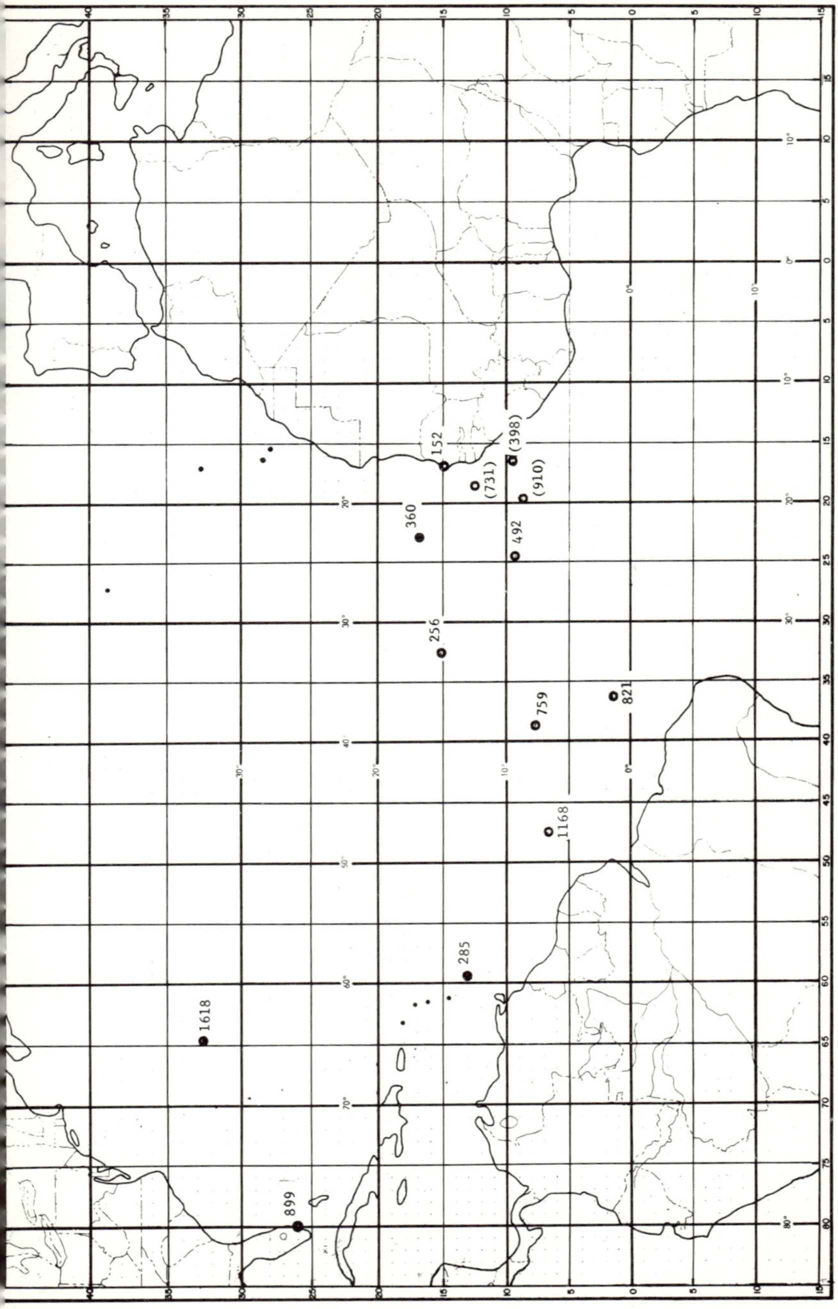

Figure 8.2 Alpha, arithmetic mean values (x 10^3) for GATE

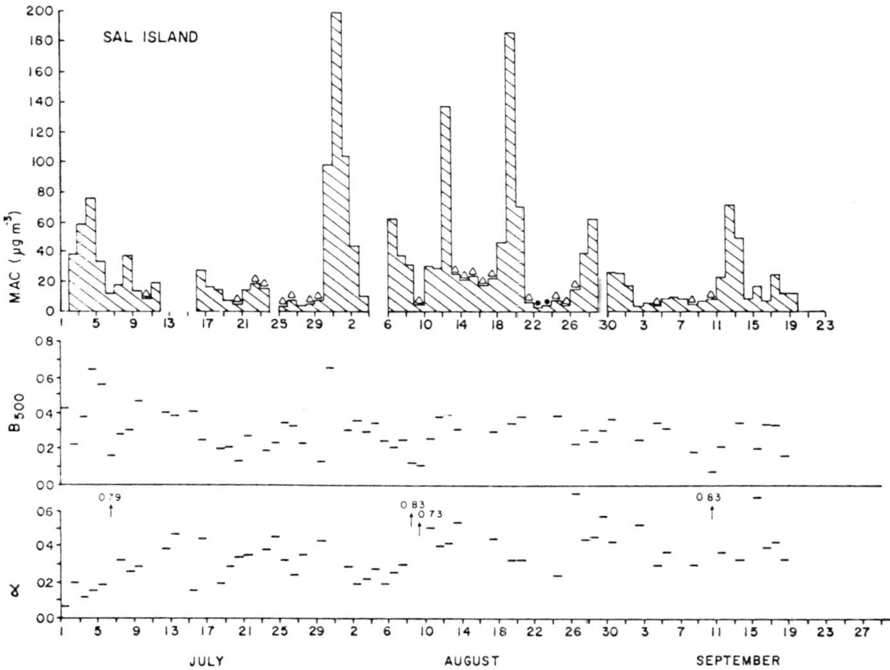

SAL ISLAND
MAC (μg m⁻³)
B500
α
0.79
0.83
0.73
0.83
JULY
AUGUST
SEPTEMBER

(a)

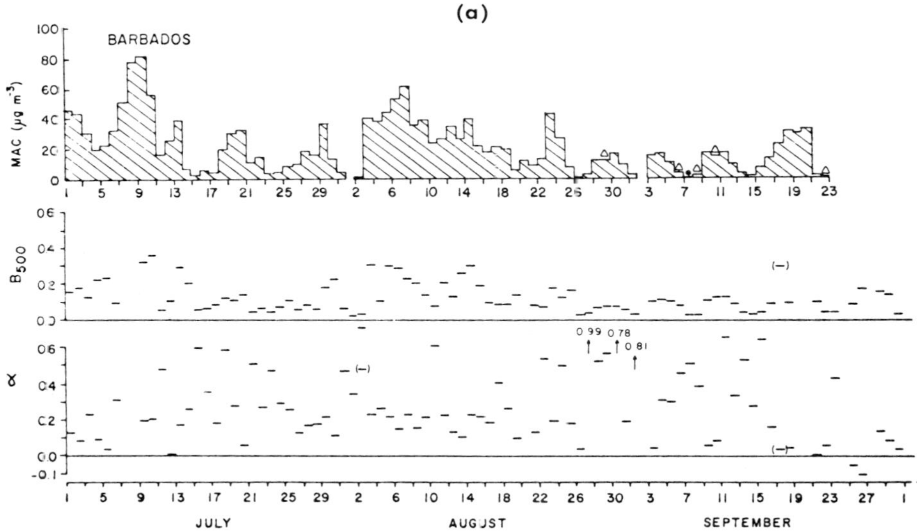

BARBADOS
MAC (μg m⁻³)
B500
α
0.99
0.78
0.81
JULY
AUGUST
SEPTEMBER

(b)

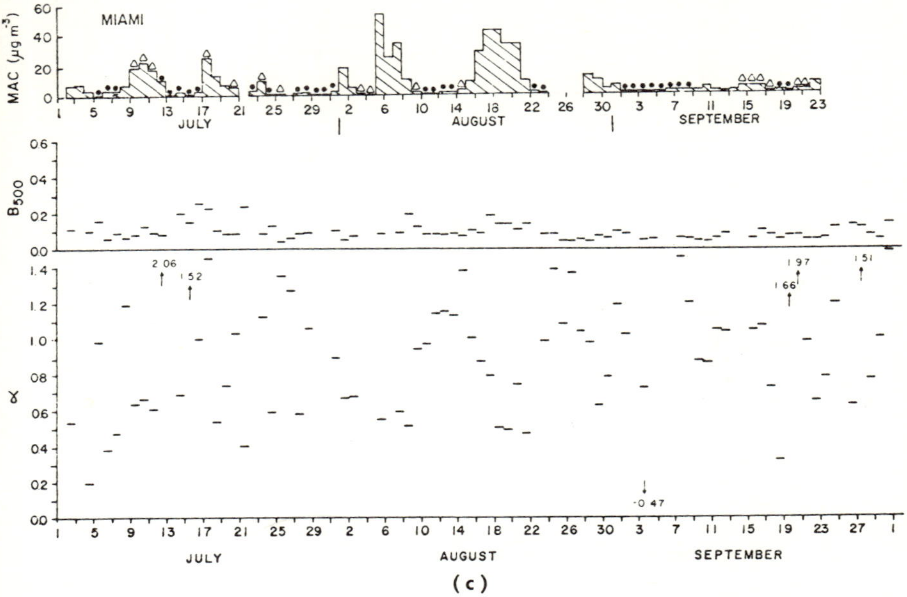

Figure 8.3 Daily values of mineral aerosol concentration (10^{-6} g m^{-3}), Volz turbidity at 500 nm and alpha

III – 20 August to 23 September. The mean values of the turbidity at 500 nm and of alpha are presented in Figures 8.1 and 8.2, respectively. As seen in Figure 8.1, the turbidity is greatest along the west coast of Africa and decreases with distance westward across the Atlantic and southward to the intertropical convergence zone (ITCZ). The high turbidities are attributable to the presence of large concentrations of mineral aerosol in the atmosphere. The daily values of mineral aerosol, turbidity (500 nm) and alpha at Sal, Barbardos, and Miami are shown in Figures 8.3a, b, c. The mineral aerosol data are coded according to filter colour, all filters being red-brown except those marked (●) indicating black or (△) indicating a mixed colour of red-brown and black. The vast majority of the Sal and Barbados aerosol samples are red-brown, a colour that is characteristic of Saharan aerosols. In contrast, Miami aerosols are predominantly characterized by a black or mixed colour. Nonetheless, on a mass basis, the dominant mineral aerosol component in Miami is Saharan dust.

The large day-to-day variations in dust load at Sal, Barbardos, and Miami are attributable to the passage of dust-laden air parcels associated with Saharan air outbreaks (SAO). The movement of these outbreaks can be observed by means of SMS-1 satellite photographs. The turbidity isopleths in Figure 8.1 are partially based on such photographs.

The turbidity values at Sal, Barbados, and Miami are not highly correlated with

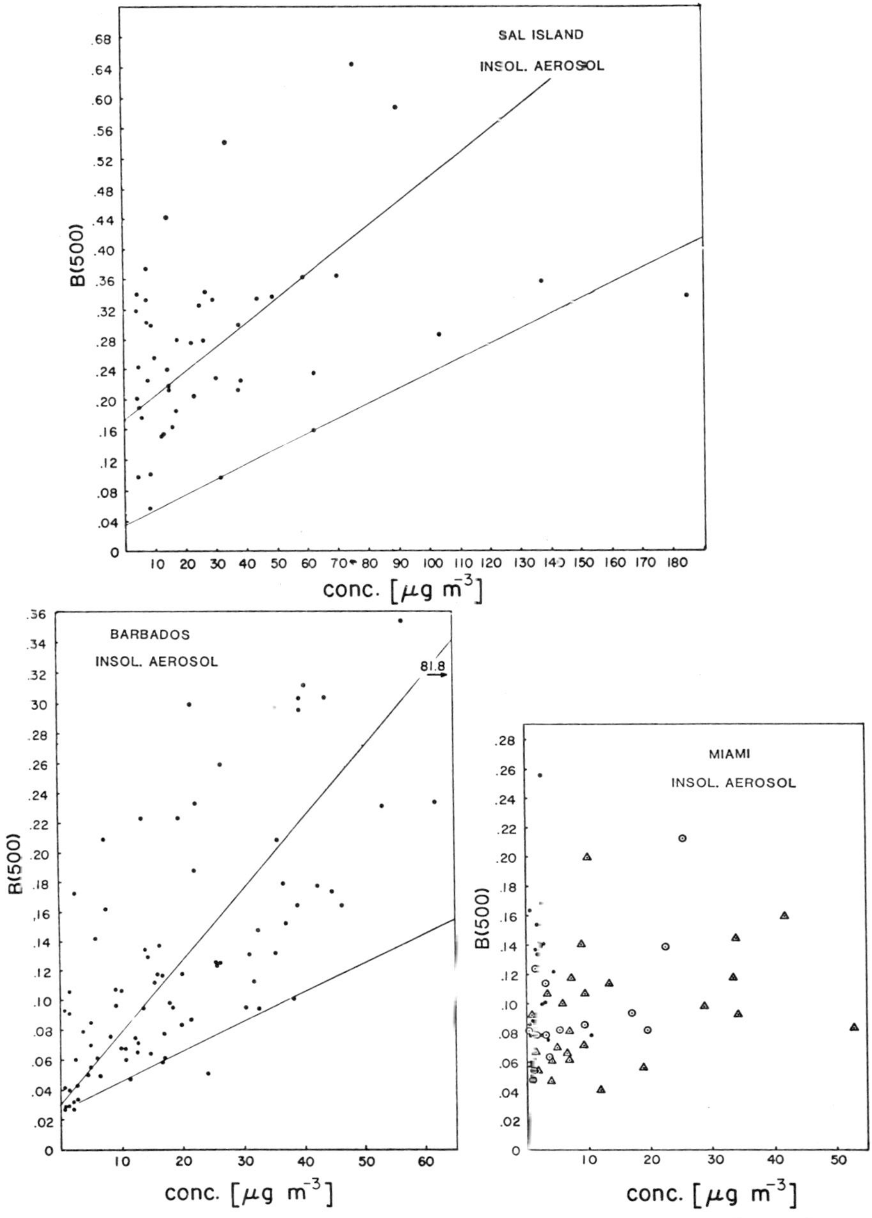

Figure 8.4 Volz turbidity *vs*. mineral aerosol concentration (10^{-6} g m^{-3}). The upper line in the Figures Sal Island and Barbados represents a least squares fit to the data points. The lower line is drawn by eye to show the trend of the lower limit of B *vs* aerosol concentration

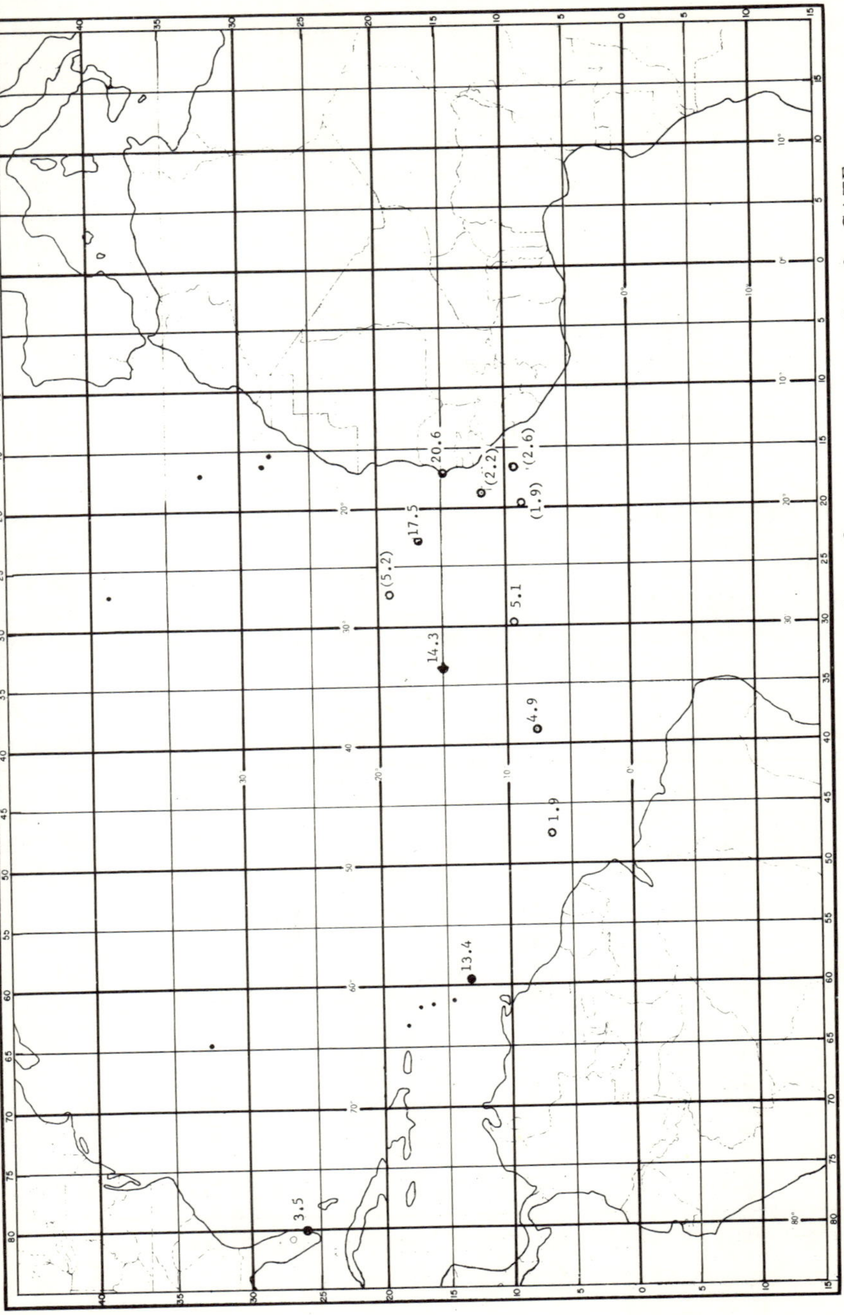

Figure 8.5 Mineral aerosol concentration (10^{-6} g m^{-3}), geometric mean values for GATE

the aerosol concentration as measured in surface level air. This can be noted from Figure 8.3; however, the lack of close correlation is best seen in Figures 8.4a, b, and c which show Volz turbidity *vs.* mineral aerosol concentration. (In Figure 8.4 the same symbols are used as in Figure 8.6.) Indeed, the mean mineral aerosol concentrations in the trade wind mixed layer at stations in the equatorial North Atlantic are not greatly different (see Table 8.3 and Figure 8.5). Concentrations drop sharply in the ITCZ and in the far western Atlantic (based on the Miami data).

The mineral aerosol and turbidity data is best understood in terms of the schematic model of Saharan dust transport proposed by Carlson and Prospero (1972). They suggest that the dust transport takes place primarily in a layer, the Saharan air layer (SAL), which has a base at 1 to 1.5 km and a top at 5 to 7 km. Dust is transported downward by gravitation (most effective for particles above *ca.* 10 μm) and by downward mixing through convective erosion of the SAL base.

The two-layer model explains a number of otherwise curious features in the data. Compare, for example, the phase I mineral aerosol data (Table 8.3) for the Endurer, which was located at 19°05′N 27°20′W, with that on Sal Island or on the Charcot at 15°00′N 35°00′W. Although the Endurer dust concentration is only a third to a quarter of those at the other two stations, the arithmetic mean turbidities are quite comparable: 0.331, 0.364 and 0.299 respectively. Thus, the SAL overlies all three stations and the vertically integrated aerosol loading in the optically active size range is relatively constant. However, the low level winds, which have a relatively strong northerly component in this region, have accumulated relatively little dust 'fallout' in the relatively short traverse under the layer.

TABLE 8.3 Mineral Aerosol Concentrations for the Equatorial North Atlantic During GAT

Station	Phase I			Phase II			Phase III			Overall	
	AM	*GM*	*n*	*AM*	*GM*	*n*	*AM*	*GM*	*n*	*AM*	*GM*
Land:											
Barbados	31.8	23.1	21	24.7	18.9	31	12.0	7.0	34	21.4	13.4
Dakar				38.3	31.1	11	17.2	9.7	6	30.9	20.6
Miami	9.2	5.9	16	11.6	4.6	26	4.1	2.0	27	8.1	3.5
Sal	28.9	23.8	13	41.7	22.1	29	18.3	12.0	29	29.8	17.5
Ship:											
Charcot	20.2	15.1	22							20.2	15.1
Charterer	5.3	2.6	14	41.8	28.3	23	15.1	7.8	28	22.4	9.7
Endurer	17.9	5.2	22	5.1	2.2	21	2.2	1.9	6	10.5	3.2
Gilliss	24.5	19.4	22	8.4	2.2	13	5.4	2.4	25	13.1	5.1
Matamoros				1.7	0.78	23	11.8	4.6	24	6.9	1.9
Sirius	19.8	16.0	21	3.7	1.4	21	9.9	5.3	19	11.2	4.9

Phase I: 26 June–18 July; Phase II: 19 July–19 Aug.: Phase III: 20 Aug.–23 Sept.

AM = Arithmetic mean; GM = Geometric mean.

The values of the Angstrom wavelength exponent, α (Tables 8.1 and 8.2, Figure 8.2) are relatively constant in the trade wind belt. Error analysis suggests that there is no significant difference amongst the mean values at Sal (0.360), at the composite ship position at 15°N 32°30′W (0.256) and at Barbardos (0.285). These low alphas, and the attendant neutral extinction, suggest that the Junge size distribution parameter, β, is close to two. This is in agreement with size distribution measurements made by Savoie and Prospero (1976) who found that the normalized geometric mean particle number-size distributions for three major dust outbreaks passing over Sal had $\beta = 2.02 \pm 0.29$ in the size range 0.5 to 1.0 μm diameter. In contrast, non-arid region aerosols have β values in the range 3 to 3.5 which should yield α values of 1 to 1.5. Alpha values in the latter range are seen in Miami during those periods when Saharan aerosols are not present in significant concentration; conversely, there is a tendency for low α's to obtain when Saharan air parcels are in the area. The relationship of α to Saharan aerosol concentration is best seen in Figure 8.6; here, the data points are coded according to colour with the symbol (△)

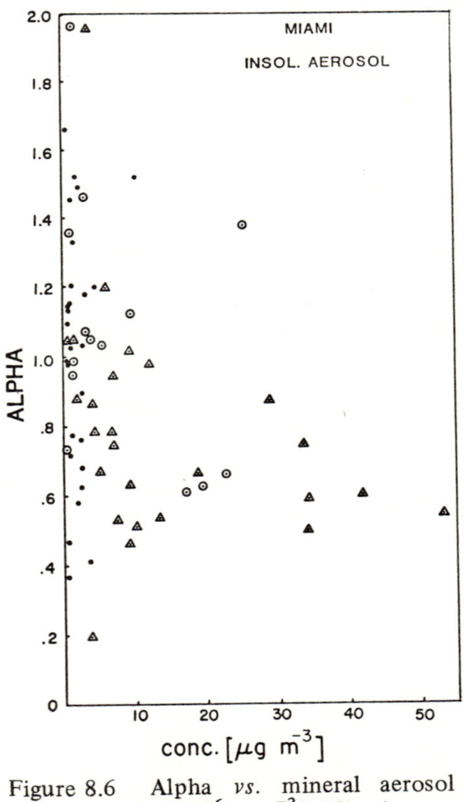

Figure 8.6 Alpha *vs.* mineral aerosol concentration (10^{-6} g m^{-3}), Miami

representing red-brown aerosol, (○) representing black and (●) representing mixed black and red-brown. There is a clear tendency for low α's to occur in association with red-brown aerosol and high α's in association with black aerosol. There is a certain degree of ambiguity in the relationship because of the fact that the surface level air may not be very representative of the vertically integrated aerosol at the site. Nonetheless, of the 28 days when α was greater than 1, black aerosols dominated on 15 days, mixed colour on eight and red-brown on five. Of the 39 days when α was less than one, black occurred on 12, mixed on six and red-brown on 21.

The persistence of low α's in Barbados and Miami in the presence of Saharan air parcels is consistent with the size distribution measurements at these locations; although the mean β's for the size range 0.5–1 μm were somewhat higher than that at Sal, the differences were not significant and the combined mean for all three stations was 2.18 ± 0.13 (Savoie and Prospero, 1976). Low α values appear to be a characteristic feature of air masses over, or emanating from, arid regions (Roosen *et al.*, 1973).

8.7 ACKNOWLEDGEMENT

Research supported by the Global Atmospheric Research Program, National Science Foundation (grant ATM76-01195) and the GATE Project Office, NOAA.

REFERENCES

Carlson, T. N., and Prospero, J. M. (1972). The large-scale movement of Saharan air outbreaks over the northern equatorial Atlantic. *J. Appl. Meteorol.*, **11**, 283–297.

Roosen, R. B., Angione, R. J., and Klemck, C. H. (1973). Worldwide variations in atmospheric transmission: I. Baseline results from Smithsonian observations. *Bull. Am. Meteorol. Soc.*, **54**, 307–316.

Savoie, D. L., and Prospero, J. M. (1976). Saharan aerosol transport across the Atlantic Ocean: characteristics of the input and the output. *Bull. Am. Meteorol. Soc.*, **57**, p. 145.

Savoie, D. L., and Prospero, J. M. (1977). Aerosol concentration statistics for the northern equatorial Atlantic, *J. Geophys. Res.*, in press.

CHAPTER 9

East Mediterranean Trajectories of Dust-carrying Storms from the Sahara and Sinai

D. H. Yaalon and E. Ganor

ABSTRACT

A detailed study of desert dust reaching and settling in Israel indicates that some 25 million tons of dust reaches annually the East Mediterranean Basin mostly settling into the Mediterranean Sea. Computed trajectories of the dust particles showed that the dust originates from the Libyan, Egyptian, Sinai, and Negev deserts. Quantity and mean particle size of the deposited dust in Israel decreases with distance from the desert as dust washed out by rainfall is finer grained than dry dust settling out in stable atmospheric conditions in the desert fringe regions.

9.1 INTRODUCTION

Atmospheric dust penetrates into Israel in the rainy and late spring seasons when barometric cyclones cross the Eastern Mediterranean Basin. Measurements of suspended and deposited dust were carried out over a period of 6 years, mainly from 1968 to 1973, in Jerusalem and at various climatological stations of Israel. Over 1,200 dust samples were obtained by different instruments and methods (Ganor, 1975).

Dust concentration of suspended dust with height during two dust storms in Jerusalem is indicated in Figure 9.1. The grain size distribution of deposited dust at several stations collected during a single dust storm is shown in Figure 9.2. The stations are arranged from South (Avdat and Beer-Sheva) to North (Kefar Gileadi and Mt. Hermon) indicating a decreasing mean size, mainly due to a gradually increasing clay content as the coarse silt population settles out on the fringes of the desert. A similar trend is obtained when total annual deposition is calculated, as has been reported elsewhere (Yaalon and Ganor, 1975). The total annual accretion decreases from the semi-arid to the subhumid Mediterranean region and amounts to some 50 to 200 ton/km^2. This constant accretion is identifiable in the soils and has had a significant effect on the nature of the soils of Israel (Yaalon and Ganor, 1973).

The usual composition of the clay mineral-free fractions in the dust was

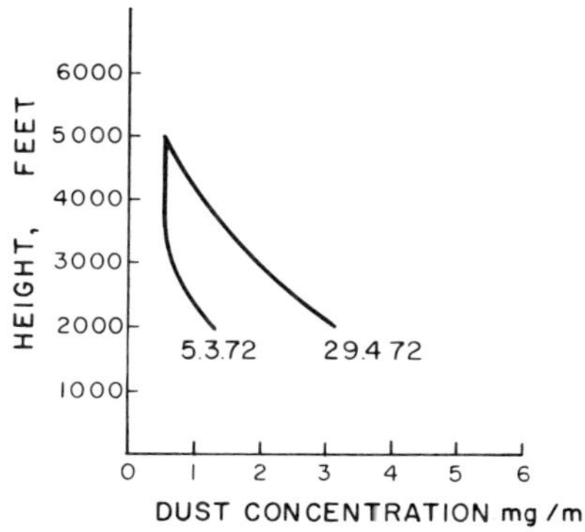

Figure 9.1 Dust concentration at various heights during two dust storms in 1972 in the coastal plain of Israel. The modal near surface concentration during dusty days is 1 mg/m^3 or 10 to 400 times more than during clear days (One foot equals 0.3048 metres)

35%–45% quartz, 30%–40% calcite, 10%–20% dolomite, 5%–10% feldspar and less than 1% halite. Carbonates were common both in the silt and in the clay fraction, some of it as forams and calcareous nanoplankton. Among the clay minerals montmorillonite and mixed layer minerals are dominant, followed in

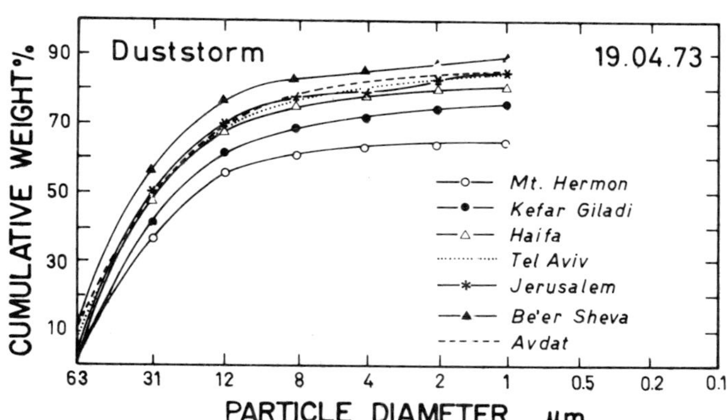

Figure 9.2 Cumulative grain size curves for dust collected during the storm of April 19, 1973. Note the decreasing coarse silt content going from the desertic South (Avdat, Beer Sheva) to the Mediterranean North (Kefar Gileadi, Mt. Hermon)

TABLE 9.1 Estimates of Suspended and Deposited Dust Quantities During Selected Dust Storms in the Eastern Mediterranean Basin

Date of dust storm	Wind direction D (azim.)	Visibility L (km)	Wind speed V (km/h)	Dust concentration C (ton/km^3)	Width of storm W (km)	Height of storm H (km)	Cross section of storm A (km^2)	Duration t (h)	Suspended dust T (10^6 ton)	Deposited dust Q (10^3 ton)	Ratio Q/T (%)
21-3-67	250	0.4	50	(3.0)	1200	(4.0)	4800	48	(34.5)	300	(0.8)
21-3-70	270	4.0	40	(1.0)	800	1.5	1200	25	(1.2)	220	(1.8)
30-3-71	200	0.3	35	10.0	1200	3.0	3600	20	25.2	460	(1.8)
7-11-71	260	4.0	20	0.6	1000	2.5	2500	30	0.9	60	6.6
5-3-72	270	2.0	30	2.0	1200	2.5	3000	36	6.5	120	1.8
29-4-72	270	0.8	40	3.0	1000	4.0	4000	40	19.2	160	0.8
5-5-72	270	7.0	60	1.0	1000	5.0	5000	43	14.4	16	0.1
Short duration storm		3.0	50	1.0	1000	1.5	1500	10	0.75	20	2.7
Long duration storm		0.3	60	1.0	1000	3.0	3000	50	9.0	400	4.4

D = Wind direction of dust storm in Jerusalem
L = Minimum horizontal visibility during dust storm
V = Maximal wind speed during dust storm
Q = Estimate of deposited dust quantities over Israel based mainly on measurements in Jerusalem
Q/T = Proportion of dust deposited over Israel in relation to total estimated quantity of suspended dust over the Eastern Mediterranean Basin during the dust storm

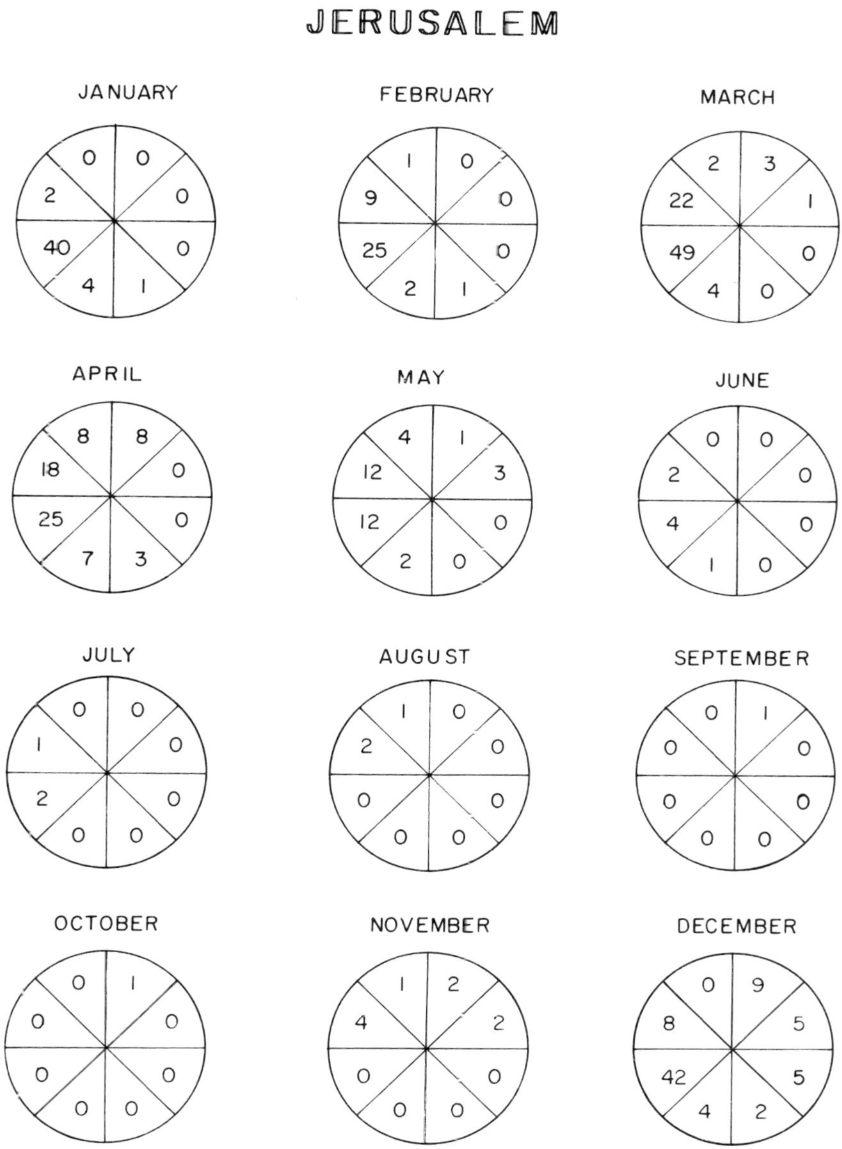

Figure 9.3 Frequency of wind directions during duststorm days (visibility $\leqslant$ 5 km) in Jerusalem, 1957–71. Numbers are total number of observations

abundance by kaolinite, some illite and palygorskite. The angular shapes and rough surface morphologies clearly indicate a desert weathering and origin of the mineral particles.

9.2 TOTAL QUANTITIES TRANSPORTED

Estimates on the total quantity of dust carried during selected dust storms have been made on the basis of observations of their areal extent, visibility, dust concentration, wind speed and duration of the storm, as indicated in Table 9.1.

It can be seen that during a short duration storm (less than one day) about one million tons of suspended dust is brought into the East Mediterranean Basin. During the more exceptional long duration storms (several days) the amount increases to about 10 million tons of dust and exceptionally to two or three times this amount.

The monthly frequency of 368 dusty days in Jerusalem, by wind direction, is shown in Figure 9.3, clearly indicating the dominance (80%) of the south-western and western dust storms during the four months December to March. Easterly dust storms are very exceptional (Yaalon and Ginzbourg, 1966). The average number of dust storms per year is 10 to 12 (Ganor, 1975; Katsnelson, 1970).

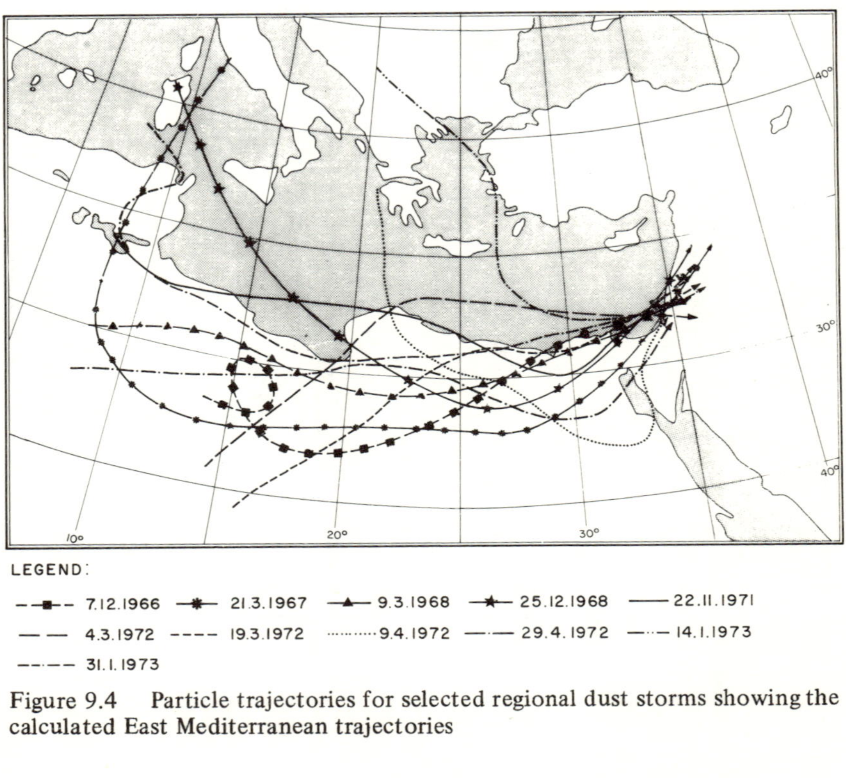

Figure 9.4 Particle trajectories for selected regional dust storms showing the calculated East Mediterranean trajectories

Combining these data of the frequency of dust storms and the amount of dust carried by them (Table 1) we calculate that the total yearly amount of dust brought into the Eastern Mediterranean Basin is some 20 to 30 million tons of dust. By far the largest part of this is deposited in the Mediterranean Sea. Assuming a transport of several hundreds of kilometers from the source area, it must be assumed that a much larger amount must have been detrained in the Saharan and Sinai source areas. This overall figure is then comparable, though probably still smaller than the estimated annual westward transport from the Sahara of some 200 million tons (Junge, 1979).

9.3 THE DUST TRAJECTORIES

The actual trajectories of the dust carrying storms were calculated (Djuric, 1961; AWS, 1968) and are shown for a number of storms in Figure 9.4.

Earth satellite photos were also used to observe the source area and extent of the dust plumes (Figure 9.5). Both the calculated trajectories and a meteorological analysis of synoptic pressure-patterns clearly indicate that the Libyan, Egyptian, Sinai, and Negev deserts are the main source areas of the dust reaching the Eastern Mediterranean.

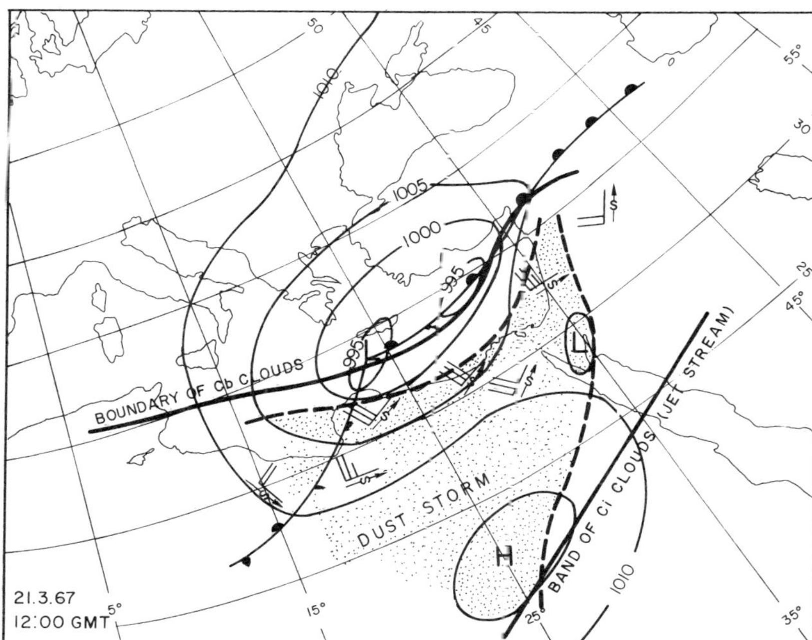

Figure 9.5 Aereal extent of the March 21, 1967 dust storm, as traced from satellite photos, superimposed over the synoptic map

The meteorological analysis of different synoptic pressure-pattern types showed two characteristic causes of dust storms. Western dust storms were brought about by barometric cyclones passing the Eastern Mediterranean Basin (Cyprus and 'Sahara' type depressions), while the rare eastern dust storms occurred when a Red Sea trough penetrated from the south or when a small depression developed in the Jordan Rift Valley and northern Negev. Dust particles become air-borne when atmospheric conditions are unstable and when strong turbulent winds are in existence. These conditions occur when the westerly winds pass over the wide flat wadis with their accumulated flood deposits (Yaalon, 1969). Part of the dust is redeposited locally after a relatively short transport but most is carried to the fringes of the desert (Yaalon and Dan, 1974). Settling of the dust is effected by precipitation and also in stable atmospheric conditions.

REFERENCES

AWS. (1968). The use of trajectories in terminal forecasting. U.S. Air Force, Air Weather Service, *Tech. Rep.*, **210**, 34 pp.

Djuric, D. (1961). On the accuracy of air trajectories computations. *J. Meteorol.*, **15**, 597–605.

Ganor, E. (1975). Atmospheric Dust in Israel-Sedimentological and Meteorological Analysis of Dust Deposition. Unpubl. *Ph.D. Thesis*, Hebrew Univ., Jerusalem, 224 pp. (in Hebrew).

Junge, C. (1979). *The Importance of Mineral Dust as an Atmospheric Constituent.* In this report.

Katsnelson, J. (1970). Frequency of duststorms in Beer Sheva. *Israel J. Earth Sci.*, **19**, 69–76.

Yaalon, D. H. (1969). Origin of Desert Loess. *Etudes Quater.* Monde, Congress, INQUA, Paris, **2**, p. 755.

Yaalon, D. H. and Dan, J. (1974). Accumulation and distribution of loess-derived deposit in semi-desert and desert fringe areas of Israel. *Z. Geomorph. Suppl.*, **20**, 91–105.

Yaalon, D. H. and Ganor, E. (1973). The influence of dust on soils during the quaternary. *Soil Sci.*, **116**, 146–155.

Yaalon, D. H. and Ganor, E. (1975). Rates of aeolian dust accretion in the Mediterranean and desert fringe environments of Israel. *Intern. Congr. Sedimentology, Nice,* **2**, 169–174.

Yaalon, D. H. and Ginzbourg, D. (1966). Sedimentary characteristics and climatic analysis of easterly dust storms in the Negev (Israel). *Sedimentology*, **6**, 315–332.

SECTION 2

(d) Monitoring and Deposition

CHAPTER 10

Saharan Dust Sedimentation in the Western Mediterranean Sea

K. G. ERIKSSON

ABSTRACT

The seas bordering North Africa receive yearly a varying amount of wind-borne material, which will be incorporated in the bottom sediments. In the Western Mediterranean Sea a sediment core has been collected at a depth of about 2,800 metres between Spain and Algeria and investigated concerning its amount of eolian material. A sequence of distinct layers of such material is described and approximately dated by the C-14 method. Comparatively great amounts of wind-borne material of silt-size have been brought to the investigated area especially during late Middle Würm and early Post Glacial time.

10.1 INTRODUCTION

Wind-borne materials constitute an important part of the marine sediments in the oceans and seas bordering North Africa; locally they may be the main part. Depending on the topography and the main wind directions certain sea areas receive more dust than others. The amount of dust is always highest near the coast and the grain size as well as the quantity is decreasing in a distal direction. During the last 30,000 years the frequency of strong dust storms have varied considerably according to the investigation of some sediment cores from the Western Mediterranean Sea; in this report only the results from one core, called No. 210, is described. This sediment core was collected from the research ship 'Albatross' by the 1947–1948 Swedish Deep-Sea Expedition led by Professor Hans Pettersson (1957). The core site is centrally located in the southern part of the Algiers–Provençal Basin (Figure 10.1).

10.2 RECENT WIND-BORNE MATERIAL

As an introduction to the description of some sediment layers of wind-borne material in the Western Mediterranean Sea a couple of examples of such dust found in Europe and originating from North Africa will be briefly described.

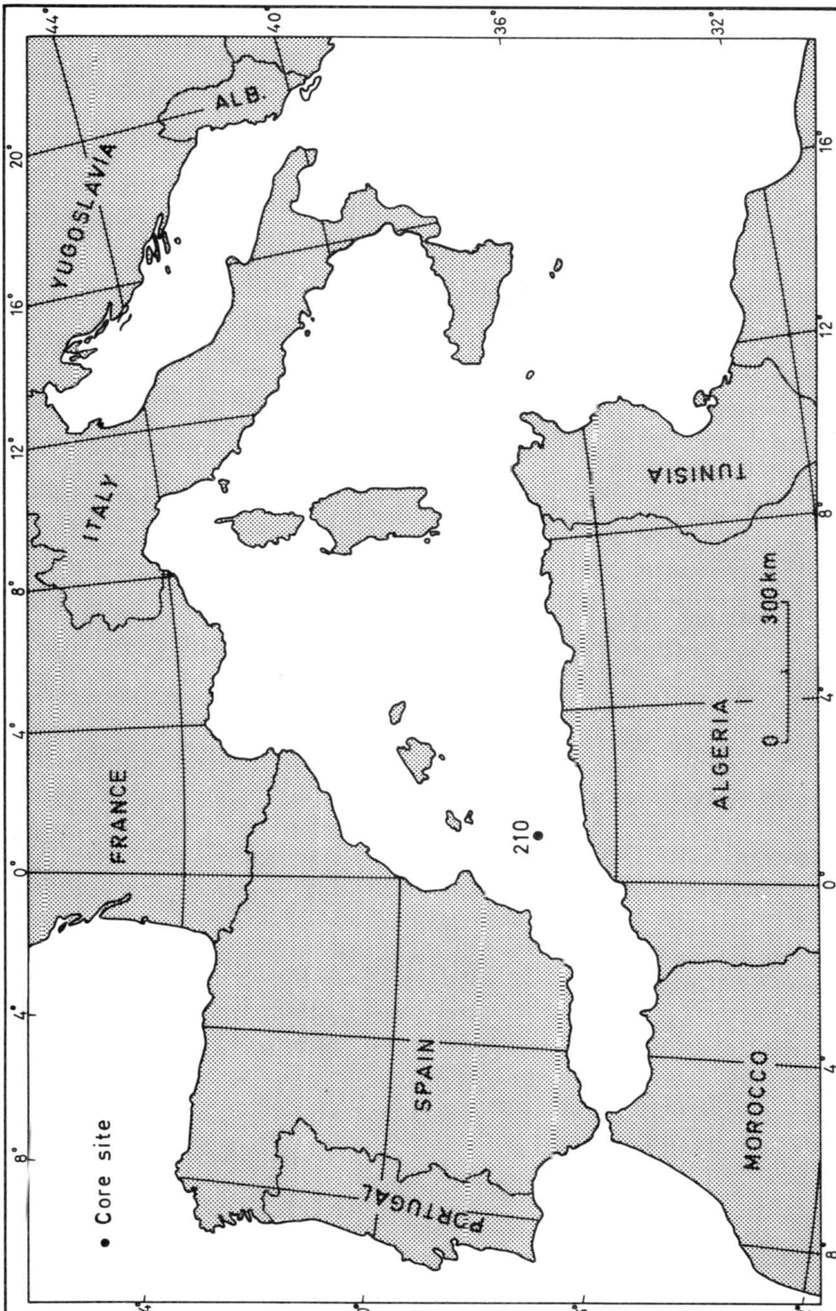

Figure 10.1 Western Mediterranean Sea and the location of Core No. 210

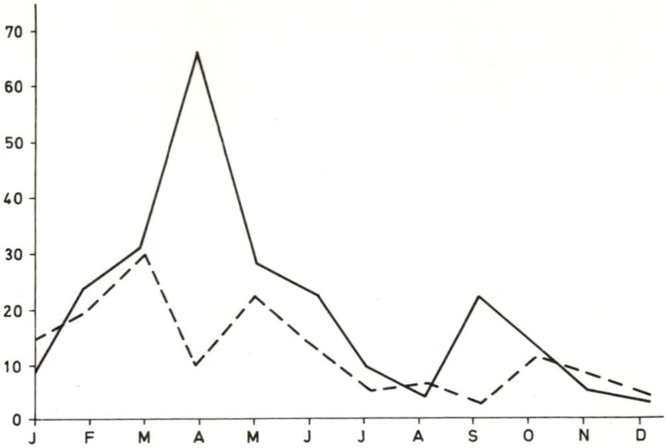

Figure 10.2 ——— Number of sand storms in the Sahara in 1936

– – – – Number of dust falls in Europe in 1936
(after Glawion, 1939)

Studies of modern sand dunes show that there is a definite relationship between the direction of dune migration and the general atmospheric circulation (Dubief, 1963; Opdyke, 1961). In the Sahara Desert the dunes are moving predominantly southwest and parallel to the trade winds. In the Mediterranean area, on the other hand, the dune fields are moving mainly eastwards because there the westerlies are the dominant winds. Those two facts plus the fact that the Western Mediterranean Sea is surrounded by high mountain ranges suggest that the major part of the sand-sized material transported to the sea by wind is derived from these mountain areas or the coastal regions (North Africa, Spain and France) and not from the Sahara Desert, However, during almost the entire year, especially in autumn and spring when there are changes between the winter and summer atmospheric circulation, there are many sand storms in the Sahara and North Africa. At times these sand storms transport the finest material, the dust, northward over the Mediterranean Sea, (Figure 10.2). Also, on rare occasions, winds are so extraordinarily strong and persistent that dust from the Sahara can be transported as far north as Scandinavia.

Investigations of many of the great dust storms in central Europe during this century have shown that the mineral assemblage of the dust arriving from North Africa and the Sahara Desert is fairly uniform (Thoulet, 1908; Radczewski, 1937; 1939; Gaubert, 1937). As early as 1901, Hellman and Meinardus showed that the principal mineral in the Sahara dust was quartz, many of the grains being well-rounded, frosted, and covered with a film of limonite, with subordinate amounts of mica, calcite, felspars, and limonite. The accessory minerals were usually gypsum,

hornblende, tourmaline, garnet, magnetite, epidote, titanite, rutile, and zircon. None of the samples examined at that occasion contained volcanic glass, but many samples contained abundant plant remains and diatoms, the latter of marine as well as of fresh water origin. The grain size of the material investigated ranged between 0.002–0.007 mm.

Götz (1936, p. 227) described an important dust fall over Arosa in Switzerland in 1936. This dust consisted mainly of calcite fragments with subordinate amounts of mica, chalcedony and quartz and most of the particles were covered by limonite. The grain size ranged between <1 and 100 μm; 5 μm was the median grain size.

Stumper and Willems (1947) have examined the great dust fall of March 29th, 1947, at Luxembourg in which the material < 8 μm (33%) consisted mainly of quartz (39%) and plant remains. The mean calcium carbonate content was 9% and the limonite content 9.5%. An approximately similar composition of the same dust deposit was found in Marseille.

Wind-blown dust, probably derived from the Spanish Sahara, collected on a ship in 1962 west of the Canary Islands has been described by Game (1964). This dust consisted mainly of 'aggregate' particles (mostly quartz, mica and clay minerals, determined by X-ray), calcite and rounded to semi-rounded, uncoated quartz grains. More than 80 per cent of the dust particles were between 5 and 30 μm in size.

These and other mineralogical investigations of the modern dust from North Africa and the Sahara Desert show that its principal component is quartz, part of which is rounded and frosted and sometimes coated by a thin film of iron oxide (Eriksson, 1961). Next in abundance is calcite or mica, then plant remains and then accessory heavy minerals, but volcanic glass is unrecorded. The most fine-grained dust usually is reddish and the coarse-grained dust yellowish (Radczewski, 1939; p. 497). Grain-size analyses of dust particles collected on ships off the West African coast show that the particles generally are less than 15 μm in diameter (Radczewski, 1939; p. 498), whereas the median values of particles collected in central Europe are usually smaller.

The dune and sand fields on the coastal areas of the Western Mediterranean Sea are fairly limited in size today, but during the stages of lower sea level they were more extensive. Quartz is usually the dominating mineral. In Late Pleistocene and recent sand dunes near Bone in Tunisia the quartz content is 60–80% followed by calcium carbonates, micas, felspars and heavy minerals (Moussu and Moussu, 1952). Locally in Tunisia, however, some beds of sand-sized material consists mainly of calcium carbonate fragments and it is called calcareous sand (Castany, 1952, p. 83; Norin, 1958, p. 95).

10.3 MICROSTRATIFIED SEDIMENTS

Three types of microstratified sediment have been observed in core 210. Only one type is of importance and this is designated 'microstratified wind-borne material' (Eriksson, 1965).

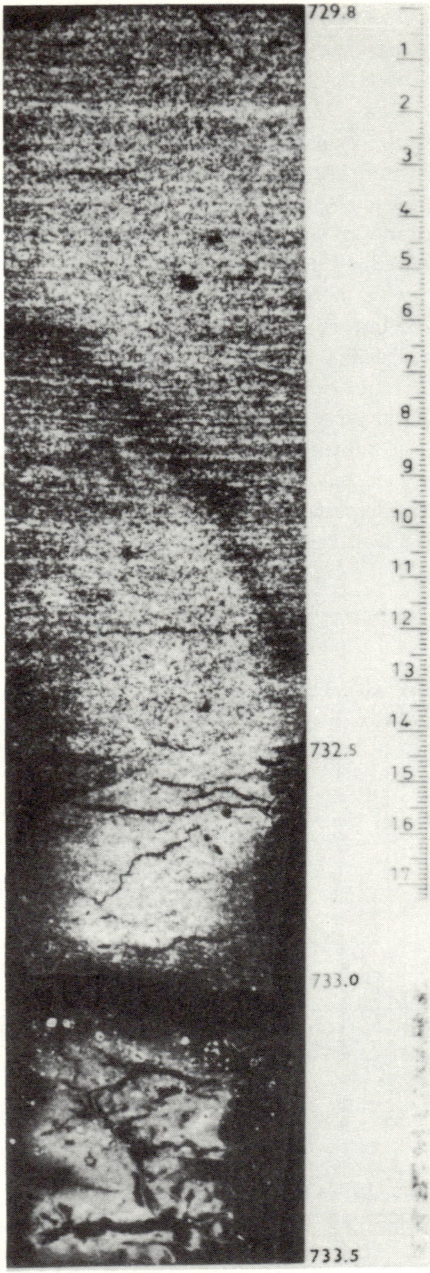

Figure 10.3 Microstratified wind-borne material. Thin section 733.5–729.8 cm in Core No. 210 (10 times)

To the naked eye the most conspicuous feature of the principal type of micro-stratified sediment is the thinly and regularly laminated structure in which coarse silt alternates with silty clay, in the form of laminations, a fraction of a mm to a few mm in thickness. This is repeated through sequences, up to 1/2 m in thickness, with remarkable regularity. The sediments are classified as marly, clayey silt or silt, usually highly micaceous. This type constitutes about 12% (107 cm) of the core and is observed in 7 beds, having an average thickness of 15 cm; a less well-developed lamination has also been noticed in a few small sections of the ordinary heterogeneous sediments. A bluish-grey shade dominates in the beds of a Würmian age, while yellow-grey predominates in the Postglacial sediments.

In the bed 733–723 cm about 200 laminae were observed in a sediment thickness of 7 cm and in the bed 590–541 cm the same thickness contained between 10 and 100 laminae. However, in the bed examined only the main part is laminated and other parts are non-stratified. No signs of erosion were observed within the beds, but the base usually corresponds to a surface of discontinuity (Figure 10.3).

The amount of silt-sized material averages 70 per cent, clay-sized about 29 per cent and sand-sized 1 per cent only. The low sand-content is noteworthy in view of the high precentage of silt. Two grain-size curves of this sediment type are shown in Figure 10.4.

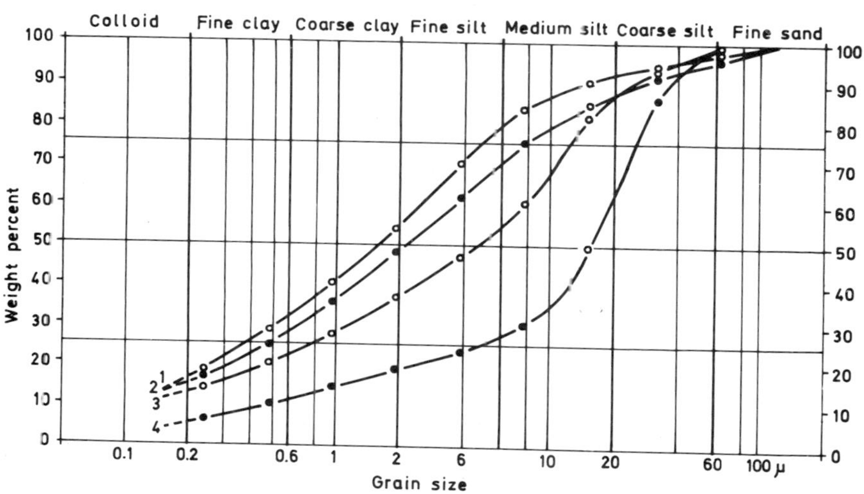

Figure 10.4 Cumulative curves of sediments from Core No. 210
(1) Ordinary, homogeneous sediment; mean of 77 pipette analyses.
(2) Ordinary, heterogeneous sediment; mean of 22 pipette analyses
(3) Microstratified wind-borne sediment; mean of 8 pipette analyses from bed 312.0–284.0 cm.
(4) Microstratified wind-borne sediment; sample 309.5 cm

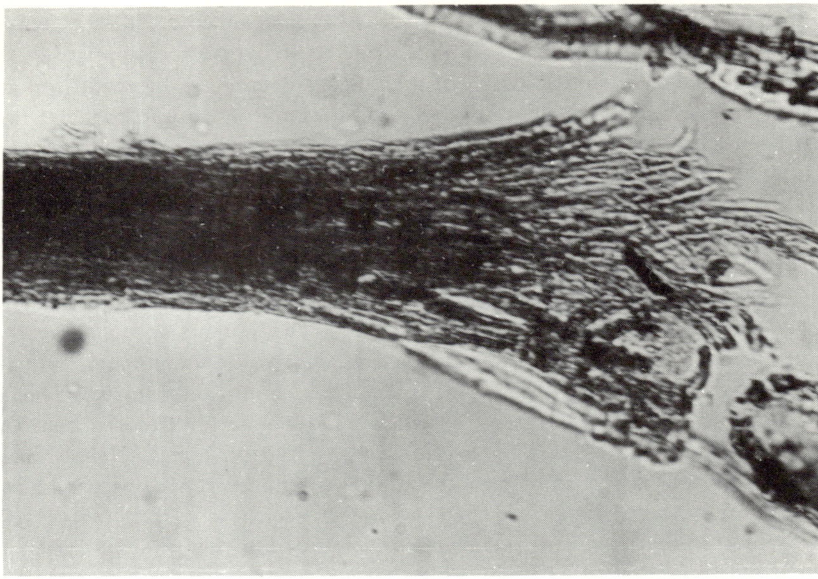

Figure 10.5 Plant fragment which occurs with 'coaly rods' (120 times)

A very conspicuous feature of the microstratified beds at 814.5, 733.0, 331.0, 312.0 and 7.5 cm is the very high content of minute mica flakes, probably mainly sericite, deposited parallel to the plane of sedimentation, sometimes giving the sediment an almost 'shaly' appearance; these beds are also characterized by a high content of 'coaly rods', which has been interpreted as being the ash of burnt grass (Figure 10.5). Fragments of volcanic glass are more frequent in these beds than in the other sediments. The fact that volcanic glass and 'coaly rods' are deposited together is considered as evidence of their continental, wind-borne origin.

Although a similar microstratified structure is exhibited in the beds 590 and 65.5 cm, the content of mica is nevertheless low and 'coaly rods' do not occur. The peculiar microstratification is therefore a feature independent of these components.

The component which predominates in the micro-stratified beds is carbonate which attains a relatively high level in spite of the low content of foraminiferal tests. The mean value, calculated from 64 analyses, is 40%, due mainly to the presence of a large amount of microcrystalline calcite. In the fraction $> 31\ \mu m$ the content of terrigenous minerals, organic detritus, opaque particles and fragments of siliceous organisms has increased relatively to that of the ordinary sediments.

The principal non-calcareous component in the terrigenous assemblage is quartz, which, however, occurs at a lower percentage than in the ordinary sediments. In the heavy mineral assemblage the amphibole/pyroxene minerals are the most abundant (average amount: 0.3 v %). Next in order comes garnet, tourmaline and zircon.

Most of the sanidines observed in core 210 occur in this sediment type. The content of micas is actually higher than that shown in the analyses because many of the particles classified as 'opaque' consist of mica flakes densely coated with iron and manganese oxides or hydroxides. High concentrations of opaque particles have mostly been observed in the micro-stratified sediments, e.g. at the levels 811.0, 730.0, 309.0, 305.5, and 4.5 cm. Moreover, studies of thin sections of the above mentioned mica-rich beds, revealed that the amount of opaque particles seems to be more abundant here than elsewhere in the core, a feature not observed in any other sediment type in the core.

In these sediments are also found the largest amounts of organic detritus (average value: 11 v %) ever measured in core 210. The amount *of organic matter*, as determined chemically is, on the other hand, relatively low compared with the core-mean, which is partly associated with the coarse grain size of this component. The mean value of Fe_2O_3 attains 3.5%, which is fairly close to the core-mean. This is probably due to the coating by iron oxide observed on many of the mineral grains. In addition many laminae in the ordinary sediments, rich in wind-borne material, show a marked increase in the iron-oxide content.

The structure of the micro-stratified sediment is such that the deposition by water currents appears unlikely. Thus, (1) its basal contact is a fairly sharp surface of discontinuity lacking conspicuous signs of erosion; (2) the structure is one of alternating, but indistinctly delimited, very thin laminae, many of which contain 'coaly rods' and volcanic glass, and is therefore atypical of current-deposited sediments in core 210; (3) almost all elongated or flat particles are oriented parallel to the plane of sedimentation which suggests that calm water conditions obtained during deposition. In view of the rather uniform grain-size and the high content of 'coaly rods' the present author has interpreted this structure as arising from a more or less rhythmic supply of wind-borne material. The mineral assemblage with its high content of microcrystalline calcite is suggestive of a wind-borne material of loessic character.

10.4 PERIODS OF HEAVY DEPOSITION OF WIND-BORNE MATERIAL

The dust-material is derived from three principal regions adjacent to the Western Mediterranean Sea, (1) the Sahara Desert, which seems to be the most common source of dust today, (2) the 'tablelands' in Spain, Morocco and Algeria, and (3) the coastal areas, which during the glacial periods were more extensive than today. The investigation gives no reliable evidences of the provenance of the dust material in the microstratified sediments. The relatively low content of microfossils and high amount of 'desert sand' indicate, however, that the source area is a desert or a semidesert.

The time-periods mentioned below are based on dating of selected beds in the core by the ^{14}C method calcareous microfossils $>44\ \mu m$ (Olsson, 1959; Olsson, 1960; Olsson *et al.*, 1961; Olsson and Blake, 1961–1962; Olsson and Kilicci, 1964; Olsson and Eriksson, 1965). The datings are compared with changes in the number

of the most temperature-sensitive species of the foraminiferal assemblages and with other features of stratigraphic importance. Of many reasons it has not been possible to date the different beds with microstratified wind-borne material, and their age has therefore been extrapolated from datings of suitable layers below and above (Figure 10.6).

In the sediment core 210 appears a few periods when the deposition of wind-borne material has been more pronounced than normally.

10.4.1 Würm

The oldest micro-stratified bed in core 210 (814.5–809.0 cm) is rich in micas, 'desert quartz' and volcanic glass. This bed was probably deposited during an early part of the Paudorf–Arcy Interstadial; it might also represent an oscillation caused by an ice advance in the Loess récent II sequence according to Bonifay (1962).

The second micro-stratified bed of mainly wind-borne material is 10 cm thick (733.0 to 723.0 cm). This bed is also rich in mica and, in addition, it has the largest amount of 'coaly rods' and volcanic glass found in core 210.

The chronological position of this bed is uncertain due to the lack of ^{14}C dating; an extrapolation from a dating made on level 670 cm gives a value with a magnitude between 24,000 and 25,000 years corresponding to an early part of the Würm III (Brandenburg ice advance), i.e. the deposition of the oldest sediments of Loess récent III. During this substage of severe cold and rather dry climatic conditions, a steppe-vegetation on loess and sand deposits is known to have existed in Europe as well as in North Africa (Quézel, 1963).

The core-section, 675.0–663.0 cm, has been dated by the ^{14}C method. The resulting age coincides with the Frankfurt ice advance. An extrapolation based on this dating suggests that these sediments were deposited approximately between 24.000 and 18.000 years ago. This very cold and fairly dry substage of the Middle Würm is considered as being a principal time for loess deposition in central Europe, and it is therefore somewhat surprising that the content of loess is not more prominent in this part of core.

Next follows the thickest (49 cm) microstratified bed in core 210, 590.0–541.0 cm. This bed contains no 'coaly rods' and the amounts of organic detritus and volcanic glass are comparatively moderate; the terrigenous mineral assemblage is dominated by micas, chiefly biotite; quartz and heavy minerals are rather common. The sediment is considered to be loessic in character. Due to a low amount of foraminiferal tests this bed has not been ^{14}C-dated, but according to an extrapolation from the dated level 347–337 cm the bed may have been deposited about 18,000 years ago, or during the later part of the maximum extent of Würm III ice advance. During this time of cold and dry climate, the Loess récent III was deposited in southern France. The extension of bed 590.0–541 cm is tentatively correlated with the extensive deposition of loess which took place during the later part of the Würm glaciation (Gross, 1962/63, p. 62).

WESTERN MEDITERRANEAN SEA, Core 210

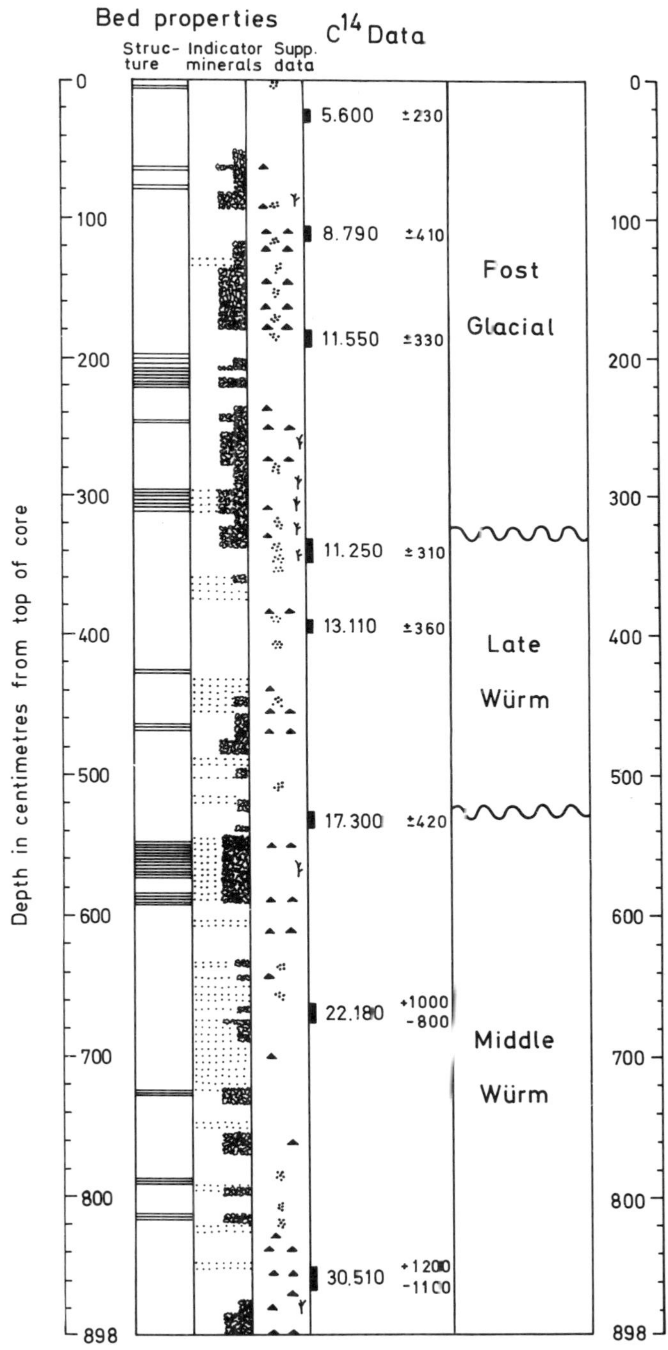

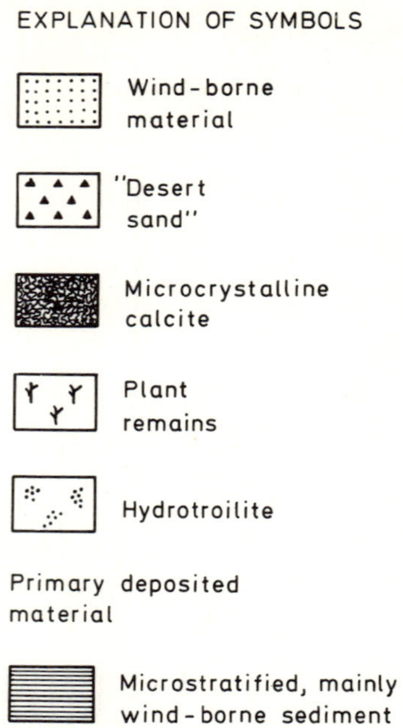

Figure 10.6 Stratigraphy and approximative chronology of the most conspicuous beds of wind-borne material in Core No. 210

10.4.2 Late Würm

During Late Würm or from about 17,000 to 10,000 years ago the deposition of dust in the Algiers–Provençal Basin seems to have been comparatively limited. Only one microstratified bed in the Late Würm period is recognised in core 210 namely the one between 331.0 and 326.0 cm. The sediment is rich in micas and 'coaly rods'. According to an extrapolated dating, this bed was probably deposited during Younger Dryas (11,000–10,000 B.P.). This proposal seems plausible since at that time the deposition of Loess récent IV in southern France continued and in the Sahara the arid climate of Steppe time still persisted.

10.4.3 Post-Glacial

Above a surface of erosion at 312.0 cm there is a bed of microstratified material, 28 cm thick (Figure 10.7). It consists mainly of microcrystalline calcite and calcareous particles of silt-size (mean carbonate content = 45%) and the predominant terrigenous components are quartz, micas, organic detritus including

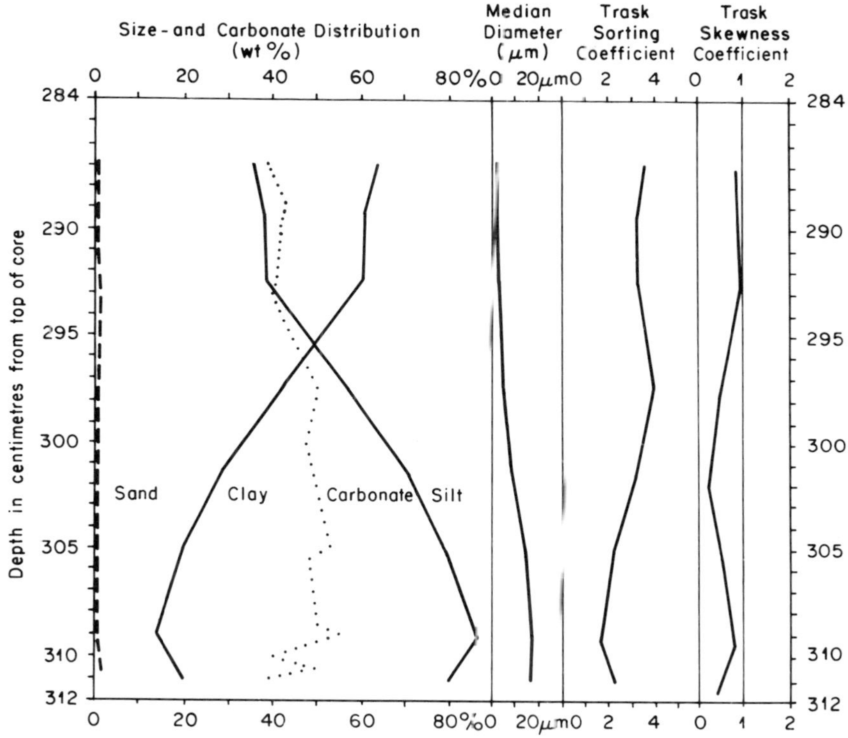

Figure 10.7 Diagram of the grain size and carbonate content distribution in the division 312.0–284.0 cm. The sediment is rich in wind-borne material

'coaly rods' and opaque particles. This bed is probably deposited in the very beginning of the Post-glacial time.

A 5 cm thick microstratified sediment bed (65.5 to 60.0 cm) is deposited around 6,000 years ago or in the beginning of Late Atlantic time. This bed is also rich in calcite grains of silt-size together with quartz and micas; volcanic glass is also present as well as a fair abundance of organic detritus.

10.5 CONCLUSION

This investigation is too limited to allow tenable conclusions concerning the variations in supply of wind-borne material to the Western Mediterranean Sea. It seems, however, rather clear that the periods of transition from one climatic period to another have been extremely stormy bringing comparatively very great amounts of relatively coarse-grained wind-borne material to the investigated area.

REFERENCES

Bonifay, E. (1962). Quaternaire et préhistoire des régions méditerranéennes françaises. *Quaternaria,* **6**, 343–370.

Castany, G. (1952). Atlas Tunisien oriental et Sahel. *19th Int. Geol. Congr. 1952: Monogr. Reg. 2*, Ser. Tunisie-6, Alger, 152 pp.

Dubief, J. (1963). Contribution au problème des changements de climat survenus aux cours de la période couverte par les observations météorologiques faites dans le nord de l'Afrique. In *Changes of climate. Proc. Rome Symp. org. by UNESCO and WHO*, Paris: 75–79.

Eriksson, K. G. (1961). Granulométrie des sédiments de l'île d'Alboran, Méditerranée occidentale. *Bull. Geol. Inst. Univ. Uppsala,* **40**, 269–284.

Eriksson, K. G. (1965). The sediment core No. 210 from the Western Mediterranean Sea. *Rep. Swed. Deep-Sea Exped.*, **8**, 395–594.

Game, P. M. (1964). Observations on a dust-fall in the eastern Atlantic, February, 1962. *J. Sediment. Petrol.,* **34**, 355–359.

Gaubert, M. P. (1937). Etude mineralogique de la pluie de boue du 27–28 Novembre, 1930. *Mem. Off. Nat. Météorol. de France*, No. 27.

Glawion, H. (1939). Staub und Staubfälle in Arosa. *Beitr. Phys. Frei. Atmos.,* **25**, 1–43.

Götz, F. W. P. (1936). Staubfälle in Arosa im Spätwinter 1936. *Meteorol. Z.,* **6**, p. 227.

Gross, H. (1962/63). Der gegenwärtige Stand der Geochronologie des Spätpleistozäns in Mittel- und Westeuropa. *Quartär,* **14**, 49–68.

Hellmann, G., and Meinardus, W. (1901). Der grosse Staubfall vom 9. bis 12. Märtz 1901 in Nordafrika, Süd- und Mitteleuropa. Abh. Königlich Preuss. *Meteorol. Inst.,* **2**:**1**, Berlin, 93 pp.

Moussu, P. and Moussu, H. (1952). Etude hydrogéologique des dunes de Bone. *19th Int. Geol. Congr. 1952,* 2: Données sur l'Hydrogeologique Algérienne: 112–129, Alger.

Norin, E. (1958). The sediments of the central Tyrrhenian Sea. *Rep. Swed. Deep-Sea Exped.,* **8**, 1–136.

Olsson, I. U. (1959). Uppsala natural radiocarbon measurements 1. *Am. J. Sci. Radioc. Supp.,* **1**, 87–102.

Olsson, I. U. (1960). Uppsala natural radiocarbon measurements 2. *Am. J. Sci. Radioc. Supp.,* **2**, 112–128.

Olsson, I. U., Cazeneuve, H., Gustavsson, J. E. and Karlén, I. (1961). Uppsala natural radiocarbon measurements 3. *Radiocarbon.,* **3**, 81–85.

Olsson, I. U., and Blake, W., Jr. (1961–1962). Problems of radiocarbon dating of raised beaches, based on experience in Spitsbergen. *Norsk Geogr. Tidskr.,* **18**, 47–64. Oslo.

Olsson, I. U. and Kilicci, S. (1964). Uppsala natural radiocarbon measurements 4. *Radiocarbon,* **6**, 291–307.

Olsson, I. U. and Eriksson, K. G. (1965). Remarks on C^{14} dating of shell material in sea sediments. *Prog. Oceanogr.,* **3**, 253–266.

Opdyke, N. D. (1961). The paleoclimatological significance of desert sandstone. In Nairn, A.E.M. (ed.) '*Descriptive Paleoclimatology*', 45–60. Intersci. Publ. Ltd., London, New York.

Pettersson, H. (1957). The voyage. *Rep. Swed. Deep-Sea Exped.,* **1**, 1–123.

Quézel, P. (1963). De l'application de techniques palynogiques à un territoire désertique Paléoclimatologie du quaternaire récent au Sahara. In *Changes of climate. Proc. Rome Symp. org. by UNESCO and WMO*. Paris: 243–249.

Radczewski, O. E. (1937). Die Mineralfazies der Sedimente des Kapverden-Beckens. *Wiss. Ergeb. Dtsch. Atl. Exped. a.d. 'Meteor'* 1925–1927, 3, 43–134, Berlin.
Radczewski, O. E. (1939). Eolian deposits in marine sediments. In Trask, P. D. (ed.) *Recent Marine Sediments. A Symposium*. Am. Assoc. Petrol. Geol., 496–505.
Stumper, R. and Willems, A. (1947). La pluie de boue du 29 mars 1947. Inst. Grand-Ducal Luxemb., Sect. sci. Nat. Phys. *Math. Arch.*, **17**, 105–106.
Thoulet, M. (1908). Origine éolienne des minéraux fins contenus dans les fonds marins. *C.R. Acad. Sci.*, **146**, 1346–1348.

CHAPTER 11

The Use of Moss-bags in Aerosol Monitoring

G. T. GOODMAN, M. J. INSKIP, S. SMITH, G. D. R. PARRY, and M. A. S. BURTON

ABSTRACT

Ideally, it would be desirable to characterize the atmospheric aerosol as follows:

(i) identify the aeolian dust component (derived from arid regions)
(ii) locate the areas of its origins
(iii) estimate its rate of deposition to the ground (including interception by vegetation)

Such information would assist in understanding the input–output dynamics in and around deserts including regional dust transport and indicate any dry-flushing effects on the productivity of vegetation receiving dust inputs.

Trace-element analysis has been used in elucidating items (i) and (ii) above, and for all those items above various kinds of traditional dry deposit gauges (e.g. filter paper pads) and dustfall collectors have been used.

This paper explores the possible use of specially cleaned samples of moss contained in flat, nylon-mesh envelopes (moss-bags) as substitutes for these gauges.

Moss-bags are very inexpensive to produce, can be sent easily by postal service to analytical laboratories and thus can be used in large numbers to make wide-scale synoptic surveys. They collect aerosol in terms of particle sizes and quantity per unit area with an efficiency closely similar to that of short grass sward, without the 'bounce-off', 'blow-off' or turbulence effects shown by certain types of dry-deposit gauge, which may tend to undersample, especially at higher wind-speeds.

Mosses have a high cation-exchange capacity and there is virtually no loss of metallic elements even under periods of intense rainfall. They can thus be used as total (wet and dry) collectors. This usefully avoids the need to carry out separate rainfall analyses to allow for the leaching of soluble trace-elements in dust-fall gauges following rain-episodes.

Evidence so far collected indicates that moss-bags orientated vertically to face various compass directions act somewhat like 'dipole aerials' so that, for pollutants at least, distant emission sources can be located. The possible utility of this for identifying the origins of transported dust is also discussed.

11.1 INTRODUCTION

Although dustfall buckets and various other kinds of dry-deposit (DD) and total-deposit (TD) collection devices have been widely used in the past to sample the atmospheric aerosol, recent improvements in direct air-filtration equipment, including isokinetic and high-volume samplers have proved capable of providing more unbiased estimates of atmospheric constituents. Additionally, the greatly increased sensitivity of modern chemical analytical methods has reduced the need for the large samples formerly required for analysis, thus further decreasing dependence on deposition samplers.

However, although the difficulties of relating the results obtained from deposit gauges to the actual concentrations of airborne constituents in the ambient atmosphere are well known, there are three principal reasons why deposit gauges still have a useful part to play in our understanding of aerial transport phenomena:

(1) They do not depend on the availability of electrical power supplies and although by no means a substitute for the direct measurement of air concentration (AC), they do at least give some idea of airborne constituents in remote places.

(2) Unlike AC gauges, they are not fixed volume samplers. Thus, in this respect they do not resemble the human lung and hence cannot be so conveniently related to medical studies as AC gauges can. However, they do reflect the nature and quantities of airborne constituents likely to be intercepted by the ground (including surface waters, soils and vegetation). Hence they are approximate measures of the atmospheric input of nutrients and pollutants to the food-chain. In this respect they relate usefully to agricultural, forestry, ecosystem and water-quality studies of chemical distribution.

(3) When investigating the fluxes of airborne materials from series of measurements in time and space, deposit gauges can often be more informative than AC gauges as they can show the net effects of aerial transport from one place to another.

When these features are seen in the context of aerosol-transport studies in arid zones, deposit gauges may be useful in studying: (a) the dry flushing effect on agricultural soils of airborne macronutrients and trace elements (K, Ca, Mg, N, S, Mn, B, Co, Cu, Zn, etc.) derived from deserts; (b) the genesis of aeolian soils and (c) the input/output dynamics of ground–surface materials in and around deserts.

The ensuing paper highlights some of the above-mentioned differences between AC, DD and TD methods and examines how far flat, nylon-mesh packets containing epiphytic or ombrotrophic moss (moss-bags) might be worth trying out as a convenient and inexpensive substitute for the more traditional type of deposit gauge and their possible use in future arid zone studies.

Unless otherwise stated, the type of deposit gauges being compared are as

follows:

(a) *Total deposit gauge* (TD). This is like a rain gauge with a polythene funnel (the mouth is screened with 0.5 mm terylene mesh to exclude insects). It delivers deposited dust and precipitation into a polythene collecting bottle (Cawse and Peirson 1972).

(b) *Dry deposit gauge* (DD). This is essentially a flat horizontal pad of filter-paper protected above from rain by a sheet of perspex, but open to winds on all sides so that blown particles can deposit on the filter paper surface (Cawse and Peirson 1972).

(c) *Moss-bag* (MB). This is a flat 10 x 10 cm envelope of 2 mm nylon-mesh containing specially cleaned moss (usually *Sphagnum acutifolium* aggregate), evenly packed to occupy the whole surface area of the packet which is orientated horizontally or vertically to face a particular direction or hung freely suspended from a nylon thread.

All the gauges are normally positioned 150 cm above the ground surface. The air concentration sampler with which they are compared sucked air through a Whatman filter-paper at the rate of approx. 5 1/min (Cawse and Pierson, 1972).

11.2 MOSS-BAGS

11.2.1 Development of the method

Epiphytic mosses receive all their mineral nutrients from the air (Tamm, 1953), and therefore any metallic substance in the moss must come from this medium. Following this premise, Rühling and Tyler (1970) showed how varying metal concentrations in naturally growing epiphytic mosses in Sweden might reflect prevailing aerial metal burdens.

Investigations in the Lower Swansea Valley area (Goodman and Roberts, 1971) showed abnormally high levels of several heavy metals in the epiphytic moss *Hypnum cupressiforme*, both from natural populations and also in moss collected from clean areas and hung out in nylon mesh bags at atmospherically polluted sites. Elevated metal levels first appeared 10 km WSW (upwind) from Swansea. These rose to about 4–20 times the 'normal values' found in and around the city of Swansea and steadily fell off again to 'normal values' 25–30 km NNE (downwind of Swansea). This was the first study using moss-bags and the method was developed further and has been used extensively by the present authors and also by other workers.

Clean Sphagnum moss (*Sphagnum acutifolium* aggregate) is collected from rural areas in Wales, away from industrial activity. Following three washes in 0.5N HNO_3 for three days, the moss is soaked and washed three times in double-distilled water, surplus water removed and the moss mixed to obtain as homogenous a material as possible.

Approximately 10–15 g (equivalent to about 2.0–2.5 g dry weight) is then sewn into flat 10 x 10 cm nylon mesh bags of approximately 2 mm mesh size (obtained from Henry Simon, type 18 GGN) so as to occupy the whole internal area of the bag. The bag is positioned in the centre of a circular loop of plastic-covered wire, held at the corners by four nylon threads each of which is tied to the loop.

After exposure to the air at the test site (usually on a bamboo cane at 150 cm above ground for 2–4 weeks) the bags are dried at 40 °C to constant weight. The surface area of moss is accurately measured and the moss then removed carefully into Kjeldahl flasks and weighed. Following wet digestion in 15 ml of a mixture of 'Aristar' grade nitric and perchloric acids (4 : 1 ratio), the solutions are boiled gently down to a few millilitres and made up to 25 ml volume.

The solutions are then analysed for metals by atomic absorption spectrophotometry using a Techtron model AA.3 (Mercury is analysed using a flameless technique).

After correction for the original metal content of unexposed moss, the amount of metal deposited and retained on the bag is calculated in terms of μg metal per cm^2 per month.

Further development work showed that the moss-bag appeared to be a useful, inexpensive, semi-quantitative indicator of metal pollutant levels and can be used to monitor relative changes in airborne metal burdens in space and time.

11.2.2 Using moss-bags to locate sources (April 1971, Isopleth survey)

In an experiment financed by the Natural Environment Research Council and the University of Wales, an array of moss-bags was set up in the Lower Swansea Valley, UK, at a network of sites, for a period of one month (10 April–11 May 1971). At each site, four bags were freely suspended 2 m above the ground-surface; two bags contained *H. cupressiforme* and the other two contained *Sphagnum acutifolium* aggregate. At the end of the period, bags from 80 sites were analysed in order to discover how far:

(a) meaningful spatial patterns of metal distribution would emerge;
(b) the results from the *Hypnum* and *Sphagnum* bags were comparable.

The results showed that regular patterns of metal distribution could be clearly discerned, allowing sites with similar metal values to be joined by isopleth lines giving a type of 'contour map' of intercepted airborne metal levels.

Magnesium isopleths showed a tendency to increase parallel to the coast as the sea is approached, an example of the, by now, well documented marine source of airborne magnesium. Other metals fell into two distinct groups. One group, nickel and cobalt showed a roughly concentric zonation of isopleths with the highest values centred close to a nickel refinery. The other group zinc, lead, and cadmium showed a broadly similar pattern with the highest values centred some 4.5 km farther south, close to a zinc smelter. The copper isopleths had a double centre,

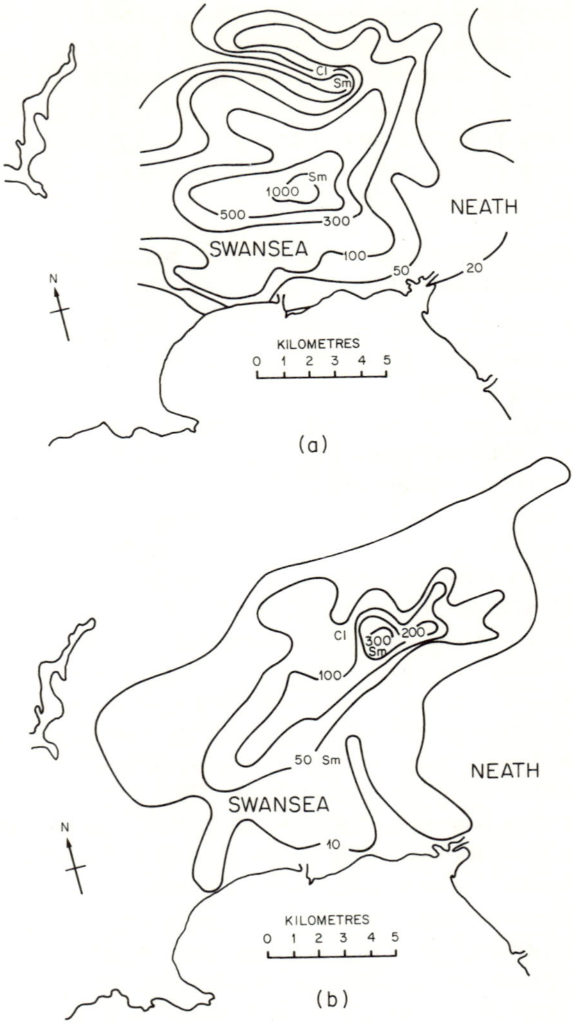

Figure 11.1 (a) Lead and (b) Nickel deposition onto Sphagnum moss-bags exposed during April 1971 (expressed as ng cm^{-2} day^{-1})

indicating some copper output by both works. The results clearly showed that metal emissions as: (a) wind-blown dusts from the old contaminated soils and waste-tips of former industry; (b) urban smoke; (c) vehicle exhausts and (d) other industries; were of far less significance than anticipated when compared with output from the two ongoing industries. Figures 11.1(a) and 11.1(b) show isopleth maps for Pb and Ni respectively.

The above conclusions emerged irrespective of whether the moss-bags contained *Hypnum* or *Sphagnum*. The level of each metal at each site, obtained from a *Hypnum* filled bag, compared closely with that obtained from a *Sphagnum* filled bag. All *Hypnum* data, taken site by site, correlated in a highly significant way with the corresponding *Sphagnum* data indicating that closely similar isopleth pictures of metal distribution are obtained irrespective of which moss is used.

11.2.3 Using moss-bags to follow airborne-metal changes with time (December 1971, Isopleth survey)

The use of moss-bags as a relative measure of the geographical distribution of metals around an emission source – changes in space – can be paralleled by their use to follow changes with time.

A good example was provided when in mid-May 1971 the zinc smelter in the Lower Swansea Valley mentioned above, closed down following industrial restructuring by the company concerned. The nickel works, 4.5 km north of the zinc smelter, continued to operate as usual. The effect of this shutdown on metal burdens in the area was monitored by regular sampling. In particular a second large-scale moss-bag survey using the same sites as in April 1971 was carried out in December (10 December–7 January 1972).

Although December patterns for the geographical distribution of all the elements were basically similar to those of April 1971, i.e. centred around the two metal works concerned, there were marked quantitative differences.

The average retention of nickel and cobalt by moss-bags doubled during the December period. There appeared to be a 'heaping-up' effect of Ni and Co isopleths around the nickel works (Figure 11.2(b)). This may have been caused by: (a) increased production at the works (considered unlikely); (b) the prevailing winter weather being less favourable for the dispersion of aerial particulates; and/or (c) the damper weather causing increased interception on the moss-bags because they stayed moist for longer periods. This latter phenomenon, i.e. wet moss-bags being more efficient interceptors, was demonstrated by Chamberlain and Clough in wind-tunnel experiments with moss-bags and is similar to what would happen with moist ground vegetation. Changes in wind speed and direction were largely discounted as being of any importance because apart from a slight increase in westerly and north-eastern winds in December, wind-rose data for the April period were substantially the same as those for the December period.

In sharp contrast to this increase, zinc, lead, and cadmium levels fell in December, by factors of × 4–10, to much less than the April levels (Figure 11.2(a)). According to these results, the area was no longer affected by appreciable fall-out of zinc, lead and cadmium and the removal of the swamping effect of this emission source enabled subsidiary emission sources to be located, e.g. wind-blown dusts from the old metal-rich wastes of former industry now contributed discernibly to the isopleths. This can be seen as a 100 ng isopleth 1 km North of Swansea (Figure 11.2(a)).

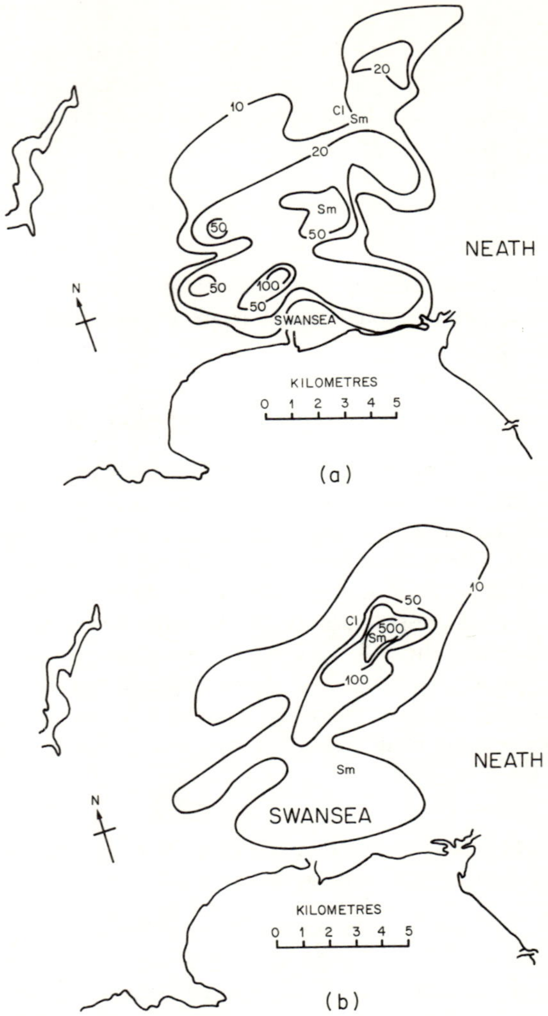

Figure 11.2 (a) Lead and (b) Nickel deposition onto Sphagnum moss-bags exposed during December 1971 (expressed as ng cm^{-2} day^{-1})

The copper isopleth map now lost is double centre; the single effective centre now being close to the nickel works.

A further feature of great interest was that the moss used had a very high cation exchange capacity. Thus, elution of solutions of metallic elements through columns of moss retained virtually all the dissolved metal on the moss. Likewise, moss-bags exposed to drenching rain (both natural and simulated) retained both the metallic

elements already deposited on the bags and also the soluble metals in the rain itself.

In view of their cheapness, and the likelihood of their behaving like a grass sward as regards particle interception (Clough, 1974), moss-bags might be suitable for routine network sampling and monitoring. A special advantage is that these can easily be sent in polythene bags by letter post to a central analytical laboratory. This makes them more convenient to handle than TD bottles. However, further information on the behaviour of moss-bags as a deposition gauge is required. It is natural to ask:

(a) How they relate to the existing more traditional methods (TD, DD and AC)?
(b) How they and the traditional gauges relate to interception by ground vegetation?

11.3 COMPARISON OF MOSS-BAGS WITH OTHER GAUGES

During the period 1 June 1972–31 May 1973 a study of metal deposition on to moss-bags (MB), total deposition (TD), dry deposition (DD) gauges and measurements of air concentration (AC) were made at nine sites in South Wales sponsored by the Welsh Office. The Natural Environment Research Council sponsored a study of a further seven sites (six in the UK and one in the Netherlands). MB measurements were made by the present authors and the other measurements (TD, DD, and AC) by AERE Harwell. The following seven metals were analysed for all gauges: Co, Cu, Fe, Mn, Ni, Pb, and Zn. This enables comparisons to be made between airborne metal deposition on to moss-bags and the other three gauges. In addition, the relationship between metal retention by moss-bags and a grass sward was examined.

11.4 FACTORS INFLUENCING AIRBORNE METAL DEPOSITION ON GROUND VEGETATION AND SAMPLING GAUGES

Extensive work has been carried out by Chamberlain and Clough (Chamberlain, 1966a and b; Chamberlain and Chadwick, 1972; Clough, 1973, 1974) using a wind tunnel to investigate the behaviour of airborne particles during their deposition on receptor surfaces (trays of grass sward, trays of moss, moss-bags and the filter paper surfaces of dry deposit gauges). This showed that the proportion of airborne particles retained by the receptive surfaces of the grass sward and the three types of gauge depends upon:

(a) The retentive characteristics of the receptor surface ('hairiness', wetness, or stickiness).
(b) Factors affecting the delivery of particles to the receptor surfaces (wind speed, turbulence).

In the following summary it must be borne in mind that, as indicated in the Introduction 11.1 above, AC gives a rough estimate of lung burden whilst TD, DD and

MB all provide approximate guides to food chain burden *via* vegetation interception.

It should be emphasized at the outset, however, that the deposit gauges (TD and DD above) were not originally designed to simulate a grass-surface, but to provide a useful measure of deposition behaviour of the various elements in the atmospheric aerosol (Cawse and Pierson, 1972). Similarly, the moss-bag gauge was designed to simulate deposition to natural populations of moss growing epiphytically on walls or trees (Goodman and Roberts, 1971). However, since these are the *only* convenient, inexpensive methods available to us for estimating deposition to ground vegetation, it is clearly desirable to determine how they relate to one another and, as far as possible, to discover how well they reflect interception by grass swards or other ground vegetation.

(1) Despite the apparent design differences in these three gauges (TD, DD, MB), the physical principles governing deposition on their retentive surfaces are basically the same as those controlling interception by a grass sward.

(2) The physical properties which make deposition surfaces efficient interceptors of airborne particles are 'hairiness', wetness or stickiness. The DD filter paper, which is kept dry and is relatively smooth, would not be expected to trap particles as efficiently (V_g)* as grass which is 'hairy', and periodically wet. Moss-bags, which get periodically wet to about the same extent as the grass at the site where the moss-bag is exposed, and are relatively far more hairy than grass, would be expected to trap particles more efficiently than grass. Because of its terylene mesh cover, the filter funnel of the total deposit gauge may possibly be more efficient than the filter paper of the dry deposit gauge. It is usually assumed that dry deposition on the filter paper of the dry deposition gauge (DD) is equivalent to that on the rain funnel of the total deposition gauge (TD) for equal projected horizontal areas.

(3) Particles of different sizes are not retained on any of these deposition surfaces with equal efficiency. Thus, for filter-paper, larger particles (c. 10 μm) and the smallest (c. 0.1 μm) are retained most efficiently (high V_g), with intermediate sized particles showing a sharp minimum in the proportion retained at c. 0.5 μm diameter (low V_g). This gives a pronounced V-shaped curve for the proportion of particles retained with increasing particle size. Grass swards and moss-bags have a closely similar particle size retention curve which is somewhat the same as that for filter paper but more U-shaped, i.e. there is a flattened trough at the minimum between particles of 0.1–2.0 μm – an important size fraction in air.

*The proportion of airborne particles retained is normally expressed as velocity of deposition V_g:

$$V_g = \frac{\text{quantity } (\mu\text{g or ng}) \text{ deposited per cm}^2 \text{ of ground surface/second}}{\text{quantity } (\mu\text{g or ng}) \text{ contained in 1 cm}^3 \text{ of ambient air}}$$

Velocity of deposition is the summation of the processes of sedimentation, impaction and diffusion (i.e. dry deposition).

This differential retention of the various particle sizes by TD, DD and MB means that there can be no simple fixed relationship between air-concentration and deposition to ground vegetation or deposition on a gauge. In this respect the gauges would be a better reflection of vegetation retention than would be provided by the air-concentration gauge which draws an unbiased sample of airborne metals independent of prevailing wind speeds.

(4) The proportion of particles in the aerosol retained by the deposition surface in a wind tunnel is strongly increased with increasing wind speed. Since wind speed increases logarithmically with height above ground, the gauges placed at 150 cm height will be exposed to more wind than the grass sward.

Wind speeds above 500 cm/sec. (11 mph, Beaufort 3, gentle breeze), at 150 cm, may cause particles c. 30 μm to bounce off the filter paper of the DD gauge or to blow off again if already deposited earlier under less windy conditions. This effect is not so important at normal sites where particles are usually $<10\ \mu$m, but could be very important at breezy sites close to an emission source. It is negligible for moss-bags but could reduce the deposition on to DD. By contrast, the air-concentration gauge sampling at a constant rate (c. 5 1/min.) is independent of wind speed.

(5) Although it is very difficult to quantify the integrated effects of the physical features of the receptor surfaces and wind speed upon retention efficiency, it is possible to make a very rough estimate from these wind tunnel observations as follows.

The filter paper dry deposit gauge would most likely *underestimate* input to a grass sward by a factor of 2–10. The *vertical* moss-bag placed at right angles to an emission source would be likely to *overestimate* input to a grass sward by a factor of 5–8. Thus, the methods might be expected to differ by a factor within the range x 10–80 as between moss-bags and the AERE dry deposition gauge, with a central tendency of *c.* x 30.

(6) There is evidence that doubling the wind speed in a wind tunnel approximately doubles the proportion of 0.5–2.0 μm particles retained by a filter paper pad, moss-bag or grass sward. Because the physical characteristics of the moss-bag or grass sward make them highly retentive compared with filter paper (x 20–30), doubling the metal input to grass or moss-bags represents a very large arithmetic increment compared with a relatively small increment to filter paper.

(7) Both moss-bags and the total deposit gauge suffer from the disadvantage that relative to a grass sward, although there is a scaling factor [see (5) above] operating for dry deposition, it is most likely that no scaling factor operates for metal deposition in rain, i.e. the quantity of metals received in rain/cm^2 of receptor surface is about the same for filter paper, moss-bags, and grass sward. Thus, if metal input from washout is rather variable and large, both gauges could give very misleading results. Since washout of industrially derived metals is believed to contribute only a small and relatively constant percentage to total deposition (10–20%), this problem is not thought to be a severe one in

practice, particularly for the moss-bag which overestimates dry deposition. However, the only way to overcome it thoroughly would be to use dry and total gauges together or pairs of moss-bags, side by side, which are exposed to and sheltered from rain.

11.5 FIELD TRIALS

These experimental findings in the wind tunnel need verification by field trials and in particular suggest the following questions:

(a) Do the results in the field indicate that all gauges are sampling the prevailing airborne metals much as they do in the wind tunnel, i.e. AC drawing an unbiased sample of all particle sizes and thus giving metal results somewhat different from MB, TD, and DD (all three of which undersample in the intermediate particle size range 2.0–0.1 μm)?
(b) Do the MB, TD, and DD metal results suggest that they each sample the various particle sizes to about the same extent, i.e. similar degrees of undersampling of intermediate particles (similar sampling quality)?
(c) Do moss-bags collect more metal/unit area than DD or TD, and TD more than DD in a given time period (sampling quantity)?
(d) If so, is this a fixed ratio (MB/TD; MB/DD; TD/DD, etc.) irrespective of month or site?
(e) Does metal deposition in a moss-bag bear any relation to deposition on to a grass sward in the field?

11.5.1 Comparing the qualitative characteristics of different gauges

This can be done most conveniently by correlating the quantities of all seven metals (Co, Cu, Fe, Mn, Ni, Pb, and Zn) retained by any two types of gauge in any month or at any site. As an example, it is possible at Windermere in January 1973 to compare the content of the seven metals analysed for in moss-bags with the amounts of the corresponding metals retained by the total deposition gauge. If the level of each metal from the moss-bag has the same fixed ratio to the level of the corresponding metal from the total deposit gauge (every metal value from the moss-bag is, say, 3.5 times greater than the value of the corresponding metal from the total deposit gauge) all metal values will lie along a straight line when graphed. Their closeness of fit to linearity will be given by 'r' ($r = 1.0$ for a perfect fit to a straight line). This can be most easily recognized by the degree of statistical significance: xxxx ($p < 0.001$) = very good fit; xxx ($p < 0.01$) = good fit; xx ($p < 0.02$) = acceptable fit; x ($p < 0.05$) = statistically linear but not regarded as acceptable in this study. In this way, tests can be made at each site for each month comparing metal values on any pair of gauges, e.g. MB v TD; MB v DD; MB v AC; TD v DD; TD v AC; DD v AC. If the results normally fall along a straight line, this is

TABLE 11.1 Summary of Significance Levels of Correlation Coefficients (r) between Various Gauge Pairs for All Seven Metals and All Sites and Months

Significance levels	MB/TD	MB/DD	TD/DD	MB/AC	TD/AC	DD/AC
xxxx	86	90	85	25	32	24
xxx	17	12	12	50	45	50
xx	3	4	5	16	13	17
x	5	3	2	7	7	5
N.S.	6	7	12	18	19	20
N.D.	0	1	1	1	1	1
Total	117	117	117	117	117	117

Notes: xxxx; xxx; xx; x = r statistically significant at $P < 0.001$; 0.01; 0.02; and 0.05 respectively.
N.S. = not significant; N.D. = not determined

taken to mean:

(a) Each suspended particle contains one or other (more rarely two or three together) of the different metals and both gauges under comparison are thus sampling the various particle sizes and hence the metal aerosol in a conformable manner (i.e. similar amounts of over or undersampling various size fractions).
(b) The different analytical techniques used to measure the same element (in this case, atomic absorption spectrophotometry for moss-bags, by the authors, and neutron activation analysis, X-ray fluorescence and colorimetric methods for TD, DD, and AC, by AERE Harwell) are systematically compatible.

Table 11.1 summarizes the significance of correlations from the nine sites in S. Wales (seven for a 12-month period and two for six months each) and also from Windermere (grid reference SD 362974; seven months) Styrrup (grid reference SK 606989; eight months) and Petten (Netherlands; six months).

The results show a high degree of linearity in all comparisons involving MB, TD, and DD all of which are distinctly more linear than comparisons of any of these with AC.

This is what would be expected if MB, TD, and DD are sampling in a similar qualitative manner but not in quite the same way as AC. It is assumed that this result is caused by the unbiased sampling of AC as contrasted with the other gauges which tend to undersample at intermediate particle sizes to about the same extent.

11.5.2 Quantitative sampling characteristics of the gauges

What the above results do not show is whether the actual ratios of metals retained by each pair of gauges is the same from month to month and site to site. As an example, if metal levels on the moss-bag are, say, 3.5 times greater than those on the TD gauge for February 1974 for Styrrup, will this factor be the same for all

other months and all other sites? Or will it change from month to month and site to site? Thus, although we know that gauge comparisons between MB, TD, and DD are linear, we do not know whether the slope of the line stays constant or changes from site to site or month to month, or depending which types of gauge are being compared. The following factors are thought to be most likely to influence the slope:

(a) Sampling rates of the two gauges being compared, especially as influenced by increasing wind speeds.
(b) Proximity of sampling site to main metal emission sources.
(c) Relative importance of aerial diffusion processes versus wind transfer for the various particle sizes of the metallic aerosol.
(d) Frequencies and strength of the various wind directions.
(e) Angle of moss-bag orientation in relation to main metallic emission sources.
(f) Frequency of rain showers.

These may interact in such a complex way that it seems unlikely that ratios will be constant.

Table 11.2 summarizes information on slopes (given as geometric means of monthly linear regression coefficients) for eight of the sites in Wales and shows that ratios vary widely from site to site and depending on which gauge pair is being compared. The fall in ratios generally observed when reading from left to right of the table are expected from the different effective sampling rates of the gauge, which in decreasing order, are: MB $\gg$ TD $\geqslant$ DD $\gg$ AC.

TABLE 11.2 Geometric Means of Regression Coefficients for Metal Content of Various Gauge Pairs

SITE \ GAUGE COMPARISON	MB/AC	MB/DD	MB/TD	TD/AC	TD/DD	DD/AC
Port Talbot	48.86	13.74	7.12	6.86	1.93	4.80
Trebanos	7.87	4.50	2.90	3.08	1.55	2.17
Kidwelly	6.95	4.74	4.57	1.88	1.43	1.31
Skewen	7.65	5.41	4.26	1.78	1.37	1.30
Penmaen	4.02	4.21	1.62	2.64	2.67	0.97
Mount Pleasant	3.24	2.48	0.93	3.54	2.70	1.29
Llansamlet	2.61	2.65	2.14	1.44	1.35	1.06
Clydach	2.60	2.14	0.91	3.16	1.91	1.17

TABLE 11.3 Regression Coefficients for Metal Content of Various Gauge Pairs

	VMB/HMB	VMB/AC	HMB/AC	VMB/DD	HMB/DD	VMB/TD	HMB/TD	TD/AC	TD/DD	DD/AC
Port Talbot	3.2	74.6	24.1	16.4	5.2	8.5	2.7	8.6	1.9	4.6
Trebanos	2.1	13.3	6.1	5.3	2.5	4.7	2.1	2.7	1.4	2.4
Penmaen	1.7	9.4	5.2	15.4	9.1	5.4	3.2	1.7	2.9	0.6

Note: Vertical moss-bags (VMB) were orientated at right angles to prevailing metal sources to give maximum 'dipole' effect. Receptor surface area is the same for all wind directions in horizontal moss-bags (HMB).

The ratios also tend to fall from top to bottom of the table. The eight sites were arranged partly subjectively, in decreasing order of windiness (no wind data were available for several sites) as follows: Port Talbot ≥ Trebanos ≥ Kidwelly > Skewen > Penmaen ≥ Mount Pleasant ≥ Llansamlet ≥ Clydach. The decreasing slope with decreasing windiness is in accordance with expectation for all gauge comparison, involving AC which is a constant rate sampler whereas MB, TD, and DD are all expected to be wind sensitive. *This emphasizes the fact that MB, TD, and DD gauges cannot be used to determine air-concentration directly.*

The ratios MB/DD and MB/TD appear to show sensitivity to wind speeds, especially at the windiest site (Port Talbot) and one explanation for this may be that larger particles are being bounced off or blown off DD and TD gauges under the very breezy conditions of this more polluted site where large particles are most likely to be present.

Another important factor is that a number of experiments have shown that vertical moss-bags may seriously undersample all particle sizes when not orientated at right angles to the main metal emission source. This 'dipole' quality may be used to locate emission sources but the experiments showed that less-biased estimates of metal deposition can be obtained by using horizontally orientated bags. An indication of this is given (Table 11.3) by a comparative study of vertical and horizontal moss-bags which was made together with total deposition, dry-deposition and air-concentration during November 1972–May 1973, at some of the sites listed in Table 11.2 above. All regression coefficients shown in Tables 11.2 and 11.3 were calculated using the following units: VMB; HMB; TD and DD – μg metal/cm^2/month; AC – μg metal/m^3 air.

11.6 METAL ACCUMULATION ON MOSS-BAGS AND GRASS SWARD IN THE FIELD

A number of plastic seed trays containing clean (unpolluted) soil were sown with the grass *Festuca rubra* and allowed to grow in an uncontaminated environment.

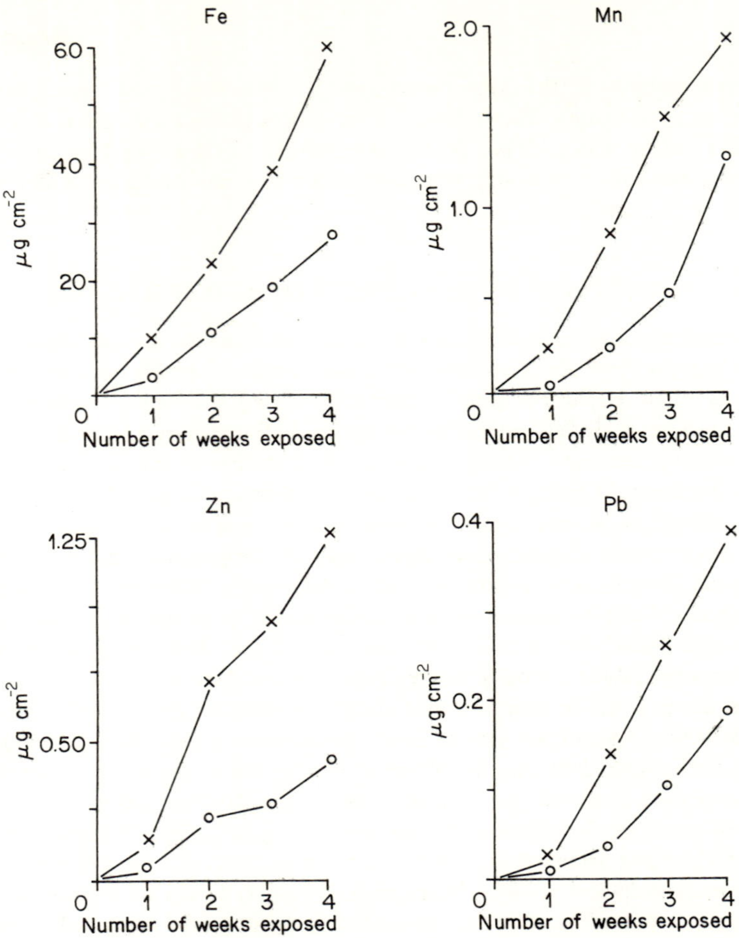

Figure 11.3 Weekly deposition of Fe, Mn, Zn, and Pb ($\mu g\ cm^{-2}$) as measured by moss-bag (x——x) and grass sward (o——o), exposed from 1–4 weeks at Port Talbot, South Wales. (Note vertical scale differences)

When the grass was about 6 cm high the trays were placed in a relatively polluted site together with a number of moss-bags suspended at the same height above the ground as the trays. Each week a fixed standard area of grass was cut from the centre of each tray to within 1 cm of the soil surface and the whole analysed. At the same time, moss-bags were removed and analysed. Figure 11.3 shows a comparison of weekly accumulation of Fe, Mn, Zn, and Pb per unit area of grass sward and moss-bag.

The moss-bags are roughly twice as retentive as the grass sward for these metals. This is in good agreement with the work of Clough (1974) who found that in wind

tunnel experiments at the same wind speeds, moss-bags were about twice as retentive as grass swards. Further work over longer periods indicated clearly that grass tends to lose about 50 per cent of its intercepted trace-metals to the ground. This makes horizontal moss-bag retention about the same as grass under similar conditions.

11.7 EFFECTS OF WIND DIRECTION AND STRENGTH

If a metal is being emitted from a particular point source, winds blowing onto the sampling site from the direction of the source will deliver the airborne metal to the gauges. This could be quite important for larger airborne particles where diffusion processes are less important for transport but where the turbulent air in strong winds could carry the particles long distances from the source. Such considerations would indicate that the amount of metal accumulated by a gauge each month will be proportional to the number of miles of wind (i.e. frequency or duration each month x speed of wind) passing over the gauge from the direction of the emission source each month. No such correlation would exist for winds from other directions and a negative correlation for winds from the opposite direction, which would blow the metal away from the gauge. This situation should be particularly applicable to wind-sensitive gauges and to vegetation.

It was not possible to monitor wind direction and strength at each of the nine sites in S. Wales. Instead the present study relied upon observations of wind speed and direction made daily at the Penmaen site by Mr J. Powell and upon three hourly wind data collected on the coastal plain near the Port Talbot site (OS Ref. SS 789867) and kindly supplied by the Meteorological Office, Bracknell. In both cases, the data was arranged to give the number of nautical miles of wind passing over the meteorological stations Penmaen and Port Talbot from various points of the compass (reported as compass bearings in intervals of degrees) each month for the 12 month period. Each set of wind data was then correlated with the monthly metal accumulation on the gauges at the corresponding site. Correlation coefficients for moss-bags, total deposit and dry deposit gauges at Port Talbot and Penmaen are summarized in Tables 11.4 and 11.5. At Port Talbot, both moss-bag and total deposit gauges show that the higher the amount of metal deposited, the greater the number of wind miles received from the NW (260–340°). Winds from this direction pass over two large iron and steel works (lying at 290° and 310° respectively) and copper and nickel works (lying 310–320°). Negative correlations for moss-bags are observed with wind miles from the opposite direction. There appears to be a lead source in the SE (140–160°) at present unidentified. These results are in good agreement with expectation.

Preliminary analysis of the gauges at Penmaen showed activity in the SE (70–140°) and NW (320–360°). Only these directions are reported in Table 11.5. Significant correlations for Fe and Mn, particularly for moss-bags, are seen from 70–140° and for Cu at 320–360°. The two iron and steel works referred to above

TABLE 11.4 Summary of Statistically Significant Correlations (r) between Monthly Wind-miles from Stated Compass Bearing and Monthly Metal Content of Gauges at Port Talbot

	Compass bearing (0° = N; 90° = E; 180° = S; 270° = W)											
	350-010°	020-040°	050-070°	080-100°	110-130°	140-160°	170-190°	200-220°	230-250°	260-280°	290-310°	320-340°
MOSS BAG												
Co	.	.	.	–N.S.	.	.	.	.	.	.	.	xxxx
Cu	.	.	.	–x	.	.	.	.	.	.	x	.
Fe	.	.	–N.S.	–x	.	.	.	.	.	.	xxxx	.
Mn	.	.	.	.	.	.	.	.	.	x	.	.
Ni	.	.	.	.	.	.	.	.	.	.	xx	.
Pb	.	.	.	.	.	xx	.	.	.	.	N.S.	.
Zn	.	.	.	–N.S.	.	.	.	.	.	.	.	.
TOTAL DEPOSITION												
Co	N.S.	xx	.	.	.	.	.	.	.	–x	x	N.S.
Cu	.	.	.	.	N.S.	x	.	.	.	.	.	.
Fe	.	.	.	.	.	N.S.	.	.	.	.	.	.
Mn	.	.	.	–N.S.	.	.	.	.	.	.	x	.
Ni	.	.	.	.	.	N.S.	.	.	.	.	xxx	.
Pb	.	.	.	.	.	.	.	.	.	.	xxxx	N.S.
Zn	xx	.	.	.	.	.	.	.	.	.	x	N.S.
DRY DEPOSITION												
Co	.	.	.	.	.	.	–N.S.	–x	–N.S.	.	.	.
Cu	.	.	.	.	.	.	.	.	.	.	.	.
Fe	.	.	–x	–xx	.	.	.	.	.	.	x	.
Mn	.	.	.	–N.S.	–N.S.	.	–N.S.	.	.	.	.	.
Ni	.	.	.	.	.	.	.	.	.	.	.	.
Pb	.	.	.	.	.	.	.	.	.	.	.	.
Zn	.	.	.	.	.	.	.	–N.S.	.	.	.	.

Note: x; xx; xxx; xxxx = r statistically significant at $P < 0.05$; < 0.02; < 0.01; < 0.001 respectively

N.S. = $r > 0.500$ but not statistically significant; – = negative correlation

TABLE 11.5 Summary of Statistically Significant Correlations (r) between Monthly Wind-miles from Stated Compass Bearing and Monthly Metal Content of Gauges at Penmaen

	Compass bearing (N = 0°; E = 90°; S = 180°; W = 270°)			
	30-70°	70-110°	110-140°	320-360°
MOSS BAG				
Co	.	.	.	.
Ni	.	.	.	.
Cu	.	.	.	xxxx
Fe	.	xxx	xx	.
Mn	.	xxx	xx	.
TOTAL DEPOSITION				
Co	.	.	.	.
Ni	.	.	.	.
Cu	–x	.	.	.
Fe	.	.	x	.
Mn	.	x	.	.
DRY DEPOSITION				
Co	.	.	.	.
Ni	.	.	.	.
Cu	.	.	.	.
Fe	.	.	x	–x
Mn	.	.	.	.

Note: x; xx; xxx; xxxx = r statistically significant at $P < 0.05$; < 0.02; < 0.01; < 0.001 respectively; – = negative correlation

lie between 70–90°. The Cu source to the NW is not known. The interesting feature of this site, selected as an uncontaminated 'background' site is the extent to which it is affected by industrial sources.

From the correlations presented above it may be concluded that the moss-bag is somewhat more responsive to the quantity of wind from the direction of an emission source of airborne metals than TD, and TD more than DD.

11.8 POSSIBLE USE OF MOSS-BAGS AS TOTAL PARTICULATE GAUGES

The results so far indicate that moss-bags may be useful in obtaining reasonably quantitative estimates of the total amounts of airborne metallic elements intercepted by a grass sward. It is of interest to consider whether they could be used to estimate total particle deposition. This cannot be done by direct weighing before and after exposure because in windy sites, small fragments of moss may become detached from the moss-bag and blow away during exposure.

TABLE 11.6 Observed Monthly Weights (μg) of 'Dust' in TD Gauge Compared with Estimates Based on Content of Various Metals respectively on Moss-bags (All Figures are per dm^2 Horizontal Deposition Surface)

Month 1976–77	Obs. 'Dust' in TD	Estimated 'dust' based on metal analysis					
		Fe	Mn	Pb	Ca	Mg	Mean
Aug.	40.7	47.0	68.0	34.0	22.0	70.0	48.2
Sept.	—	—	—	—	—	—	—
Oct.	54.0	95.0	52.0	67.0	54.0	65.0	66.6
Nov.	63.0	70.0	41.0	78.0	85.0	144.0	83.6
Dec.	48.5	60.0	42.0	46.0	35.0	60.0	48.6
Jan.	53.9	44.0	49.0	81.0	30.0	47.0	50.2
Feb.	38.9	60.0	40.0	37.0	26.0	31.0	38.8
March	30.4	46.0	37.0	39.0	39.0	34.0	39.0
Mean	47.0	60.0	41.0	55.0	42.0	64.0	52.4

However, if an array of moss-bags were exposed during the same period of time as one or two total deposit gauges, the TD gauges could be used to determine the concentration of various trace metals in the total-deposit material and by working backwards, the weight of dust intercepted by the moss-bag might be estimated from the total metal content of the bag itself. This of course assumes that the composition of the dust intercepted by moss-bag and TD gauge remains unaffected by the nature of each type of sampler. Data summarized in Table 11.1 above show this to be largely correct.

In work carried out at a very sheltered site, the Chelsea Physic Garden, London, UK, the weight of total deposit, sampled monthly for 8 months (including dissolved material in the precipitation collected by the TD gauge – without a terylene covered funnel) correlated significantly ($P < 0.001$) with Ca and Mg values respectively in moss-bags correspondingly exposed under both wet (completely open to the atmosphere) and dry (sheltered from precipitation by a horizontal sheet) conditions.

This suggested that it might be possible to estimate total particulate interception using moss-bags, as suggested above.

At the same site, the amount of particulate interception on moss-bags was calculated, based on a monthly comparison of the concentrations of Fe, Mn, Pb, Ca, and Mg respectively in the TD gauge and the total amount of each of these metals deposited on a moss-bag for each month. Table 11.6 compares these estimates based on each metal for the period August 1976–March 1977 with the total weight of material received by the TD gauge.

The first results look promising. The moss-bag gives deposition values approximately equal to the TD gauge. This is what would be expected in a very sheltered site. More results are needed before drawing firmer conclusions as to whether the

approach might be worth trying under arid-zone conditions. As indicated earlier, it is unlikely that the moss-bag will lose dust in wind or rain-episodes and so it might be expected to perform adequately even under arid or semi-arid conditions.

11.9 CONCLUSIONS

Considerable research work has been done on interception of airborne particles by grass surfaces and other obstacles and despite their different designs, the physical principles governing deposition on all gauges (MB, TD, DD) are the same.

Laboratory work by Chamberlain (1966a,b) and Clough (1973) using wind tunnels has shown that 'hairiness' and wetness are important properties making objects efficient interceptors of airborne particles. Because grass sward and other ground vegetation, MB, TD and DD are all different for these characteristics, each will tend to retain airborne particles with different 'efficiencies' in terms of the proportion of airborne particles retained and also exhibit a tendency to retain particles of different sizes in different proportions compared with their occurrence in the air. Doubling wind speed was found roughly to double the proportion of particles retained on vegetation, MB, TD, and DD.

Because AC is quite independent of wind speed and extracts an unbiased sample in terms of the different particle size fractions, there can be no simple relationship between AC and deposition to ground vegetation or to MB, TD, and DD. *Thus, these latter gauges cannot be used to determine air-concentration directly.*

Metal retained on MB correlated well with metal retained by TD and DD at each site for each month. Correlations between AC and MB, TD, or DD were less good. This is taken as evidence for the fact that MB, TD and DD are all sampling the chemical composition of the aerosol in similar qualitative ways but not absolutely representatively, as done by AC. This may be caused by differential retention of different particle sizes. It is also evidence for the quantitative compatibility of the different chemical analytical techniques used by the authors for MB and AERE Harwell who analysed for TD, DD, and AC.

The metal ratios obtained from comparing gauge pairs showed that comparison of MB or TD or DD with AC gave ratios that were highest for windiest sites. This is as expected, the gauges being wind dependent whereas AC is a fixed rate sampler. Other comparisons gave ratios MB/DD, MB/TD, and TD/DD as expected from the greater proportion of airborne particles retained by MB than by TD and DD. Each of these last three ratios varies depending on month and site. Higher ratios were generally associated with the most exposed (and polluted) site (Port Talbot). One explanation might be that larger particles (*c.* 30 μm) of similar chemical composition to the smaller particles were being retained more efficiently on MB and to a lesser extent on TD or DD where they may bounce off or blow off. The way the moss-bag is presented to the air is also important, horizontal bags giving less biased results than vertical bags. This 'dipole' effect can be used to pinpoint metal emission sources.

Strong winds and rain episodes do not remove particles already deposited on the moss-bags. Soluble cations are held on the exchange-surfaces of the moss so that the soluble metal is not lost following rain.

Trays of the grass *Festuca rubra* placed at the same height above ground as horizontal moss-bags at Port Talbot and sampled weekly for a month accumulated about the same quantity of metals per unit surface area as did moss-bags. The grass leaves lost half of this amount of metal to the ground surface making moss and grass results similar to those found in the wind tunnel by Clough.

The results indicate that the moss-bag is a useful gauge providing informative metal deposition figures in terms of space and time. It is relatively inexpensive and can be used for extensive surveys. It is wind-sensitive like ground vegetation and other deposit gauges. Further critical studies, which would include wind direction and strength, and rainfall measurements are desirable to quantify further the moss-bag as a reliable index of metal input to a grass sward. The results obtained from the study described above indicate that such critical experiments would further validate the horizontal moss-bag as a reasonably accurate deposition gauge and the vertical bag as an indicator of emission sources.

Preliminary evidence indicates that it might be possible to use moss-bags as total 'dust' monitors in arid-zone conditions but this needs to be verified experimentally.

11.10 ACKNOWLEDGEMENTS

We wish to thank the Welsh Office (Cardiff) and the Natural Environment Research Council for financial support throughout this study. We are indebted to University College of Swansea and Chelsea College, London for laboratory facilities and also the Royal Society for providing items of research equipment. We also wish to thank Dr D. H. Peirson and his colleagues of the Environmental and Medical Sciences Division of AERE Harwell for their continued collaboration and the provision of all the data on total deposition, dry deposition and air-concentration.

REFERENCES

Cawse, P. A. and Pierson, D. H. (1972). An analytical study of trace elements in the atmospheric environment. *AERE – R7134* (HMSO Lond.).

Chamberlain, A. C. (1966a). Transport of *Lycopodium* spores and other small particles to rough surfaces. *Proc. R. Soc. Lond. (A)*, **296**, 45–70.

Chamberlain, A. C. (1966b). Transport of gases to and from grass-like surfaces. *Proc. R. Soc. (A)*, **290**, 236–265.

Chamberlain, A. C. and Chadwick, R. C. (1972). Deposition of spores and other particles on vegetation and soil. *Ann. Appl. Biol.*, **71**, 141–158.

Clough, W. S. (1973). Transport of particles to surfaces. *J. Aerosol Sci.*, **4**, 227–234.

Clough, W. S. (1974). The deposition of particles on moss and grass surfaces. *AERE – M2612* (HMSO Lond.).

Goodman, G. T. and Roberts, T. M. (1971). Plants and soils as indicators of metals in the air. *Nature, Lond.*, **231**, 287–292.

Ruhling, Å. and Tyler, G. (1970). Sorption and retention of heavy metals in the woodland moss *Hylocomium Splendens* (Hedw.), *Br. et Sch. Oikos*, 21, 92–97.

Tamm, C. O. (1953). Growth, yield and nutrition in carpets of a forest moss (*Hylocomium splendens*). *Meddn. Statens Skogsforskninginst.*, **43**, 1.

CHAPTER 12

Monitoring and Critical Review of the Estimated Source Strength of Mineral Dust from the Sahara

R. JAENICKE

ABSTRACT

A transport model is used to estimate from turbidity measurements at Sal, Cape Verde Islands, the source strength of the Sahara for mineral material. The estimated value of 260×10^6 t/yr is definitely larger than previously estimated. Because of the sea-land-circulation at Africa's west coast, fractions of this amount might return to Africa. Larger portions certainly settle out within the first thousand kilometres. Only particles in the range of 0.1 μm to 1 μm in radius are transported over longer distances. The evaluation of different types of measurements of the last decades showed no indication for a drastic increase in turbidity. Consequently, an increase in dust transport – as previously proposed – cannot be confirmed.

12.1 INTRODUCTION

The often referred to SMIC (1971) Report estimates that between 25% and 50% of natural particles production are of mineral (soil and rock debris) origin. This estimate is based mainly on the Goldberg (1971) model in which the mineral dust load in the Atlantic Ocean northeast trades enters with 2.5 $\mu g/m^3$ – first published by Prospero (1968). According to more recent papers (Carlson and Prospero, 1972; Prospero and Carlson, 1972) this value seems to be very low. Based on their measurements and on observed sedimentation rates in the Atlantic Ocean the production rate for mineral particles was estimated to be around 60×10^6 t/yr.

This paper uses direct measurements in the surface air of the NE-trades, long term optical measurements, and a transport model to estimate the strength of the Sahara as a source for mineral particles. Since this new value seems to be considerably larger than that mentioned above it will be discussed whether observations really support the idea of an increase in source strength.

12.2 ESTIMATE OF THE STRENGTH OF THE SAHARA AS SOURCE FOR MINERAL PARTICLES

Measurements of physical and chemical properties of the surface aerosol in the NE-trade wind region were carried out in two field expeditions: July 1973 on Sal, Cape Verde Islands, and November 1973 aboard the German Research Vessel 'Meteor' on a cruise from the Caribbean Sea to the African continent. These measurements focused on the size distribution of mineral particles and are discussed by Schütz (1977) and Jaenicke and Kasten (1977). In addition, Schütz (1977, 1979) discussed a transport model for mineral dust from Africa over the North Atlantic Ocean. This model uses the transport pattern recently recalled by Carlson and Prospero (1972). The surface measurements on the Atlantic Ocean and direct measurements on the African continent (Schütz and Jaenicke, 1974), led to the conclusion that the mineral aerosol leaves the African continent with an aerosol size distribution of a concentration decrease proportional to r^{-2} – quite in contrast to other typical continental aerosols with –3 as exponent. This point is of special importance because the behaviour of the dust during transport is influenced by the shape of the starting size distribution.

To support our intensive but time-limited observations optical measurements were carried out on Sal, Cape Verde Islands, during the years 1973 until 1975. Figure 12.1 shows the results in form of Linke's turbidity factor T. These measurements indicate the seasonal variation of the dust transport with maximum in summer and minimum in winter. Compared to the Rayleigh atmosphere the comparatively large values of T in the months of minimum turbidity reveal the quasi continuous character of the dust transport on a monthly basis.

Especially for these measurements Prospero (1976) pointed out that measurements with the Volz-sunphotometer might be affected by changing calibration values. A careful investigation in fact showed such variations. For the calculations in Figure 12.1 adjusted calibration values were used, obtained several times during the whole observation period.

The average Linke turbidity factor T_{500} over the whole period of observation is $\bar{T}_{500} = 3.45$ ($\bar{B}_{500} = 0.166$, $\bar{a}_{D500} = 0.382$, $\bar{\tau}_{500} = 0.382$). This turbidity is caused mainly by particles in the range 0.1 μm–1 μm in radius. Following Volz (1959) the mass of particles in a vertical column can be estimated and yield M (0.1–1 μm) = 9.1×10^{-6} g cm^{-2} for Sal. It could be shown (Jaenicke and Kasten, 1977) that this column contains practically only mineral particles. In surface air sea salt, organic material, and ammonium sulphate is of minor importance and consequently in the entire column extending up to 5000 m. Schütz (1979) model shows that particles in this size range are practically not removed from the column during transport. From the original extension in the vertical (1500 m to 5000 m altitude) at the starting point they spread over the entire column without considerable loss due to dry deposition. The above figure then is identical with the columnar mass in the source region.

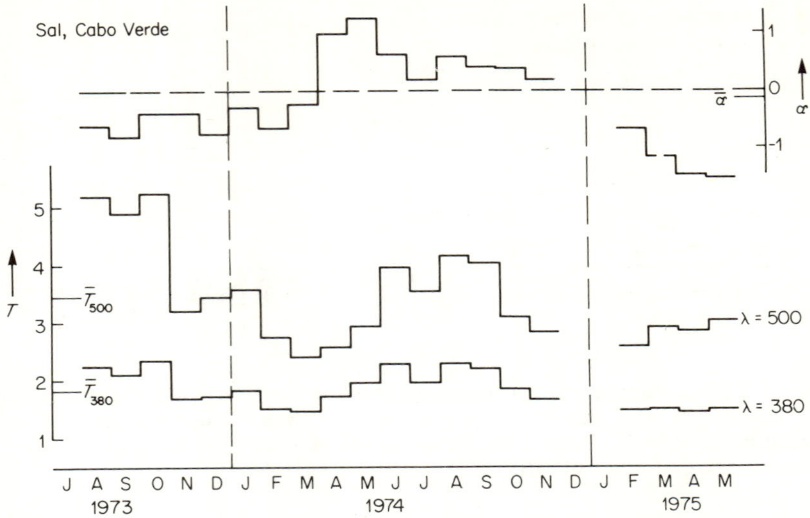

Figure 12.1 Linke's turbidity factor T for two wavelengths ($\lambda = 380$ nm, $\lambda = 500$ nm) at Sal, Cape Verde Islands. The calibration constants were monitored during the observation period and showed to be stable for $\lambda = 500$ nm, while fluctuations occurred for $\lambda = 380$ nm. Angström's exponent α, therefore, is given with all reservations

The discussion by Schütz (1979) indicated that the source aerosol exhibits a distribution with $\nu^* = 2$.* In such an aerosol the 0.1 μm–1 μm radius range covers 6.7% of the total mass, thus 1.35×10^{-4} g cm^{-2} can be expected as total columnar mass in the source area (0.1 μm to 20 μm particle radius). If we assume a dust duct of 1000 km width [from 15 °N (Dakar) to 24 °N] and the average wind velocity of 6 m s^{-1} (Newell *et al.*, 1972) a total of 260×10^6 t/yr as source strength results. This figure, of course, is only valid within a factor of 2 if the uncertainties of the assumptions are considered.

Figure 12.2 shows how this dust spreads over the Atlantic Ocean, if we follow the Schütz (1977) model. During the first thousand kilometres a rapid dry fallout occurs from which we believe considerable fractions are returned to Africa. As can be seen from satellite photographs and weather maps a sea-land-circulation up to

*Size distribution

$$\frac{dN}{d \log r} = n^*(r_0)(r/r_0)^{-\nu^*},$$

where

r = particle radius in μm;
r_0 = arbitrary chosen reference radius;
N = number of particles cm^{-3} larger than r,
$n^*(r)$ = differential number size distribution in cm^{-3};
ν^* = exponent of the distribution.

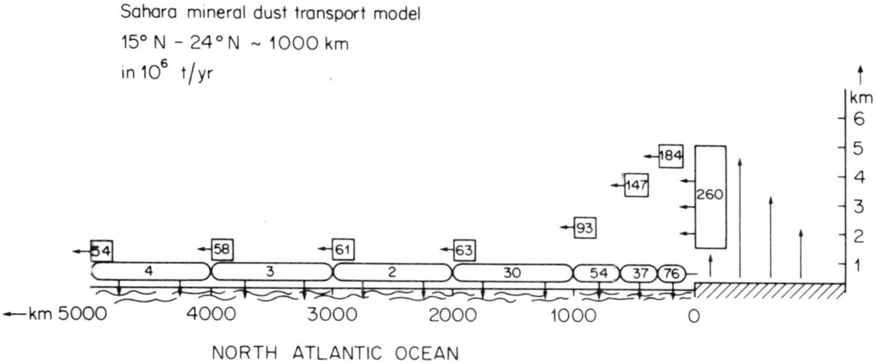

Figure 12.2 Estimate of dust transport over the North Atlantic Ocean from the Sahara. Airborne quantities are given as well as dry deposition rates. The possibility of mineral dust being transported back to the continent in the sea–land circulation is indicated

300 km off-shore developes at Africa's west coast. In that circulation dust might be returned to Africa before it settles down.

The dust does not show drastic variations beyond 1000 km and after 5000 km of transport roughly 50 x 10^6 t/yr are still airborne and carried beyond the longitude of the Lesser Antilles.

At this point, our results can be compared with those of Prospero and Carlson (1972) who used their measurements in the dust laden air to estimate 37 x 10^6 t/yr passing the longitude of Barbados during a dust transport period of 6 month. These authors feel that this value should be higher because the winter dust is carried in latitudes below 10 °N, thus not reaching Barbados. However, considering this fact and the general uncertainty of such values both estimates agree quite well.

The calculated dry deposition, which is superior to wet removal in this region of the world can be used to estimate the rate of sedimentation in the ocean caused by eolian transport. Figure 12.3 shows calculated sedimentation rates based on a packing density of 2 g cm^{-3}. This is compared to calculated values as published by Ku *et al.* (1968), based on Goldberg *et al.* (1963) measurements. In addition, one value of Rothe (1973) is given. This comparison indicates, with all reservations, that our estimate at least for distances larger than 1000 km is not unrealistic. The first thousand kilometres are difficult to compare because the sedimentation might be influenced by other processes within the continental shelf, as Rothe (1973) indicates. In addition, one has to keep in mind that the North Equatorial Current and the Counter Current might carry considerable portions of airborne sediments to more southern latitudes. The fraction of mineral dust falling on the Ocean was estimated by Prospero and Carlson (1972) as 30 x 10^6 t/yr which is quite in agreement with our estimate (39 x 10^6 t/yr, Figure 12.2) for the distance beyond 1000 km.

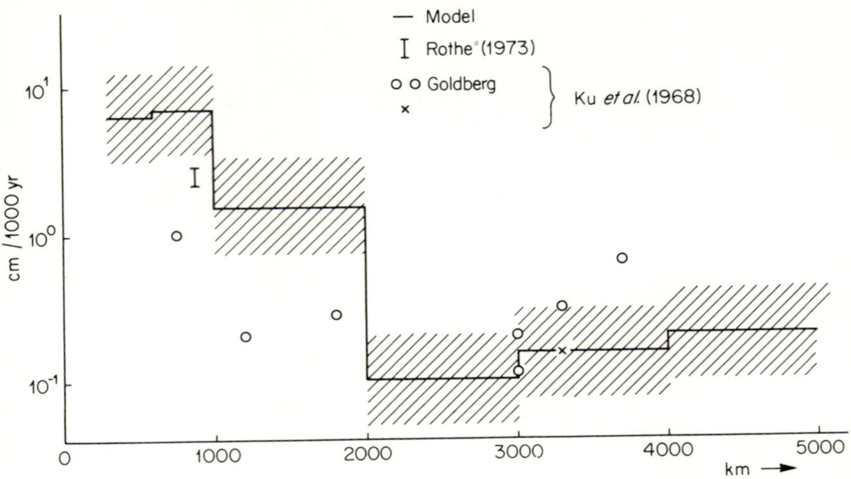

Figure 12.3 Estimate of ocean sedimentation rates due to eolian dry deposition. Measured values of Goldberg *et al.* (1963), Ku *et al.* (1968), Rothe (1973) are given for comparison. The area close to the continent is difficult to compare because of possible influences from effects within the continental shelf

Summarizing this comparison our budget calculation is not in disagreement with measured sedimentation rates at the ocean floor and with estimates of airborne material at 5000 km distance. At this distance, however, it is practically impossible to sense the rapid loss of airborne material during the first thousand kilometres of transport and, consequently, the source strength of the Sahara probably was underestimated.

The new estimate, of course, has consequences to the total world mineral production as mentioned above. The value of 100–500 × 10^6 t/yr should probably be corrected towards higher values.

12.3 RECENT VARIATION IN SOURCE STRENGTH

It was claimed recently that the transport of dust over the Atlantic Ocean has increased due to the severe drought in the Sahel zone since 1968. This is of special interest to all scientists concerned with the effects of deserts and desertification. Unsigned news reports (BAMS, 1974) disclose that Carlson and Prospero found a threefold increase in summertime dust levels at Barbados in 1973 compared to 1967. Since our estimates about source strength carried out above are based upon measurements in 1973–1975 it is necessary to ascertain whether these measurements were done in periods of increased dust transport and thus not representative for the situation in general. In this case, the comparisons made above for turbidity and sedimentation would be invalid as well.

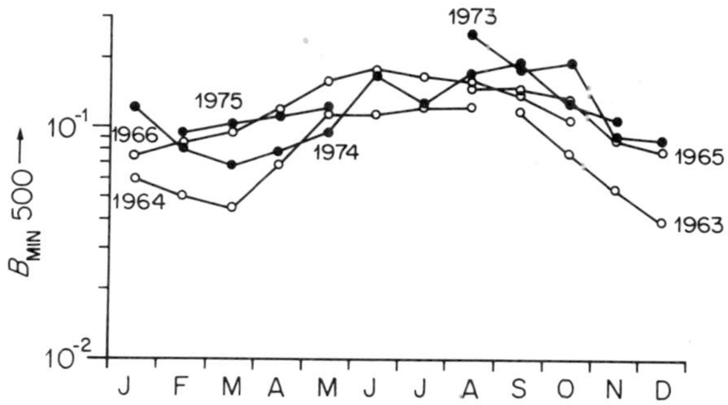

Figure 12.4 Turbidity data from the Cape Verde Islands during 1963–1966 (Volz, 1959) and 1973–1975 (Jaenicke and Kasten, 1977) $B_{\text{min } 500}$ was calculated according to Volz

The following section will look for presumptive evidence for a change in dust transport between 1968 and 1973. Since meridional shifts of the dust plume might have caused the observed increase (see below) it is necessary to obtain observations from stations comparatively close to the source to find out if the source strength varied in the past.

From 1963 until 1966 Volz (un-published draft) carried out turbidity measurements in the Cape Verde Islands region. Figure 12.4 shows monthly averages from 1963 until 1966 compared to our measurements in 1973 until 1975. Our data were calculated according to Volz's procedure, thus they might be slightly different from the data in Figure 12.1. Both sets of data show the same annual cycle and agree quite nicely. The averages of both sets of data are

$$\bar{B}_{\text{min } 500}\,(1963\text{–}1966) = 0.099$$

$$\bar{B}_{\text{min } 500}\,(1973\text{–}1975) = 0.121$$

Since B is directly proportional to the mass of 0.1–1 μm particles in a vertical column it means an increase of 22% in particle columnar mass. The statistical accuracy of both sets of data, however, is not sufficient for the statement that the dust transport has increased.

There is further evidence that the dust transport has practically not changed in recent years. Jaenicke and Schütz (1977) proposed a method to derive the atmospheric turbidity from the time difference between the recorded end of sunshine in sunshine recorders and astronomical sunrise/sunset. The method seems to be rather sensitive and comparisons with measured turbidities are quite in favour with the proposed method. Sunshine records have been available since 1880 and thus can be used to establish a climatology of turbidity. From the Sahara, data are available from a less extended period summarized in Dubief (1959). A drawback of the proposed method is that the sunshine records have to be evaluated for cloudless

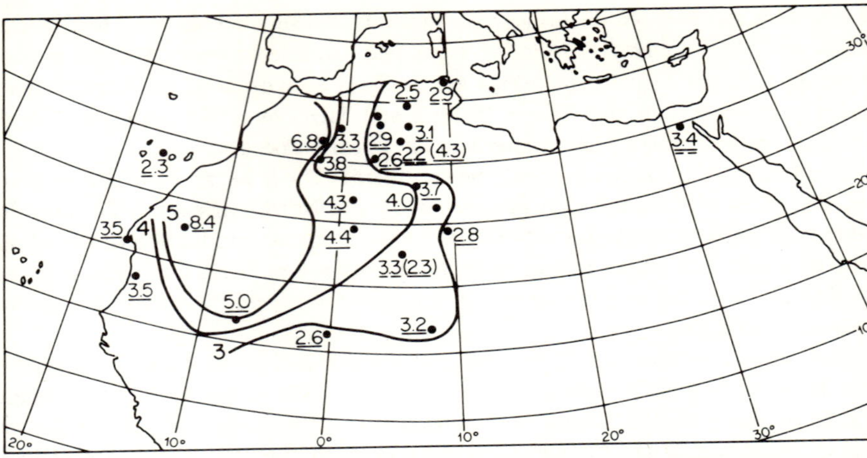

Figure 12.5 Isolines of Linke's turbidity factor calculated from suntraces (using data from Dubief, 1959) and direct measurements (Steinhauser, 1934; Dubief, 1959). The underlines indicate the source of the data
—— from suntraces Dubief (1959), $I_{cs} = 20$ mW cm^{-2}
– – – turdidity Dubief (1959), Steinhauser (1934)
—— Isolines of turbidity T_g

sunrise/sunset. At the present this has not been done for the Sahara. However, Lauscher (1957) offers a method to calculate the 'recorded' sunrise/sunset if the duration of sunshine and cloudiness is given. It would go beyond the scope of this paper to give details of this method.

I am aware of the uncertainty of this procedure, but the results obtained are surprising. Figure 12.5 shows isolines of Linke's turbidity factor T_g for the Sahara from observed data before 1959. The map shows a centre of turbidity in the West–Sahara which is practically identical to the region of increased dust storm activities, as compiled by Dubief (1953). Although the method of evaluation might give rise to objections the uniformity of the results is quite surprising. It seems to be more than coincidental that values $T_g = 3.5$ are calculated for Africa's West coast. Values equal to those were measured on Sal ($T = 3.45$).

It should certainly be permitted to summarize:

1. Evaluation of suntraces lead to turbidity values of $T_g = 3.5$ at Africa's Sahara west coast for some years before 1959
2. Turbidity measurements at Sal during 1973–1975 give $T_{500} = 3.45$ or $B_{\min\ 500} = 0.121$
3. Turbidity measurements in the Cape Verde Archipelago during 1963–1966 give $B_{\min\ 500} = 0.099$

These findings certainly do not support the idea of a drastically changed dust production and transport over the North Atlantic Ocean. This presumptive evi-

dence, however, is based exclusively on turbidity measurements. Thus the particles radius range of roughly 0.1 μm to 1 μm was considered only. It is, however, expected that the above conclusions are valid for all particle sizes. Measurements during dust storms in the Sahara (Schütz and Jaenicke, 1974) indicate that all particle sizes are activated and no particle size selection occurs.

In addition, the following should be considered. The Sahel zone certainly covers less than 50% of the total dust production area in North Africa. Should only the Sahel zone be responsible for a threefold increase in dust production of the whole area it would mean that a more than 6-fold output of the Sahel zone must be considered. It is beyond the limits of this paper to decide whether this can be regarded as realistic.

Carlson and Prospero (1972) report that the dust at Barbados during the winter months is of greyish or black colour with a possible origin from the semiarid grasslands south of the Sahara. An increased dust production in the Sahel zone should, therefore, affect the winter months, quite in contrast to the observations.

The increase in dust content in Barbados has probably other reasons than increased dust production in the Sahel zone. Hastenrath (1976) discusses the correlation between dry or wet years in the Sahel zone and modified pattern of the NE-trade winds. He reports about an increase of wind velocity, and a more southern location of the NE-trades during dry Sahel years as compared to wet years. It is recommended to investigate the influence of these large scale variations on the dust content in the West Indies.

12.4 MONITORING OF THE DUST PLUME

The above discussion indicated that two aspects of the mineral dust plume from the Sahara seem to be of interest: the radiational properties and the dry fallout. These properties then should be monitored in such a way that the whole dust plume is surveyed.

Carlson *et al.* (1977) showed the possible effects of increased dust amounts on the radiational cooling or heating of the atmosphere and its consequences on the dynamics of the troposphere. Since the particles causing optical effects are transported over long distances more or less unaffected, monitoring stations can be selected quite independently from the distance of the source so long as they are within the plume. Monitoring the turbidity of the atmosphere would give sufficient information about the radiational properties and would cover the dust plume in its entire vertical extension. A network of photometers would give quite reliable results but not sooner than some ten years from now. The same remains valid if monitoring from satellites is considered.

The use of the sunshine recorder network would give additional information for some decades in the past. It is, therefore, recommended to use a combination of a turbidity and sunshine recorder network for an optimum of information.

Optical measurements and even the determination of airborne particles mass will

give no information about dry deposition (Jaenicke, 1973). This is probably best done with simple deposition dust collectors positioned on the ground. Because of the large falling velocity of the giant particles the whole dust plume in its vertical extension contributes to the deposition. Dry deposition usually occurs in the 'vicinity' of the dust source. The giant particles are not carried over large distances as could be shown. Schütz and Jaenicke (1974) estimated that only one quarter of the giant particles which became airborne leaves the boundaries of the Sahara. Because dry deposition is observed close to the source it is affected by short time disturbances. To avoid blurring of any trend, time averaging methods are recommended. Dust deposition collectors exposed for roughly one month and weighed afterwards serve this purpose.

Summarizing this discussion:

> Turbidity and dry deposition cover two different size classes of the airborne particles and two different effects of the dust plume. Both effects give information about the whole vertical extension of the plume. Sal, Cape Verde Islands, seems to be an ideal place for a ground based monitoring station. It is located at such a distance from the source that both effects, turbidity and dry-fallout, can be observed best.

REFERENCES

BAMS (1974). *Bull. Am. Meteorol. Soc.*, **55**, 348–349.

Carlson, T. N., and Prospero, J. M. (1972). The large-scale movement of Saharan air outbreaks over the equatorial North Atlantic. *J. Appl. Meteorol.*, **11**, 283–297.

Carlson, T. N., Prospero, J. M., and Hanson, K. J. (1977). Attenuation of Solar Dust off the West Coast of Africa. *NOAA Tech. Mem. ERL WMPO-7*, 27 pp.

Dubief, J. (1953). Les vents de sable au Sahara français. *Colloq. Int. CNRS*, **35**, 45–70.

Dubief, J. (1959). *Le Climat du Sahara*, **1**. Univ. Inst. Rech. Sahariennes, Mem. (hors sér.), Alger, 312 pp.

Goldberg, E. D., Koide, M., Griffin, J. J., and Peterson, N. N. A. (1963). A geochronological and sedimentary profile across the North Atlantic Ocean. In Craig, H., Miller, S. M., and Wasserburg, G. J. (eds.), *Isotopic and Cosmic Chemistry*: 211–219, North-Holland, Amsterdam.

Goldberg, E. D. (1971). Atmospheric dust, the sedimentary cycle and man. *Comments on Earth Sciences. Geophys.*, **1**, 117–132.

Hastenrath, S. (1976). Variations in low-latitude circulation and extreme climatic events in the tropical Americas. *J. Atmos. Sci.*, **33**, 202–215.

Jaenicke, R. (1973). Monitoring of aerosols by measurements of single parameters. *Stockholm Tropospheric Aerosol Seminar. Univ.*, Stockholm, 29–31.

Jaenicke, R., and Kasten, F. (1977). The determination of atmospheric turbidity from the burned traces in the Campbell-Stokes sunshine recorder. Submitted to *J. Appl. Opt.*

Jaenicke, R., and Schütz, L. (1977). A comprehensive study of physical and chemical properties of the surface aerosols in the Cape Verde Islands region. Submitted to *J. Geophys. Res.*

Ku, T. L., Broecker, W. S., and Opdyke, N. (1968). Comparison of sedimentation

rates measured by paleomagnetic and the ionium methods of age determination. *Earth Planet. Sci. Lett.*, **4**, 1–16.

Lauscher, F. (1957). Zur Frage: 'Sonnenschein und Bewölkung = 100%?'. *Wetter Leben*, **9**, 143–146.

Newell, R. E., Kidson, J. W., Vincent, D. G., and Boer, G. J. (1972). *The General Circulation of the Tropical Atmosphere.* Vol. 1, MIT-Press, Cambridge, Mass., 258 pp.

Prospero, J. M. (1968). Atmospheric dust studies on Barbados. *Bull. Am. Meteorol. Soc.*, **49**, 645–652.

Prospero, J. M. (1976). *Pers. comm.*

Prospero, J. M., and Carlson, T. N. (1972). Vertical and areal distribution of Saharan dust over the western equatorial North Atlantic Ocean. *J. Geophys. Res.*, **77**, 5255–5265.

Rothe, P. (1973). Sedimentation in the deep-sea areas adjacent to the Canary and Cape Verde Islands. *Mar. Geol.*, **14**, 191–206.

Schütz, L., and Jaenicke, R. (1974). Particle number and mass distributions above 10^{-4} cm radius in sand and aerosols of the Sahara desert. *J. Appl. Meteorol.*, **13**, 863–870.

Schütz, L. (1977). Die Saharastaub-Komponente über dem subtropischen Nord-Atlantic. *Ph.D.-Thesis*, Univ. Mainz, FRG, 153 pp.

Schütz, L. (1979). *Sahara Dust Transport over the North Atlantic Ocean – Model Calculations and Measurements.* In this report.

SMIC (1971). Report of the Study of Man's Impact on Climate. *Inadvertent Climate Modification.* MIT-Press, Cambridge, Mass., 308 pp.

Steinhauser, F. (1934). Die mittlere Trübung der Luft an verschiedenen Orten beurteilt nach LINKEschen Trübungsfaktoren. *Gerlands Beitr. Geophys.*, **42**, 110–121.

Volz, F. E. (1959). Photometer mit Selen-Photo-Element zur spektralen Messung der Sonnenstrahlung und zur Bestimmung der Wellenlängenabhängigkeit der Dunsttrübung. *Arch. Meteorol. Geophys. Bioklimatol.*, **10**, 100–131.

Volz, F. E. (unpublished draft). *Atmospheric Turbidity and Solar Aureoles in the Azores – Cape Verde.*

CHAPTER 13

Long-range Impact of Desert Aerosol on Atmospheric Chemistry: Two Examples

K. A. Rahn, R. D. Borys, G. E. Shaw, L. Schütz, and R. Jaenicke

ABSTRACT

Approximately one-half the elements in the atmospheric aerosol occur in near-crustal proportions, and are probably soil-derived. Deserts are likely to be the main contributors of this crustal dust even to remote temperate and polar regions. Multi-elemental analyses of 7 samples of Sahara aerosol collected over the North Atlantic in 1973 revealed that its composition was surprisingly uniform over 2400 km and 8 days, with more than three-quarters of the elements in crustal proportions. Sahara dust accounted for about 80% of the aerosol mass at the surface of the NE-tradewind region of the North Atlantic during this strong outbreak. As an example of the impact of desert dust on aerosol composition even farther from the source, the recent discovery of bands of Asian desert dust in the arctic atmosphere is discussed. During April and May 1976 layers of Asian dust were observed over Alaska. These haze layers completely altered the natural chemistry and physics of the arctic aerosol. As much as one-half million tons of Asian desert dust may enter the Arctic during a 5-day episode. Desert-derived aerosol seems to be chemically different from its parent soil. Only the finest ($r < 1$ μm) soil particles, which are chemically and mineralogically quite different from the bulk soil, are capable of long-range transport through the atmosphere. Crust-air fractionation has already been noted for Si and Al, but may exist for many other elements as well, such as the rare earths, Th, and several heavy metals such as Zn, Sb, As, etc. Analysis of desert soils specifically in the aerosol size range will be important for the future understanding of desert dust in the atmosphere.

13.1 INTRODUCTION

Crustal aerosol, especially that produced from deserts, plays an important role in the physics and chemistry of the atmosphere. Recently this key role has been increasingly recognized, and studies of desert aerosol and its atmospheric effects are rapidly increasing. Nevertheless, surprisingly little is known about many aspects of desert dust.

This paper is concerned with the elemental composition of desert dust in the atmosphere and the extent to which it governs the composition of the aerosol, even

far from deserts. A general introduction will show that a large number of elements in the aerosol throughout the troposphere are essentially exclusively controlled by crustal sources. Then two examples of long-range transport of desert dust into otherwise remote areas will be given, namely Sahara dust over the Atlantic Ocean and Asian dust over Alaska. In each case the impact on the aerosol is striking. The paper will conclude with a discussion of the relations between the composition of desert dust and the aerosol derived from it. The argument will be made that there is no *a priori* reason that the two should be the same. In fact, the physical process of generation of crustal aerosol suggests that it should most likely have a different composition than its parent soil. Data from the world aerosol and from Sahara soils and aerosols in particular will be used to support this hypothesis. The importance of knowing the composition of desert soils as a function of particle size will be stressed.

13.2 GENERAL CHEMICAL IMPORTANCE OF CRUSTAL DUST IN THE ATMOSPHERE

The abundances of many elements in the atmospheric aerosol are controlled by the ubiquitous crustal aerosol. This can be seen most easily by enrichment-factor analysis, that is, by calculating the *enrichment factor* of each element in the aerosol relative to the crust and some reference element. This is usually done for an element by using the following formula:

$$\text{Aerosol Crust Enrichment Factor} = (X/Al)_{aerosol}/(X/Al)_{crust}$$

where X and Al stand for concentrations of element X and the reference element Al. In principle any of several elements (Si, Al, Fe, Ti, Sc, . . .) could be used for the crustal reference element, but Al is the most common choice because it is relatively easily determined and lacks major pollution sources. Fe is the next most common choice. The crustal reference material 'crust' ought of course to be soil, probably desert soil, but is usually taken to be average crust (i.e. rock) because so much better data are available for rock than for soil. All calculations reported here use the average crust of Mason (1966).

The enrichment factors of a large number of aerosols have been calculated and tabulated (Rahn, 1976b). It was found that each element in the world aerosol has a reasonably characteristic range of enrichment factor, with values distributed approximately log-normally. A simple geometric mean over the entire population of aerosol samples considered in this compilation provides a first estimate of the most typical enrichment factor for each element. These geometric mean enrichment factors, from more than 100 aerosols, are shown in Figure 13.1. It can be immediately seen that more than half of the elements have mean enrichments of less than about 7. These elements are therefore present in nearly crustal proportions in the atmosphere. By contrast, the other half of the elements have enrichment

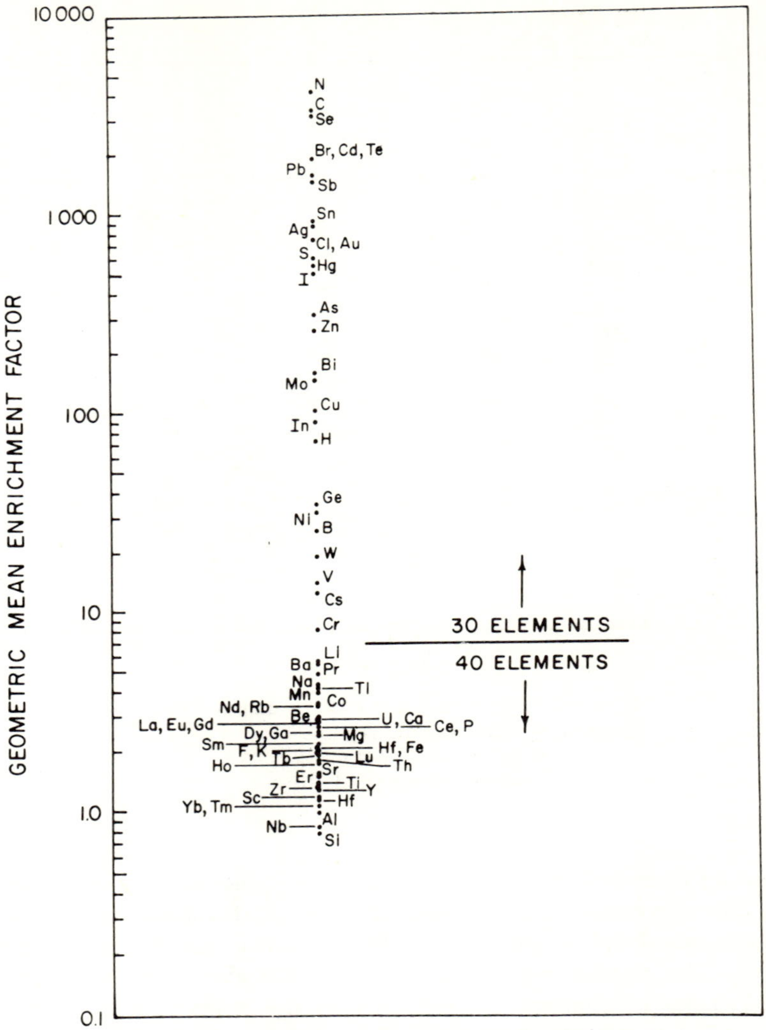

Figure 13.1 Geometric mean enrichment factors

factors between about 10 and 5000, and are therefore not crust-controlled. In general, the sources of these elements in remote atmospheres are not known.

Of the crustal elements in the atmosphere, some have important marine components in marine aerosols, whereas others have no measurable marine components even in the most remote marine regions. These two extremes of behaviour are exemplified by Na and Fe, as shown in the next two figures. Figure 13.2 shows the enrichment-factor diagram for Na (Rahn, 1976b). This type of plot is a scatter

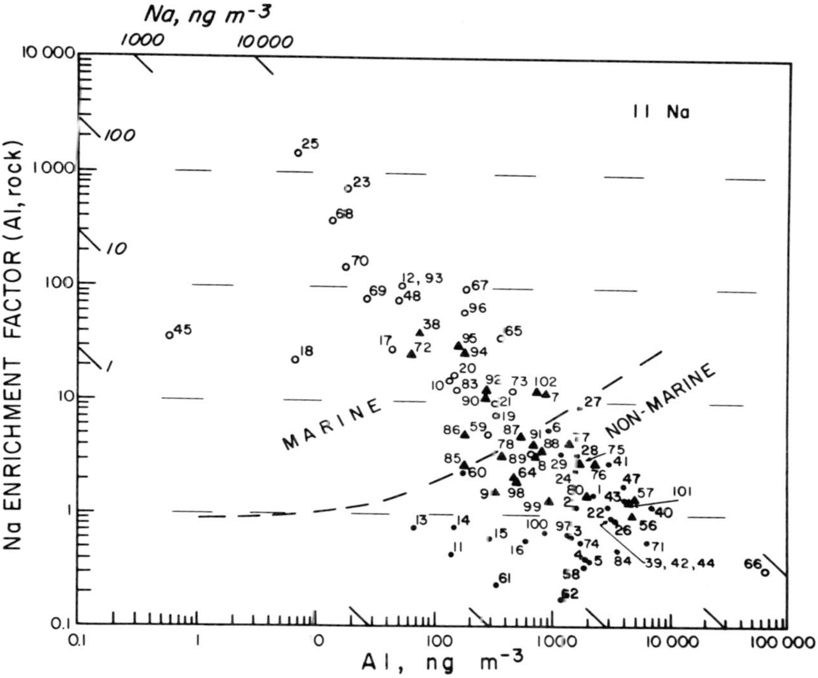

Figure 13.2 Enrichment factor diagram (Na)
Legend: ● = continental regions
▲ = semi-marine points
○ = marine regions

diagram of the aerosol-crust enrichment factor for an element *vs.* the concentration of the reference element Al, both variables being plotted logarithmically. Remote (low Al) points therefore appear on the left and urban (high Al) points appear on the right. The number next to each point is for identification only in Rahn (1976b). Lines of constant concentration of the element in question appear inclined 45° counterclockwise from the vertical. Continental regions are indicated by solid circles, semi-marine points by solid triangles, and marine regions by open circles. For Na a clear-cut marine region can be seen, as evidenced by high enrichments in remote marine and semi-marine areas. Note that in these marine areas the *concentration* rather than the enrichment factor of Na tends toward constancy, being roughly a few tenths to a few micrograms per cubic metre of air.

The opposite behaviour of Fe is shown in Figure 13.3. No matter how remote the area, and no matter whether it is marine, semi-marine, or nonmarine, the enrichment factors of Fe remain for the most part between about 0.7 and 7. Marine and semi-marine points are completely intermingled with nonmarine points. There-

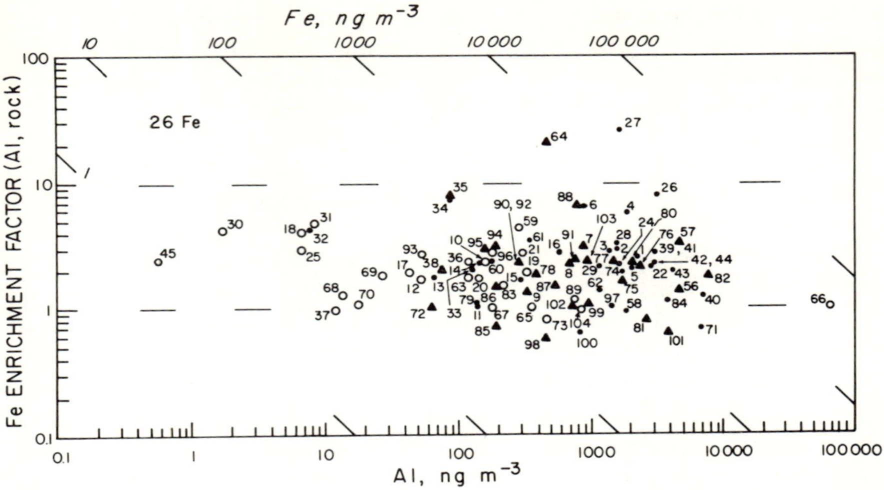

Figure 13.3 Enrichment factor diagram (Fe)
Legend: ● = continental regions
▲ = semi-marine points
○ = marine regions

fore, Fe has no measurable marine component. Even in the most marine areas Fe is crust-controlled.

Table 13.1 summarizes the behaviour of 51 elements in the atmosphere whose concentrations have been measured often enough that enrichment-factor plots can be made. As noted above, approximately half of these elements are crustal or near-crustal in abundance; the other half are enriched. Of the 26 non-enriched and

TABLE 13.1 Classification of Elements by Type of Enrichment-factor Diagram

Non-enriched elements (18)
Li, F, Na*, Mg*, Si, P, K*, Ca*, Sc, Ti, Rb*, Sr* La, Ce, Sm, Eu, Hf, Th
Intermediate elements (8)
V, Cr, Mn, Fe, Co, Ga, Cs*, Ba
Enriched elements (25)
Be, B, C, N, S, Cl*, Ni, Cu, Zn, Ge, As, Se, Br*, Mo, Ag, Cd, In, Sn, Sb, I*, W, Au, Hg, Pb, Bi

*Marine component (10)

intermediate elements, only 7, or about one quarter, show measureable marine components. Thus the study of the composition of desert soils, which make up a large fraction of crustal material injected into the atmosphere, can materially affect our understanding of nearly half the elements in the aerosol.

13.3 SAHARA DUST OVER THE ATLANTIC

During a 1973 cruise of the West German research vessel 'Meteor' across the tropical North Atlantic, a series of 15 high-volume aerosol samples were taken using 8 x 10 inch Delbag Microsorban polystyrene filters and a high-volume vacuum pump (Rahn *et al.*, 1974). Samples were 1 to 2 days in duration, with total volumes of air sampled ranging from 2500 m^3 to 7000 m^3. The cruise track is shown in Figure 13.4, along with the total concentrations of mineral matter collected. As seen in the Figure, at about the midpoint of the cruise the ship encountered a large, strong plume of Sahara dust. The seven samples numbered 9 through 15 from this plume form the basis of this section and will be considered as a unit whose composition is indicative of a single Saharan outbreak.

We felt that this suite of samples was a particularly valuable one, because it was taken on high-efficiency filters which collected all size ranges of the aerosol, as opposed to the popular nylon meshes or sails, which collect particles only larger

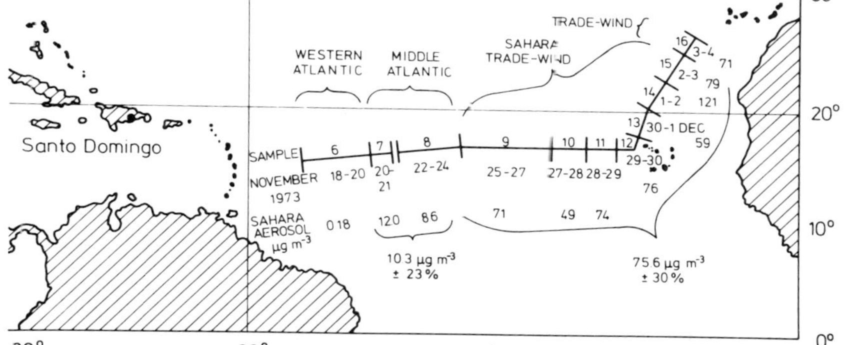

Figure 13.4 'Meteor' 1973 cruise map with dates and total Sahara dust loadings

than 1 μm or so. Furthermore, great pains were taken to avoid ship contamination and to sample as nearly isokinetically as possible. The latter was a particularly important point, because so much of the Sahara aerosol is found in the giant size range ($r > 1$ μm) which is easily missed by nonisokinetic air flow into the sampler. Lastly, the samples were important because they were taken on filters which were chemically very pure, so that they could later be analyzed chemically with a minimum of interference from the filter material itself.

Samples 9 through 15 were collected over a total distance of about 2400 km and an elapsed time of 8 days. The nearest distance to the African shore was about 600 km at the end of sample 15, and the farthest distance from the African shore was about 1300 km, at the beginning of sample 9. The 20 degrees of longitude over which this Sahara plume was sampled represents nearly one-half its total longitudinal displacement on its way from Africa to the Caribbean.

These samples were analyzed by nondestructive neutron activation for as many elements as possible (Rahn *et al.*, 1976). About 40 elements could be determined from 2 irradiations and 4 counts on each sample. In addition to these aerosol samples, 6 Libyan soil and soil-related samples were also analysed, results of which will be discussed in the last section of this paper.

The total concentrations of Sahara aerosol were higher than expected and surprisingly constant. The mean mineral concentration of the 7 samples was 75.6 μg m^{-3} ± 30%. Considering the length of the ship's track and the elapsed time during sample collection, a coefficient of variation of only 30% seems very small. The impact of these high concentrations of mineral aerosol on the region can be seen in Figure 13.5, which shows the major components of the Sahara trade-wind aerosol in the strong plume. Of the total aerosol (96.2 μg m^{-3}), sea salt accounts for 15.2 μg m^{-3}, or only 16%. By contrast, Sahara dust accounts for some 79% of the aerosol mass. In other words, the aerosol of this marine region becomes essentially continental when a strong Sahara plume passes over. The remainder of the aerosol (5%) is divided between organic matter and ammonium sulphate.

The composition of the Sahara aerosol seemed to be even more constant than was its total concentration. This is a feature of the aerosol which has not been noted in the few previous reports on its composition, perhaps because less precise analytical techniques than neutron activation were used. On the other hand, perhaps our data represent only a single outbreak, different in composition from others. Whatever its true significance, this constancy is illustrated in Figure 13.6, which is a scatter diagram of the coefficient of variation of the concentration ratio X/Fe for the various elements X *vs.* the analytical uncertainty of the same ratio X/Fe. The diagram clearly shows that the vast majority of elements show variations over the 7 samples which are just about the same as their analytical uncertainties, even down to the 2% level for Sc. In short, we were not able to detect any significant compositional variations within this particular Sahara plume. It is not yet certain whether such homogeneity is a characteristic of all Sahara outbreaks, or whether one outbreak differs measurably from another. What is certain, though, is

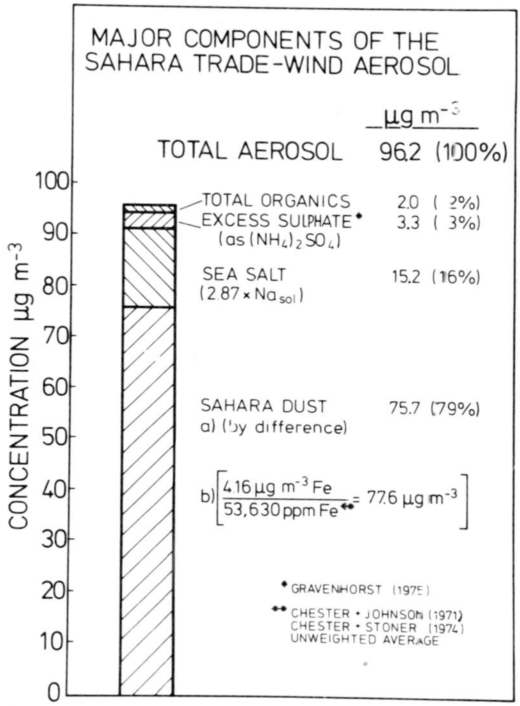

Figure 13.5 Major components of the Sahara trade-wind aerosol

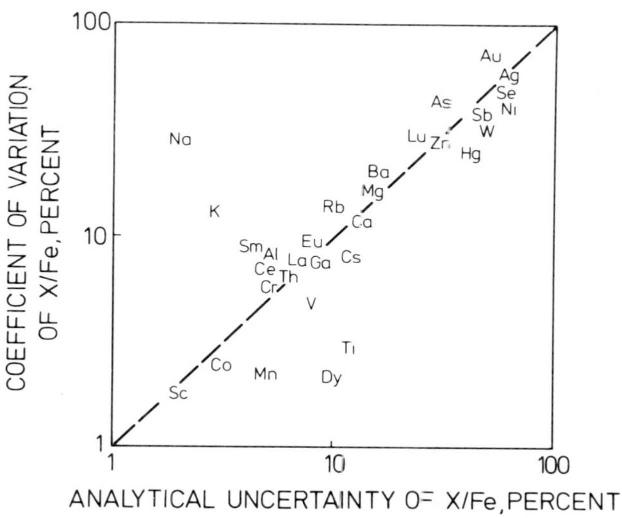

Figure 13.6 Coefficient of variation of X/Fe *vs.* analytical uncertainty of X/Fe

that the scale of this outbreak (2400 km, 8 days) was large, and that this degree of compositional homogeneity is highly unusual within environmental science. The explanation for this intriguing phenomenon must be a constancy of proportions within the Sahara sand and/or extensive mixing of the aerosol before it moves out over the Atlantic. At the moment, however, the relative importance of these two effects is not clear.

This constancy of composition for the Sahara aerosol makes its composition a somewhat more fundamental quantity than it otherwise might be. Figure 13.7 shows this composition, expressed as enrichment factors relative to Fe. The vertical bars are coefficients of variation of enrichment factor over the 7 samples, not the observed scatter directly. There are a number of features of this plot worth pointing out. First, the majority of the elements have enrichment factors of essentially unity, i.e. they are in the proportions of average crustal rock. At least 25 of the elements fit into this category, perhaps more. Only about 5 elements have enrichment factors greater than 7. This situation is to be contrasted with the world aerosol of Figure 13.1 where nearly one-half the elements have enrichment factors of roughly 7 or more. Thus from the enrichment-factor point of view the Sahara aerosol is a relatively simple one, with most elements at or very near crustal proportions. For comparison purposes, Table 13.2 shows the geometric mean ('world') enrichment factors of Figure 13.1 next to the Saharan enrichment factors. In nearly all cases the Saharan enrichments are depressed relative to the global values. The higher the global enrichment factor the more it is depressed in the Sahara aerosol.

There is, however, quite some order to the enrichments of Figure 13.7. The light Group I and II elements Na, K, Ca, and Mg have low enrichment factors, between

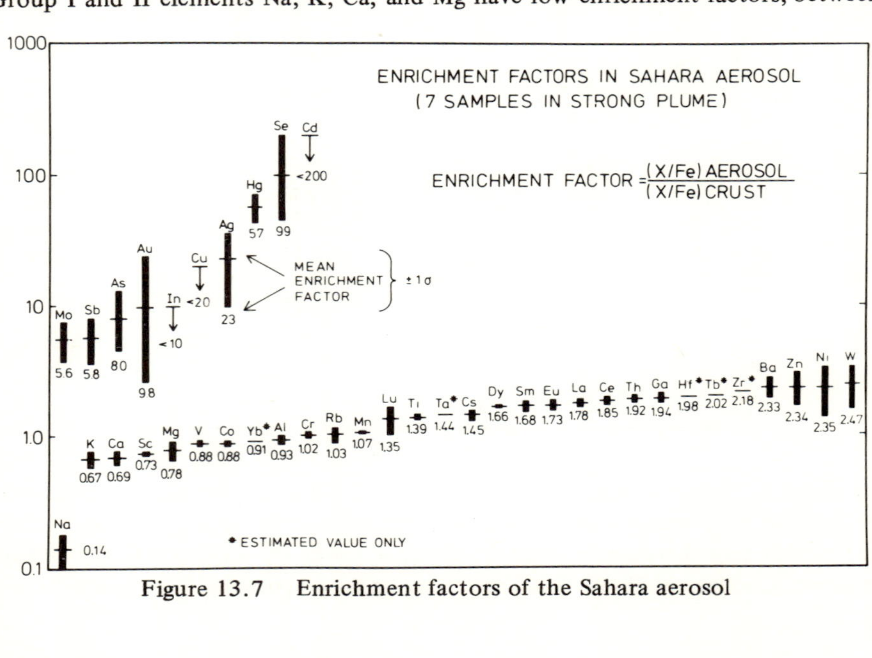

Figure 13.7 Enrichment factors of the Sahara aerosol

TABLE 13.2 Enrichment Factors in the Sahara and World Aerosols

Element	Sahara EF(Fe, rock)	World EF(A1, rock)
Na	0.14	4.4
K	0.67	1.98
Ca	0.69	2.8
Sc	0.73	1.17
Mg	0.78	2.4
V	0.88	14.0
Co	0.88	3.5
Yb	0.91	1.06
Al	0.93	1.00
Fe	1.00	2.1
Cr	1.02	8.1
Rb	1.03	3.4
Mn	1.07	3.9
Lu	1.35	1.95
Ti	1.39	1.39
Ta	1.44	1.14
Cs	1.45	12.4
Dy	1.66	2.5
Sm	1.68	2.1
Eu	1.73	2.7
La	1.78	2.7
Ce	1.85	2.6
Th	1.92	1.78
Ga	1.94	2.5
Hf	1.98	2.0
Tb	2.02	1.92
Zr	2.18	1.36
Ba	2.33	5.5
Zn	2.34	260
Ni	2.35	32
W	2.47	19.1
Mo	5.6	144
Sb	5.8	1430
As	8.0	310
Au	9.8	740
In	<10	90
Cu	<20	102
Ag	23	880
Hg	57	560
Se	99	3100
Cd	<200	1920

0.1 and 0.8. Then comes a group of heavier Group I and II elements plus transition metals with enrichments between 0.8 and 1.4. The rare earths and Th form a group with enrichments between 1.6 and 2.0. A small group of heavy metals is clustered about enrichments of about 2.4. The last group is composed of 7 heavy metals and

semi-metals with enrichments between 5 and 100. Except for Ba, all the elements in these last two groups are highly enriched in the normal aerosol.

The rare earth elements (REE) are a particularly interesting case. With the exception of Lu, they have enrichment factors between 1.6 and 2. These elements are most interesting because their enrichments have just about these same values in the rest of the world aerosol. No major pollution sources are known for the rare earths. Thus they are indisputably of crustal origin in the world aerosol but equally indisputably have enrichments markedly higher than unity. The last section of this paper addresses this point further.

In summary, then, the Sahara aerosol has a composition close to that of the average crust, i.e. low enrichment factors for most all the elements. Elements which are highly enriched in the world aerosol do show modest enrichments in the Sahara aerosol, but it is not yet clear whether this is from the Sahara aerosol itself or from a certain mixing of outside air into the Sahara plume. Certain small enrichments observed in the Sahara plume, for the rare earths in particular, are also observed in the world aerosol and are surely from the dust itself.

13.4 ASIAN DUST OVER ALASKA

The second example of long-range chemical impact of desert dust on the remote aerosol comes from an investigation of the composition and source of arctic haze over Alaska. 'Arctic haze' refers to turbid layers of air which are found often over Alaska and the pack ice to the north during periods of clear weather (Mitchell, 1956). These layers are diffuse, hundreds to thousands of kilometres in horizontal extent, and occur at various levels throughout the arctic troposphere. Their colour is grey-blue in the antisolar direction and reddish-brown in the solar direction, which suggests that they are composed primarily of small aerosol particles. Until recently the only observations of arctic haze were visual. In 1972, 1973, and 1974 however, radiative measurements on these layers near Barrow, Alaska revealed that they had a well-defined extinction, i.e. that they were affecting the radiation balance of the Arctic (Holmgren *et al*., 1974). The question of the origin of these layers became more significant in 1974 when rough trajectory analysis for a haze episode at Barrow that spring suggested that the air containing the haze may have passed over the highly polluted northeastern United States about 10 days before the episode.

In a joint venture between the University of Rhode Island and the University of Alaska, we have taken filter samples of the haze aerosol itself by flying a light aircraft within a haze layer. Chemical analysis of the aerosol collected, first results of which are reported elsewhere (Rahn *et al.*, 1977), combined with electron microscopy, has revealed that these particles are actually crustal aerosol and that their source is the deserts of eastern Asia. This programme and its results will now be briefly discussed.

During April and May 1976 a series of 15 flights with a specially equipped single-engine Cessna 180 aircraft were made from the Naval Arctic Research

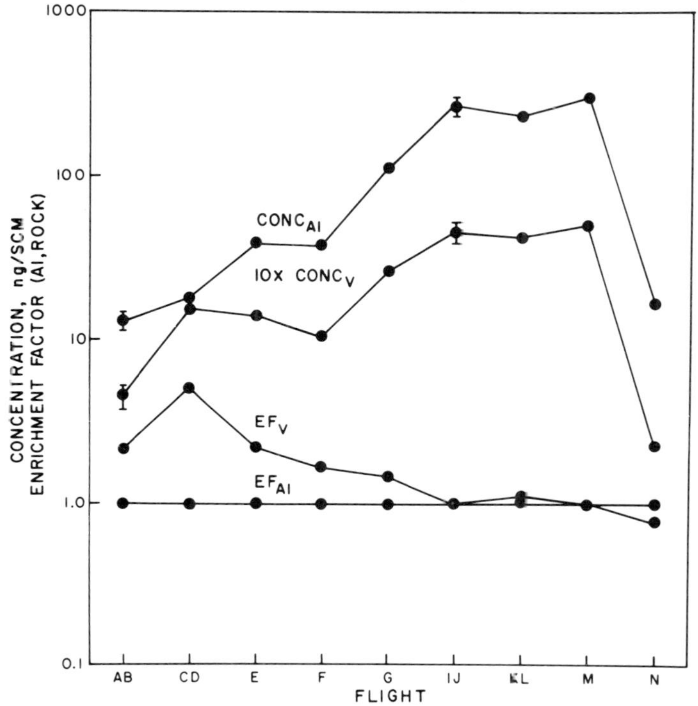

Figure 13.8 Concentration and enrichment factors of V and Al in aerosol over Barrow (Spring 1976)

Laboratory in Barrow, Alaska into the haze layers. High-volume (90 to 360 m^3) samples of aerosol were collected on acid-washed 11-cm Whatman No. 41 cellulose filter paper, low-volume and Nuclepore filter samples were taken for electron microscopy. The high-volume samples were analyzed for 20 to 25 elements by neutron activation, using essentially the same scheme as for the Sahara samples discussed above. Figure 13.8 shows the results for V and Al. This pair of elements provides perhaps the most sensitive and specific test available with neutron activation for deciding whether an aerosol is natural or pollution-derived. The major source of Al into the atmosphere is of course the crust. Vanadium has two major sources, the crust and combustion of residual (#6) oil (Zoller *et al.*, 1973). In remote areas free of significant air pollution the V/Al ratio is very close to that of the bulk crust. In other words, the enrichment factor of V is unity. In urban areas, on the other hand, the V enrichment factor is much higher, varying from 5 to 500 depending on the area and its source of residual oil (Rahn, 1976b). The northeast United States, for example, uses Venezuelan residual oils which are unusually rich in V, so has abnormally high V enrichments. Beside the crust, no other major natural sources of V are known. Thus, V enrichment factors greater than about 1.5

or so signify the presence of air pollution, and V enrichments below 1.5 mean that the aerosol is mostly crustal.

In Figure 13.8 above it can be seen that the early flights were characterized by low concentrations of both V and Al. This was in fact true for all the other elements, i.e. the atmosphere was typically 'clean' arctic air. Enrichment factors for V, however, ranged between 2 and 5, which definitely indicated that the V was pollution-derived. Flights IJ, KL, and M showed much higher concentrations of nearly all elements including V and Al, and corresponded to a visible haze event over Barrow. The elemental concentrations during this period were consistently 10 to 30 times above background levels. But the enrichment factors for V decreased sharply with the onset of the haze, and during the peak of the haze were essentially 1.0. This meant unambiguously that this arctic haze was crustal aerosol rather than pollution aerosol. In particular, it could not possibly have come from the northeast United States or from Europe. Interestingly, the composition of this haze for almost all elements was systematically skewed toward crustal material, as shown in Figure 13.9. Here the enrichment factors of this arctic haze aerosol are compared with

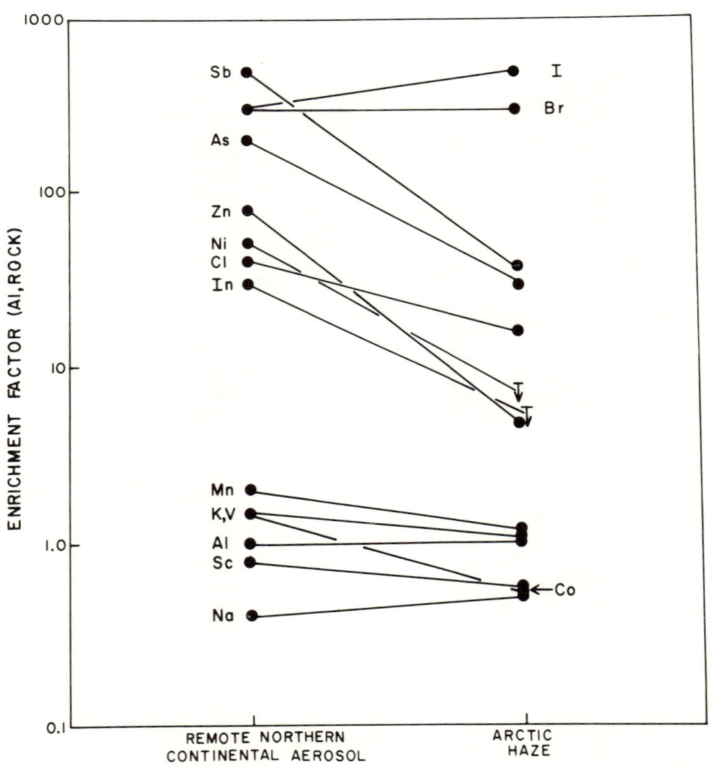

Figure 13.9 Elemental enrichment factors in arctic haze and remote northern continental aerosol

mean enrichment factors of remote northern continental aerosol (Rahn, 1976b). A general lowering of enrichment factors in arctic haze relative to the more typical northern aerosol is clearly seen. The depressant effect is greatest for the elements with the highest enrichments, similar to what was previously observed with the Sahara aerosol.

Electron microscopy of the Nucleopore filters confirmed the chemical conclusion that arctic haze was crustal aerosol. Figures 13.10 and 13.11 are electron micrographs of typical portions of filters CD and IJ, respectively. The magnification of the original 10 x 11 cm negative was 5000X; a convenient reference size for these Figures is the Nucleopore pores which are 0.4 μm in diameter. In sample CD there are many more particles smaller than the pores than larger than the pores. The largest particles are few in number and are less than 1 μm in radius. They appear to be crustal fragments, as evidenced by their angularity. In sample IJ, on the other hand, many more giant particles are seen, with radii up to several μm. Platy layers can easily be seen on several particles; all appear angular and would seem to be crust-derived.

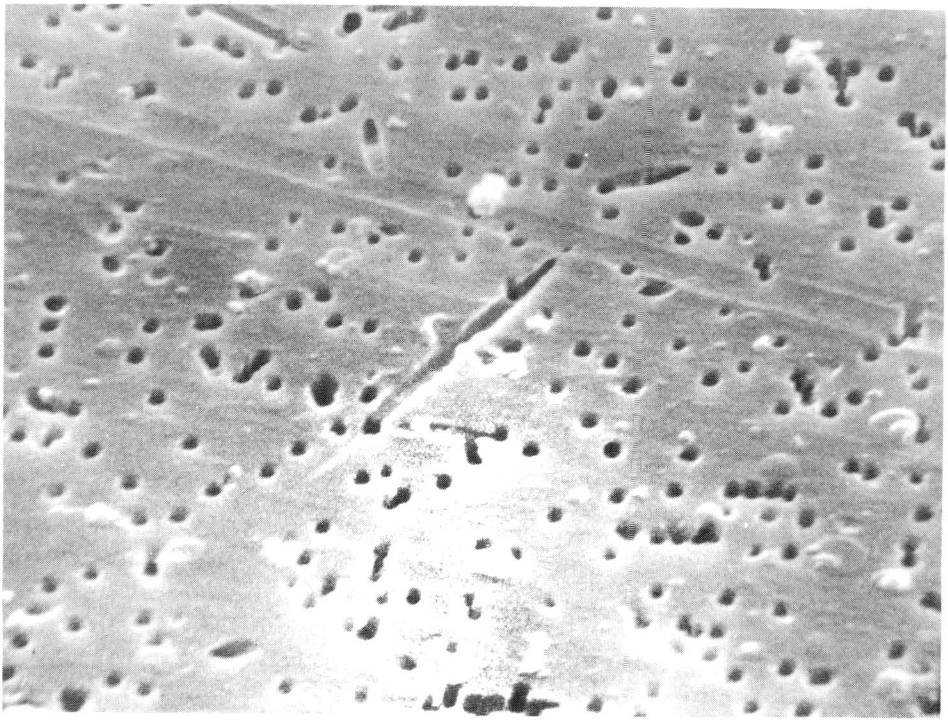

Figure 13.10 Electron photomicrograph of prehaze aerosol over Barrow, Flight CD (Spring 1976). Pore diameter is 0.4 μm

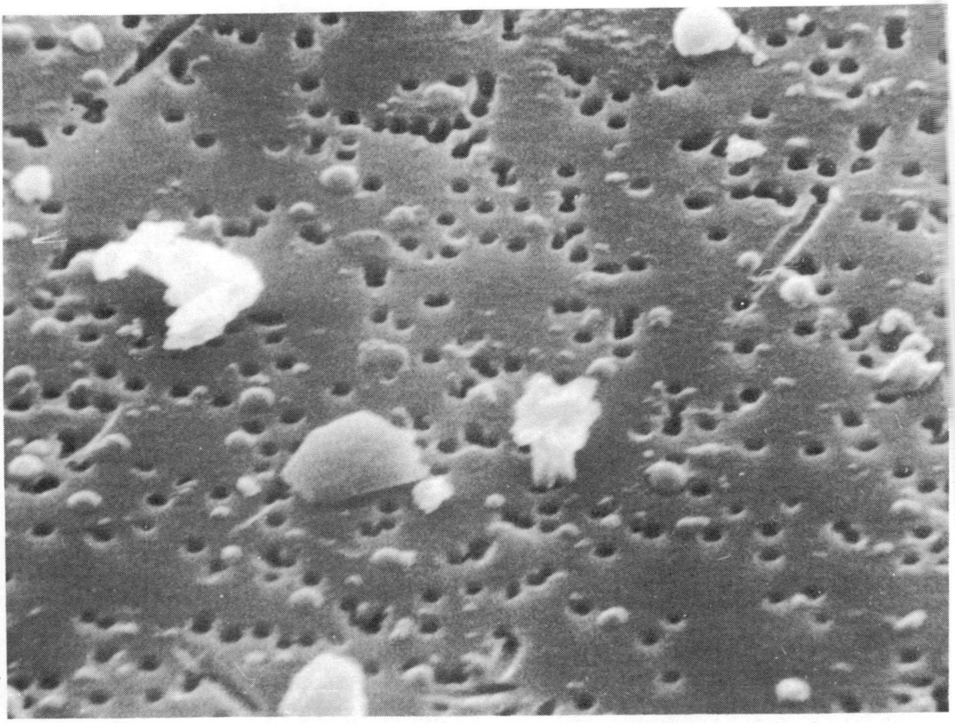

Figure 13.11 Electron photomicrograph of arctic haze aerosol over Barrow, Flight IJ (Spring 1976). Pore diameter is 0.4 μm

What could be the origin of these crustal particles? Alaska was completely snow-covered at the time; even in summer the state is only a very weak source of crustal aerosol. The haze source was clearly far distant. The first clue came from the general synoptic situation for the haze period, shown in Figure 13.12 for the 700-mb level. There was a strong flow of air into Alaska from the south, which had curved from the west and appeared to be traceable back to the Asian mainland, where the great Takla Makan and Gobi deserts are found. Actual 700-mb isobaric trajectories for the sampling period are shown in Figure 13.13; they confirm that episode air had indeed recently passed over these Asian deserts. Furthermore, April 1976 was in general a time of many dust storms in the Asian deserts, and in particular on the days when our trajectories terminated in the deserts there were dust storms. In contrast, however, trajectories for the pre-haze period did not pass over the Asian deserts. Two trajectories, from the 14th and 18th of April, came over the pole to Barrow. They corresponded to sample CD, and E, where the V enrichment factor indicated the presence of air pollution. These two air masses probably originated in Europe and/or the northeast United States, but it is not clear which source was the more important.

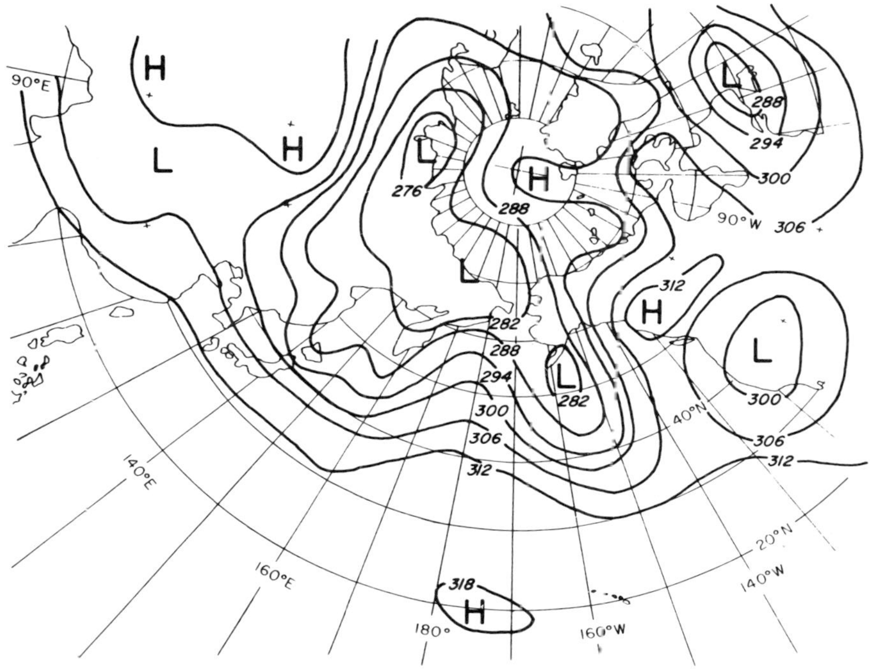

Figure 13.12 Typical 700-mb contour map for arctic haze transport to Alaska. Numbers are heights of 700-mb surface in decametres

It must be emphasized that at the time of this writing it has not yet been definitely proved that the Asian deserts are the true source of arctic haze. No haze bands have been followed from Alaska back to Asia, for example, nor has a band been observed from satellite photography. But our trajectory analysis is supported by several other factors. First, flow patterns conducive to long-range transport from Asia to Alaska do exist and routinely are strong during the spring. Second, the Asian deserts are farther north than the other deserts, lying mostly between 40° and 50° North as opposed to the more normal desert latitude of 20° to 30°. And third, the length of typical trajectories between Asia and Alaska (9000 to 12,000 km) is considerably longer than but not out of line with the well-documented 6000-km path for transport of Sahara dust to Barbados (Carlson and Prospero, 1972).

The mass of Asian desert dust transported into the Arctic seems to be very great. For example, consider some statistics from a 5-day haze episode observed at Fairbanks, Alaska 17–22 February 1976, during which visibilities were reduced from the normal 150 km to less than 30 km, and for which Asian dust seems to be the explanation. Columnar mass loadings derived from radiation measurements had values of about 45 mg m^{-2} (for aerosol density 2.5). Assuming a point source in

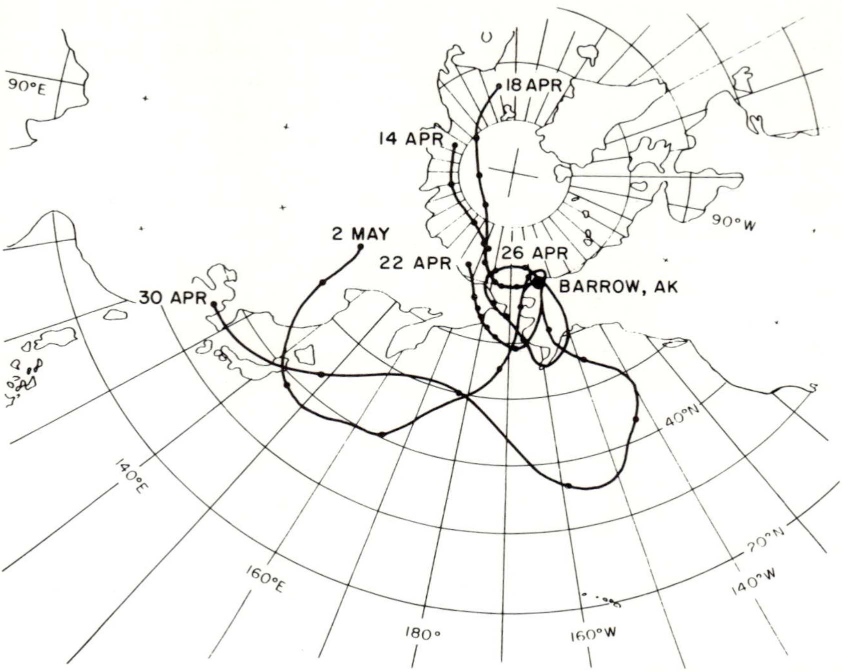

Figure 13.3 700-mb isobaric trajectories of air to Barrow. Numbers refer to date of arrival at Barrow, 1976. Solid circles each 24 hours along trajectory

Asia and a dispersion angle during transport of 5° (typical of volcanic plumes as viewed from satellite photographs), the dust cloud would be about 900 km wide by the time it reached Alaska, or about the width of the state itself. (Indeed during this episode, visibility was poor over the entire north half of the state.) Such a plume travelling at 80 km hr^{-1} would carry 4000 tons of dust into the Arctic per hour, or a half-million tons over the total 5-day episode. This corresponds to skimming off a 0.2 μm layer from a 10° x 10° desert.

Clearly, Asian desert dust must now be considered a major contributor to the overall aerosol burden of the arctic troposphere. It completely alters the aerosol chemistry of that region when it enters, and affects the radiation balance as well. Further study of this phenomenon is essential to reveal what if any climatological impact it has on the Arctic.

13.5 DESERT SOIL VERSUS DESERT AEROSOL

There is mounting evidence that desert soils cannot be directly compared with the aerosol derived from them. Recent parallel research of Gillette (Gillette and Walker, 1975; Gillette, 1976; Patterson *et al.*, 1976, for example) and Schütz and

Jaenicke (Schütz, 1971; Schütz and Jaenicke, 1974) on physical aspects of generation of soil aerosol have elucidated the high degree of physical fractionation that occurs during this process. For our purposes the important ideas can be summarized in a single diagram for the Sahara, Figure 13.14. In this Figure the mass of a typical Libyan soil and its derived aerosol is plotted as a function of particle size. In the soil itself (A) the vast majority of the mass lies between radii 20 and 100 μm. The aerosol approximately 1 μm above this soil (B) has a generally similar pattern, but shows a second mass peak at radius approximately 8 μm. By the time this aerosol has passed far out over the Atlantic (C), the faster-sedimenting particles of the large-size peak have all fallen out, and only the second, smaller-size peak remains. Even it has been shifted to a mean radius of 4 to 5 μm. When part C is compared to part A, it can be seen that virtually the entire mass of the desert aerosol which is capable of long-range transport in the atmosphere (1 μm $< r <$ 10 μm) has been

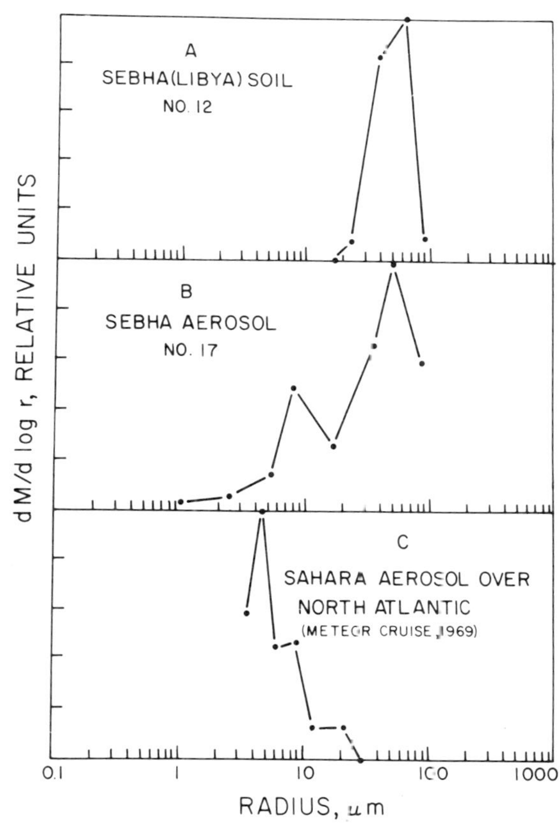

Figure 13.14 Mass distribution for parent soil and derived aerosol, Sahara Desert

derived from only a tiny mass fraction of the soil. It then seems reasonable to expect that the aerosol composition might deviate considerably from the composition of the bulk soil, especially because the lower end of the soil aerosol size range, $r = 1\ \mu m$, is the upper end of the clay-particle size range, and clay particles have a quite different structure and composition from their bulk parent soil.

We have seen above that the travel distance for Asian dust to Alaska is considerably longer than for Sahara dust to Barbados (9000 to 12,000 km *vs.* 6000 km). One might therefore expect mineral dust over Alaska to have a smaller average particle size than Sahara dust over much of the Atlantic, and this appears to be the case. For example, Figure 13.15 is a plot of volume of arctic haze aerosol over Barrow *vs.* radius. The mineral component of this aerosol, i.e. the desert-derived soil fraction, now has its volume (mass) peak at more nearly radius 2 μm. The clay-mineral contribution to this aerosol ought to be very important.

These two examples from the Saharan and Asian desert aerosols serve to demonstrate that one ought to be very careful about extrapolating from composition of desert soils to composition of the derived aerosols. But what about actuality? Is there any evidence that such chemical crust-air fractionation really occurs? There is, and it can be considered for both the general (global) mineral aerosol and for the Saharan case.

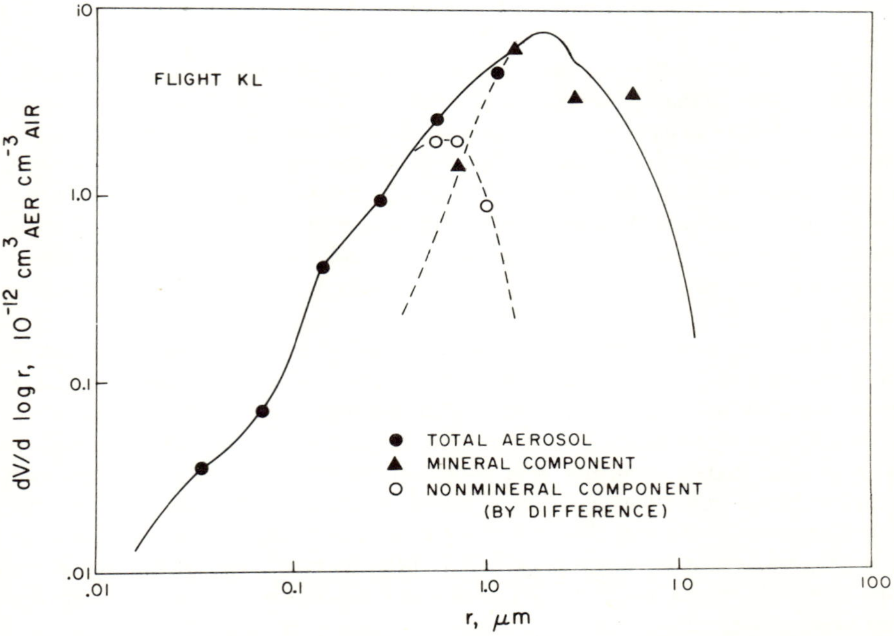

Figure 13.15 Volume distribution for arctic haze aerosol of flight KL (Spring 1976)

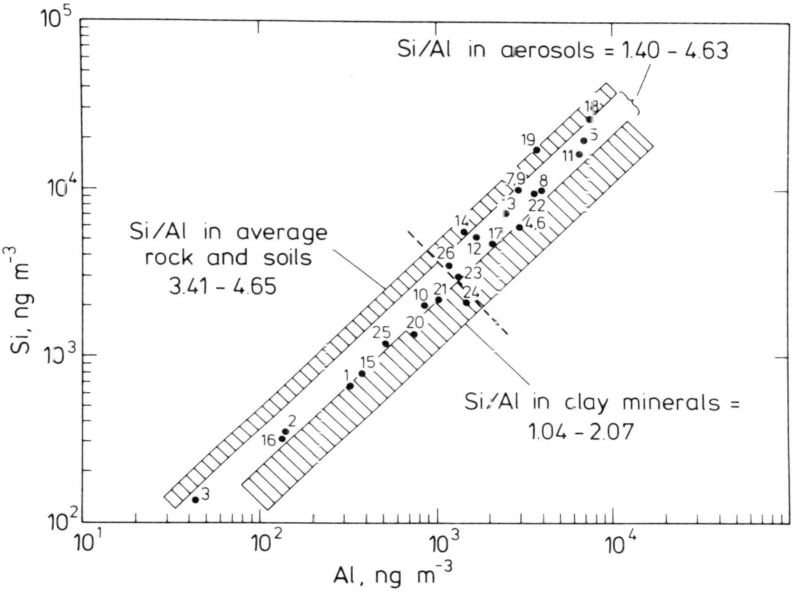

Figure 13.16 Scatter diagram of Si and Al concentration in aerosols (Numbers refer to locations as listed in Rahn, 1976a)

Crust-air fractionation has been demonstrated for Si and Al in the general world aerosol (Rahn, 1976a). This effect is illustrated in Figure 13.16, a scatter diagram of Si and Al concentrations in various aerosols of the world. From this Figure it is immediately evident that the Si/Al ratio of most aerosols, and especially those from remote areas far from dust sources, is distinctly below the average ratio of either bulk crustal rock or bulk soils. The remote-area Si/Al ratio of about 2 is nearly midway between the ratio for average crust and the ratio for clay minerals, suggesting that both of these materials contribute strongly to Si and Al in the aerosol.

Comparison of the composition of Libyan soils with the Sahara aerosol reveals compositional discrepancies of a similar magnitude to the Si–Al case, as well as a more complex soil-aerosol relationship than had been previously imagined. It was found that most elements were more enriched in the soils than they were in the aerosol, that certain elements were more enrichable than others, and that each element had a considerable range of actual enrichments over the suite of soils. Furthermore, all elements determined were more enriched in the small-size range ($r < 16$ μm) than in the larger range ($r > 16$ μm). Because this smaller range closely approximates aerosol sizes, all data considered subsequently in this section refer exclusively to it.

The different enrichments of a given element in the small size fractions of the various soils seem to depend on the percent of the soil mass in the small-size range.

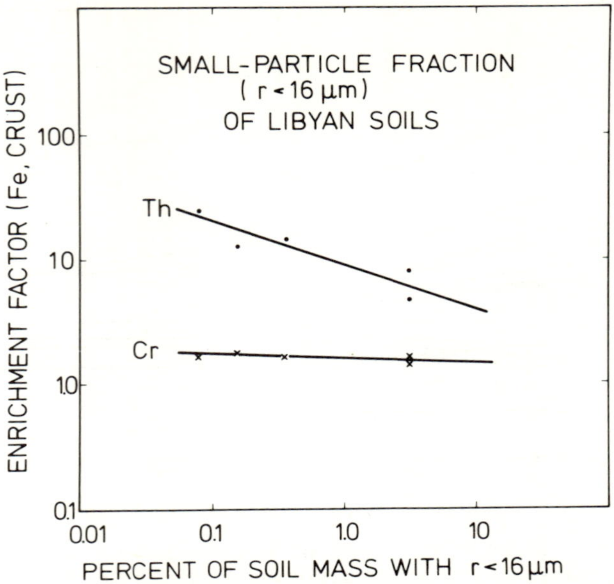

Figure 13.17 Enrichment factors of Th and Cr in $r < 16$ μm fraction of Sahara soils *vs*. percentage of soil mass in this range

When these two variables are plotted against each other, rather neat patterns emerge which bring order to much of the otherwise random scatter of the enrichments. Figure 13.17 shows this for Cr and Th. Cr represents a class of elements that show little variability of enrichment factor from soil to soil, whereas Th shows strongly increasing enrichments with decreasing soil mass in the small-size range. Note that even with Th the scatter about the best-fit line is quite small.

Figure 13.18 shows the same type of plot for more elements in the same soils, but without individual data points. The elements seem to be classifiable into 3 rough groups based on their slope, or 'enrichability': A group with near-zero slope which contains the alkali metals, alkaline earths, and heavy metals such as Zn, Ag, Sb, etc., a second group with intermediate slopes containing most of the transition metals such as Fe, Co, etc., plus other elements such as Mo, Ba, and Ta, and a third group with larger slopes composed of the rare earth elements plus Hf and Th. Note that Eu, which is known in geochemistry for its anomalous behaviour within the REE, has an anomalously low slope in this group.

At this time it is not possible to give a complete interpretation of this plot, except to note that it has demonstrated that soils have changes in composition as a function of particle size that cannot be ignored any longer in atmospheric chemistry. Furthermore, these variations seem to be regular and to depend on the

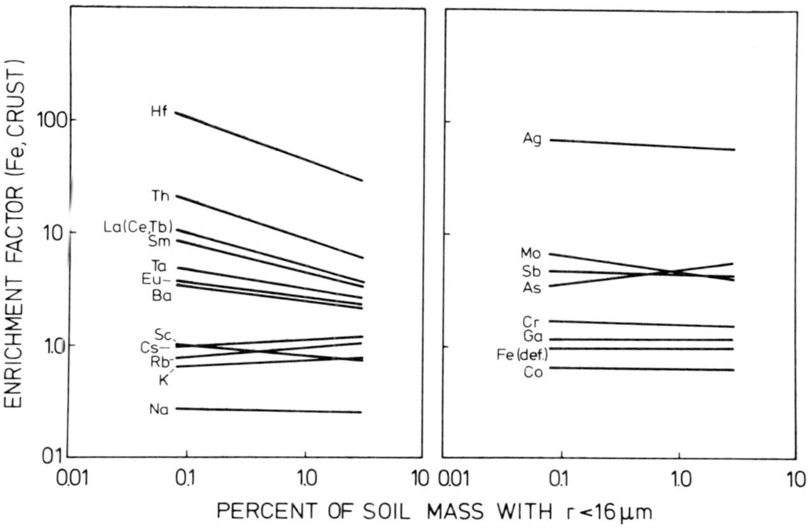

Figure 13.18 Enrichment factors of various elements in $r < 16$ μm fraction of Sahara soils *vs*. percentage of soil mass in this range

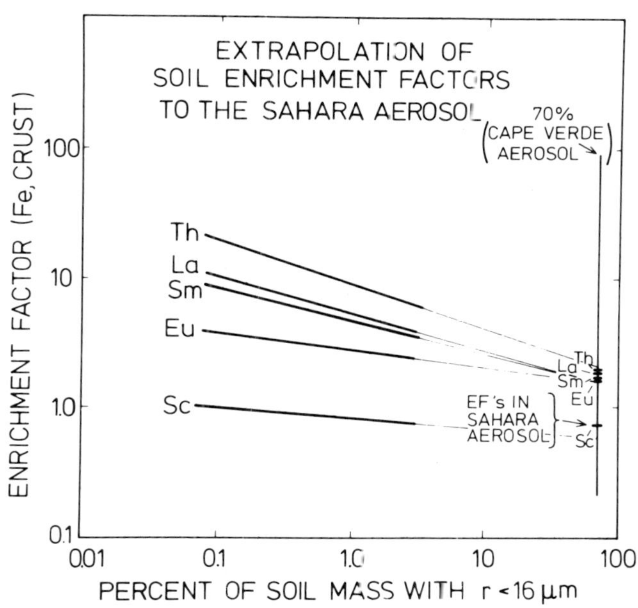

Figure 13.19 Extrapolation of Sahara soil enrichment factors ($r < 16$ μm) to Sahara aerosol

particle size distributions of the soils and the chemistry of the individual elements in some yet-to-be determined way.

In spite of the compositional differences between Sahara soils and Sahara aerosol, it did appear to be possible to extrapolate from the soils to the aerosol and reproduce some features of the aerosol composition by means of the technique shown in Figure 13.19. Here the enrichments of Th, La, Sm, Eu, and Sc are plotted *vs.* the percent of soil mass in the small-size range, as in Figures 13.17 and 13.18 above. This time, however, they are extrapolated to a value of 70% mass less than 16 μm radius, the figure for Sahara aerosol at the Cape Verde Islands. Actual enrichment factors of the elements in the Sahara aerosol are shown by the bars through the 70% line, which are in all cases close to the predicted enrichments.

This extrapolation technique is admittedly indirect and complex. Knowledge of the composition of desert soils as a function of particle size, particularly within the aerosol size range $r < 10\ \mu$m is highly desirable, and would provide a much more direct and simple means of relating aerosol to parent soil. Unfortunately, there is virtually no data available on this topic. High priority should be given to this research in the near future.

13.6 ACKNOWLEDGEMENTS

Much of the work reported here was carried out at the University of Rhode Island (Office of Naval Research contract N00014-76-C-0435) and at the Max-Planck-Institut für Chemie (Otto-Hahn-Institut), Mainz, West Germany (fellowship of the Max-Planck-Gesellschaft to K. R.). The cooperation of many staff members of both institutes, as well as the Naval Arctic Research Laboratory, Barrow, Alaska, is gratefully acknowledged. Certain samples were chemically analysed using the facilities of the Rhode Island Nuclear Science Center, Narragansett, Rhode Island.

REFERENCES

Carlson, T. N., and Prospero, J. M. (1972). The large-scale movement of Saharan air outbreaks over the northern equatorial Atlantic. *J. Appl. Meteorol.*, **11**, 283–297.

Chester, R., and Johnson, L. R. (1971). Trace element geochemistry of North Atlantic aeolian dusts. *Nature*, **231**, 176–178.

Chester, R. and Stoner, J. H. (1974). The distribution of Mn, Fe, Cu, Ni, Co, Ga, Cr, V, Ba, Sr, Sn, Zn, and Pb in some soil-sized particulates from the lower troposphere. *Mar. Chem.*, **2**, 157–188.

Gillette, D. A. (1976). Production of fine dust by wind erosion of soil: Effect of wind and soil texture. In 'Atmosphere–Surface Exchange of Particulate and Gaseous Pollutants'. *Symp. Proc.*, CONF-740921: 591–609, NTIS US Dep. Commerce, Springfield, Virginia.

Gillette, D. A., and Walker, T. (1975). Characteristics of airborne particles produced by wind erosion of sandy soil, high plains of West Texas. *Nat. Center Atmos. Res.* (Boulder, Colo.) and *Dep. Geol., Univ. Colo.* (Boulder, Colo.), preprint, 17 pp. 11 Figs.

Gravenhorst, G. (1975). Der Sulfatanteil im Atmosfärischen Aerosol über dem Nord-atlantik. *Ber. Inst. Meteorol. Geophys.*, Univ. Frankfurt, No. 30, 79 pp.

Holmgren, B., Shaw, G. and Weller, G. (1974). Turbidity in the arctic atmosphere. *AIDJEX Bull.*, **27**, 135–148.

Mason, B. (1966). *Principles of Geochemistry*, 3rd ed., Wiley & Sons Inc., New York, 329 pp.

Mitchell, M. (1956). Visual range in the polar regions with particular reference to the Alaskan Arctic. *J. Atmos. Terr. Phys., Spec. Suppl.*, 195–211.

Patterson, E. M., Gillette, D. A., and Grams, G. W. (1976). The relation between visibility and the size-number distribution of airborne soil particles. *J. Appl. Meteorol.*, **15**, 470–478.

Rahn, K. A. (1976a). Silicon and aluminum in atmospheric aerosols: Crust-air fractionation? *Atmos. Environ.*, **10**, 597–601.

Rahn, K. A. (1976b). The Chemical Composition of the Atmospheric Aerosol. *Univ. R. I. Tech. Rep.*, 1 July 1976, 265 pp.

Rahn, K. A., Schütz, L., and Jaenicke, R. (1976). *Elemental composition of Sahara Aerosol and Sahara Soil.* (Preprint).

Rahn, K. A., Borys, R. D., and Shaw, G. E. (1977). The Asian source of Arctic haze bands. *Nature*, **268**, 713–715.

Schütz, L. (1971). Messung der Grossen-verteilung von Luftgetragenem Staub und Bodenmaterial in der Libyschen Sahara. *M. S. Dissertation*, Max-Planck-Institut für Chemie (Otto-Hahn-Institut), Mainz, West Germany.

Schütz, L., and Jaenicke, R. (1974). Particle number and mass distribution above 10^{-4} cm radius in sand and aerosol of the Sahara desert. *J. Appl. Meteorol.*, **13**, 863–870.

Zoller, W. H., Gordon, G. E., Gladney, E. S., and Jones, A. G. (1973). The sources and distribution of vanadium in the atmosphere. In: Trace Elements in the Environment. *Adv. Chem. Ser., No.*, **123**: 31–47, Am. Chem. Soc., Wash., D.C.

CHAPTER 14

Sahara Dust Transport Over the North Atlantic Ocean — Model Calculations and Measurements

L. SCHÜTZ

ABSTRACT

The Sahara Desert is one of the major source areas for mineral dust particles which are transported long distances through the atmosphere. The transport mechanism for the Sahara dust itself now seems to be fairly well understood – for instance the transport which takes place in the NE-tradewind region.

This paper is concerned with a dust-transport model for the Sahara dust and measurements of the mineral dust component over the Atlantic. The model is based on the knowledge about the transport mechanism within the NE-tradewind area over the Atlantic. A steady-state transport of mineral particles will be considered which includes horizontal transport by wind and sedimentation as well as vertical turbulent diffusion. Mineral mass concentrations and particle volume distributions which had been measured across the Atlantic could be explained by the model. Both calculations and measurements indicate a very different behavior of submicron and giant particles during their long-range transport over the Atlantic.

14.1 INTRODUCTION

Investigations of the areawide distribution of Sahara dust and its chemical composition over the Atlantic were stimulated by the study of Delany *et al.* (1967). These authors as well as others studied the mineral composition and mass concentrations of mineral dust originating from the Sahara Desert. From these studies it became clear that large quanitites of mineral dust are carried from the Sahara over the Atlantic, contributing to the aerosol budget of the tradewind region and to the Atlantic deep-sea sediments.

Details of the mechanism of large-scale transport of Sahara dust over the Atlantic were reported by Prospero and Carlson (1970, 1972) and Carlson and Prospero (1972). According to their investigations the main mass transport of Sahara dust takes place above the tradewind inversion layer. There the mineral dust is transported westward within the steadily blowing 'harmattan', which is in agreement with earlier findings by Semmelhack (1934). The dust layer extends

vertically from approximately 1500 m to 5000 m altitude near the West African coast. After one week travel the dry and dust loaden air masses reach the Caribbean Sea, where the dust layer converges to 1500 m until 3000 m in altitude. The dust plume emerging from the Northwest African continent covers an area roughly between 15°N and 25°N latitude.

The relatively warm and dry continental air masses above the inversion layer are undercut by the cool and moist air of the tradewind boundary layer. This layer is originally free of dust because air mass trajectories never originate from the continent. Dust particles within this layer therefore must be supplied from the reservoir above by downward turbulent mixing and sedimentation. The inversion layer, however, suppresses vertical exchange between the dust layer above and the boundary layer beneath. One has to expect, therefore, that the vertical flux of smaller particles (radius $r < 1\ \mu m$) which are mainly subject to turbulent diffusion will be reduced. Larger particles ($r > 1\ \mu m$) remain practically unaffected by the presence of the strong inversion layer due to their high sedimentation velocities. Thus, larger particles will reach the ocean surface within relatively short distances from the source area, whereas smaller particles are deposited only after longer transport distances. Such a different behaviour of those particle-size fractions certainly has consequences for the size distribution of the mineral particles and the total suspended mineral matter found over the ocean when moving westwards under the plume.

The seasonal variation of incoming solar radiation causes seasonal dependence of both the generation of airborne dust over the Sahara Desert and of transport conditions over the ocean. The superposition of both effects produces an annual variation in the large-scale dust transport over the Atlantic ocean, as seen from Figure 14.1: Considering the sandstorm frequency for different areas of the Sahara desert as a measure of the source strength, a seasonal variation of a factor of about 2 can be seen (according to a compilation of observation data for the years 1925–1950 by Dubief, 1953). A period of increased sandstorm activity lasts from early spring until fall. Vertical wind data by SMN (1963) for the period 1951–1960 from Sal, Cape Verde Islands, clearly show favourable conditions for large scale transport especially in the summer season between May and October. During that time easterly winds prevail throughout levels up to 6000 m. For the remainder of the year the layer favorable for east to west transport is normally restricted to a shallow layer between 1500 m and 3000 m altitude. At higher levels strong westerly winds ('counter tradewinds') prevail. The transported mass of mineral dust is large during summer, as indicated by the turbidity coefficient B (for $\lambda = 500$ nm) which is a measure of the total aerosol mass in a vertical column for the particle size range between $0.1\ \mu m \leqslant r \leqslant 1.0\ \mu m$. These data are derived from a two-year series of measurements carried out at Sal by Jaenicke and Schütz (1977).

For a better understading of the transport mechanism, a model of the Sahara dust transport over the Atlantic was developed. For comparison of the calculations with actual data the total aerosol size distribution with special regard to the mineral

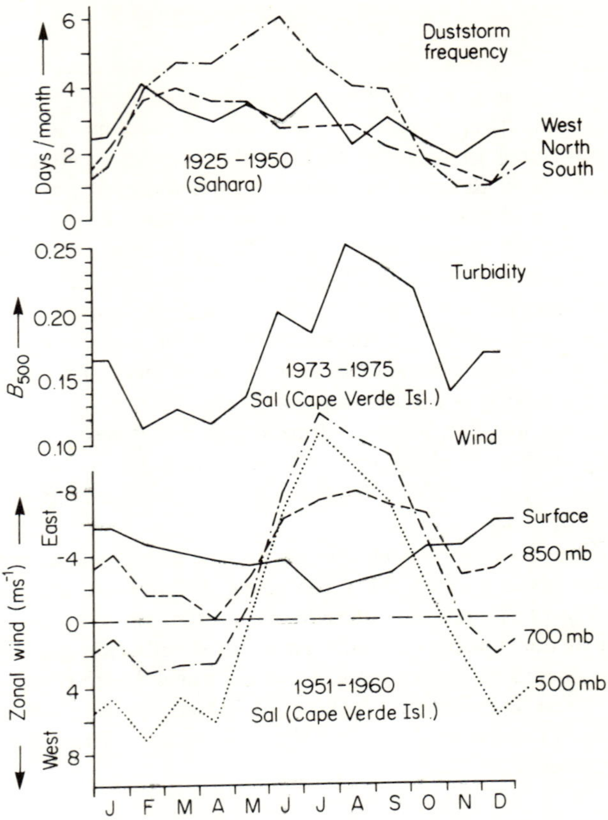

Figure 14.1 Comparison between mean annual frequency of sandstorms according to Dubief (1953), for different regions of the Sahara desert, vertical wind field over the Atlantic and atmospheric turbidity derived from measurements by Jaenicke and Schütz (1977). A possible explanation for the annual variation of turbidity which is a measure of the atmospheric dust load can be the superposition of the sandstorm frequency at the source area and the vertical wind field over the ocean

component was investigated in surface air of the NE-tradewind region over the Atlantic. To study the aerosol close to the source, measurements were carried out at Sal. Further investigations were performed onboard the German research vessel 'Meteor' while crossing the Atlantic from the Caribbean Sea to the West African coast at 15°N latitude. In this way it was intended to study the transport as a function of the distance from the source area.

14.2 TRANSPORT MODEL

For simulation of Sahara dust transport a two dimensional steady-state model (Schütz, 1977) was designed. The model considers east-to-west horizontal transport by zonal winds. Vertical transport is described by sedimentation and turbulent diffusion. Other transport processes such as subsidence and convection are assumed to balance each other. Removal of particles by precipitation is also not considered.

In order to properly describe flow conditions in the NE-tradewind zone, actual profiles up to 10 km altitude were used. The diffusion coefficient for vertical exchange was calculated from temperature and wind profiles (radio-sonde data from Sal were assumed to be representative of the total transport distance over the Atlantic). The starting point of the model was fixed at the West African coast. There a dust layer was assumed between 1500 m and 5000 m altitude with a constant vertical mixing ratio. The boundary layer up to 1500 m was assumed free of dust. This was justified because airmass trajectories never originate from the continent, and surface winds close to the African shore are westerly. The upper limit of the dust layer at 5000 m is confirmed by several aircraft observations near the African coast.

The model predicts the mixing ratio for mineral particles as a function of the air mass trajectory distance from the source area, altitude and particle size. Particle radii from 0.1 μm to 20 μm were considered. In order to study the size distribution as a function of the transport distance an initial distribution has to be assumed at the coastline. According to measurements of the mineral aerosol size distribution in the Libyan desert (Schütz and Janicke, 1974) an initial power function* distribution with an exponent of $\nu^* = 2$ seemed appropriate. These assumptions led to the following results.

14.3 RESULTS

14.3.1 Vertical mass distributions

Figure 14.2 shows how the Sahara dust plume develops during transport and how the vertical mixing ratio of the particles changes as a function of altitude and distance from the source. The mixing ratio is normalized to the initial mixing ratio.

*Most aerosol size distributions for particles with radii larger than 0.1 μm can be approximated by a power function:

$$\frac{dN}{d \log r} = n^*(r_0)(r/r_0)^{-\nu^*}$$

where

r	= particle radius in μm;
r_0	= arbitrary chosen reference radius;
N	= number of particles cm^{-3} larger than r;
$n^*(r)$	= differential number size distribution in cm^{-3};
ν^*	= exponent of the distribution.

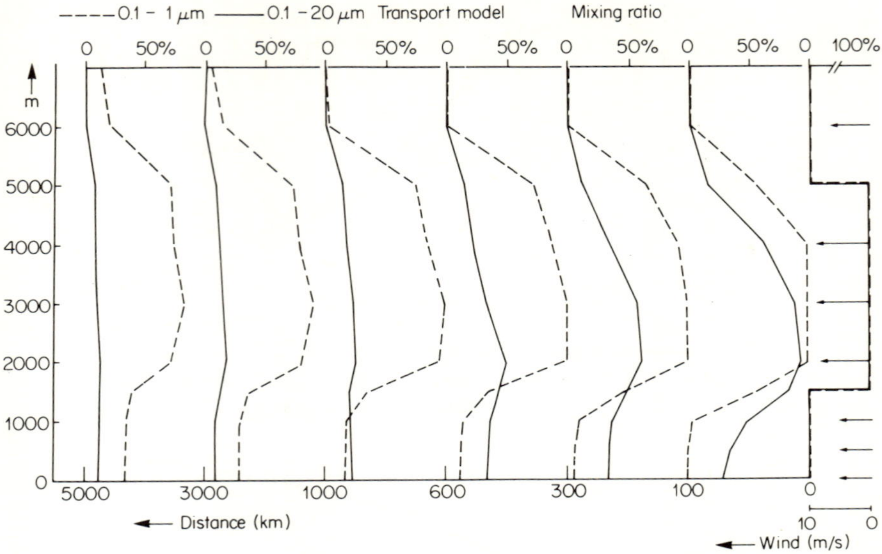

Figure 14.2 Vertical distributions of mineral dust for selected transport distances from the source. Vertical mixing ratios are plotted for the mass of mineral dust particles of the submicron range (0.1 μm $\leqslant r \leqslant$ 1.0 μm) and for the total mass (0.1 μm $\leqslant r \leqslant$ 20 μm) normalized to the initial mixing ratio at the source (x = 0 km). The vertical distributions of the submicron particles remain nearly unchanged up to large transport distances whereas those for the total mass show significant modifications up to approximately 1000 km distance. Thus most of the total mass is removed from the plume near the source but submicron particles are subject to long range transport

Obviously there is a different behavior for the submicron particles (0.1 μm $\leqslant r \leqslant$ 1.0 μm), which cause atmospheric turbidity, and those in the range of 0.1 μm $\leqslant r \leqslant$ 20.0 μm, which contain the total mineral mass. Up to 5000 km transport distance the submicron particles are only slightly removed from the plume. Their mixing ratio within the plume is only reduced to 70% of the initial value and at sea level their mixing ratio increases as the plume moves westward. These particles penetrate only very slowly from the elevated reservoir into the boundary layer by turbulent mixing and sedimentation. This clearly shows that these particles are subject to long-range transport.

In contrast, the vertical distribution of total aerosol mass (0.1 μm $\leqslant r \leqslant$ 20 μm) shows much more rapid variation with increasing transport distance. Due to the relatively large sedimentation velocities of giant particles ($r \geqslant$ 1 μm) most of the mass falls out of the plume within the first 1000 km. A constant vertical mixing ratio is nearly achieved at a distance of approximately 1000 km. Thus the dustfall area is practically restricted to the first 1000 km of transport. This agrees with earlier observations by Semmelhack (1934).

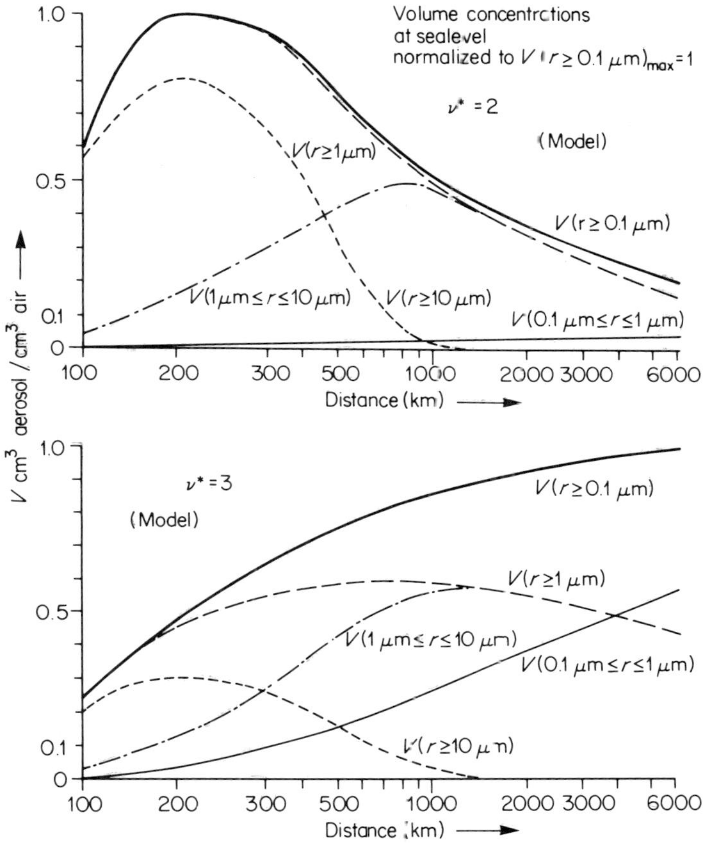

Figure 14.3 Volume concentration ratios at sea level for different particle size ranges as a function of the distance from the source. Model calculations for assumed initial size distributions with $\nu^* = 2$ and $\nu^* = 3$. Except for particles with $r \geqslant 10\ \mu m$ the graphs indicate quite different volume concentration ratios for the size ranges considered when moving westward under the plume

14.3.2 Volume and mass concentrations at sea level

In the previous chapter a general survey of the development of the plume was given. Now volume and mass concentrations at sea level for different particle size ranges as a function of the distance from the source will be discussed. Figure 14.3 shows calculated volume concentrations normalized to the maximum value of the total volume concentration $V\ (r \geqslant 0.1\ \mu m)$ for several particle size ranges with transport distances between 100 km and 6000 km. The data plotted in Figure 14.3

are for power-function size distributions with $\nu^* = 2$ and $\nu^* = 3$ assumed for initial size distributions within the dust layer at the source.

Volume concentration ratios in the submicron range increase very slowly with increasing distance from the source, especially for an initial distribution with $\nu^* = 2$. Up to a transport distance of 6000 km these particles make up only a few percent of the maximum volume concentration (at 200 km). If an initial distribution with $\nu^* = 3$ is assumed, however, considerably higher volume concentration ratios are reached with increasing distance from the Sahara. In this case the submicron particles may contribute more than 50% of the total volume for distances larger than 4000 km from the source area. But it should be pointed out that in both cases the concentrations of the submicron particles are comparatively low within the first hundreds of kilometres of transport. As mentioned above, this indicates that these particles reach the surface layer only by slow vertical mixing.

The ratios of the volume concentrations of the largest particles ($r \geqslant 10\ \mu m$) increase up to 200 km from the source and thereafter rapidly decrease with distance. At 1000 km from the source their values are roughly a few percent. At further distances their contribution to the total concentration is practically negligible. Independent of the assumed initial distribution, these particles, which sediment quickly, are deposited within 1500 km of the Sahara.

Giant particles in the intermediate size range ($1\ \mu m \leqslant r \leqslant 10\ \mu m$) increase in relative values up to 1000 km, where a maximum of approximately 50% of the total volume is reached. For distances larger than 1000 km their volume concentration ratios decrease but still account for most of the mineral volume which is transported over larger distances. For $\nu^* = 2$, 90% of the total volume transported over distances larger than 1500 km consists of these particles. For $\nu^* = 3$ the contribution of this particle range is smaller, because of rapidly increasing volume fractions of submicron particles at distances greater than 2000 km from the Sahara.

For verification of the model calculations a comparison with actual mass concentrations (M) measured at sea level over the Atlantic ocean will be given. Data were collected during field studies of the NE-tradewind aerosol over the Atlantic as mentioned above. Figure 14.4 shows measured and calculated mass concentrations for the submicron and giant particles as a function of the distance from the source. As above, initial distributions with $\nu^* = 2$ and $\nu^* = 3$ will be distinguished. Calculated mass concentrations are normalized to M ($0.1\ \mu m \leqslant r \leqslant 20\ \mu m$) $= 100\ \mu g\ m^{-3}$ at a distance of 1000 km from the source. This value was selected from mean mass concentrations of our Atlantic data. From Figure 14.4 some important conclusions can be drawn about the shape of the initial assumed size distribution. Measured total mass concentrations agree reasonably well with the calculations using $\nu^* = 3$. For the submicron range, however, the model with $\nu^* = 3$ predicts concentrations approximately one order of magnitude higher than observed. A better assumption for the initial distribution within the size range of $0.1\ \mu m \leqslant r \leqslant 20\ \mu m$ would be to use $\nu^* = 2$. In that case the measured mass concentrations for both size ranges are better approximated by the model, as seen

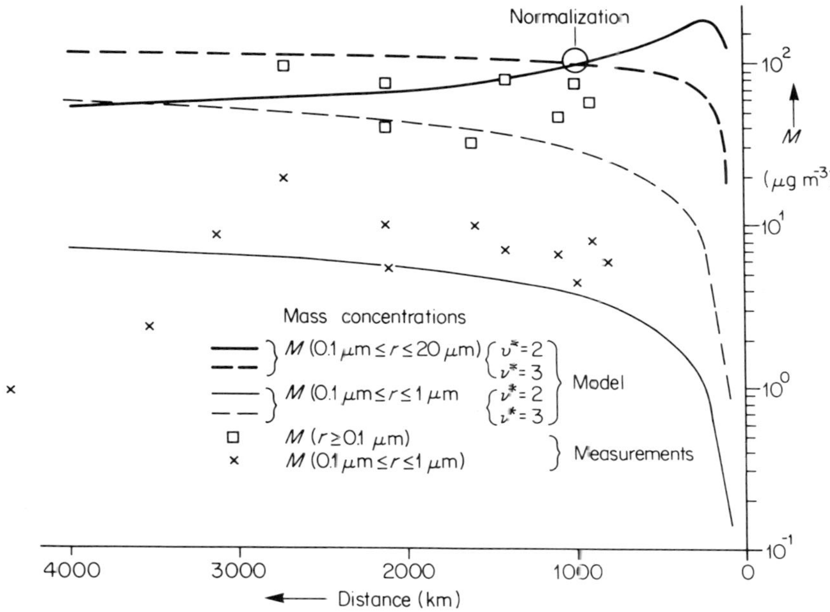

Figure 14.4 Mineral mass concentrations at sea level as a function of the transport distance from the Sahara. Comparison between calculated data [normalized to M (0.1 μm $\leqslant r \leqslant$ 20 μm) = 100 μgm^{-3} at a distance of 1000 km] and data measured by the authors for the submicron and the total mass range. The measurments can best be described by the calculations when assuming an initial distribution at the source with $\nu^* = 2$

in Figure 14.4. The assumption of $\nu^* = 2$ at the source is confirmed by measurements of mean mineral size distributions in the size range $r \geqslant 1$ μm over the Libyan Sahara by Schütz and Jaenicke (1974). A similar value of $\nu^* \sim 2$ has been obtained by Gillette *et al.* (1972) for soil-derived aerosol in Texas and Nebraska, which suggests that this value may be typical for soil-aerosol production in many regions of the globe. Figure 14.4 shows that neither the total sea-level mass nor its submicron fraction decrease significantly beyond 1000 km from the source. In contrast, significant variations in the mass concentrations have to be expected within the first 1000 km of transport. Unfortunately, no data are available to check this idea.

14.3.3 Volume distributions

One main goal of the investigation of the mineral dust transport in the NE-trade-wind region was the study of the development of the size distribution at sea level as

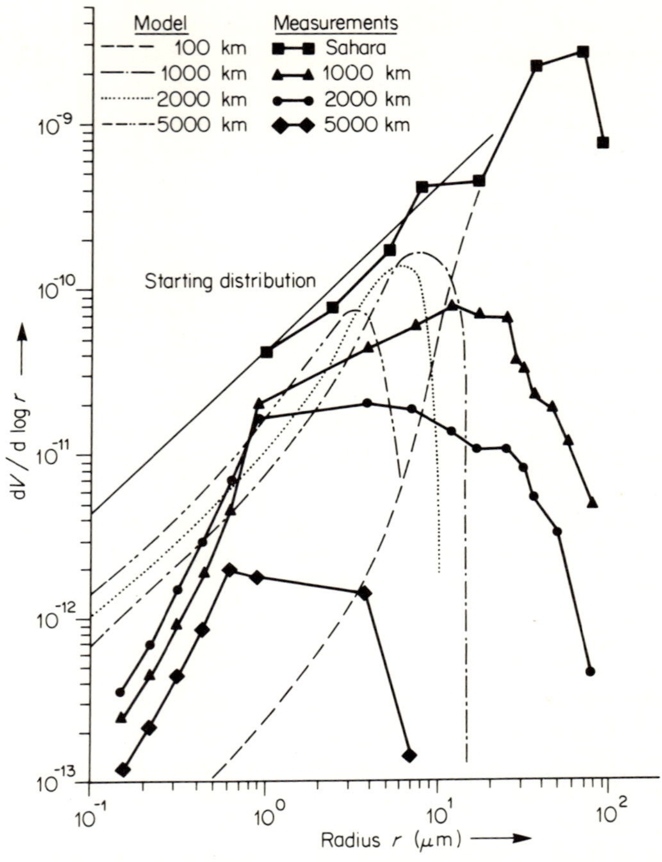

Figure 14.5 Mineral volume distributions at sea level for various distances from the Sahara. The dotted lines represent calculations assuming an initial distribution with $\nu^* = 2$ at the source. Heavy lines (measurements) show mean volume distributions at the source and for selected transport distances over the Atlantic. Both measurements and calculations indicate a shift of the maximum of the distributions toward smaller particle radii

a function of the distance to the Sahara. For a better understanding of the measured size distributions across the Atlantic, a comparison with modeled distributions will be discussed. The calculations are based on an initial distribution at the source with $\nu^* = 2$ according to the results achieved from Figure 14.4. The size distributions were then converted to volume distributions so that important variations in the giant-particle range could be more closely observed.

Figure 14.5 shows a comparison between volume distributions derived from measurements and calculations for selected distances from the source. From this Figure it is apparent that the assumed initial distribution at the source is a reasonably accurate representation of the actual distribution observed at the Sahara Desert site. When moving westward from the source, the model calculations indicate a shift of the maximum of the distributions towards smaller particle radii, and there is a general decrease in the maximum concentration. These results are in qualitative agreement with the measured size distributions. According to these data, the maximum of the volume distribution at the source is at 50 μm radius and subsequently shifts to radii of approximately 10 μm, 4 μm, and 1 μm for transport distances of 1000 km, 2000 km and 5000 km respectively. Finally the model is not able to explain the comparatively high concentrations of particles of radii larger than 10 μm observed during field measurements. These giant particles are probably agglomerates consisting of smaller particles which are formed during coalescence of precipitation and cloud droplets, followed by evaporation, which produces dry residues.

It is possible that the volume distribution observed over the Caribbean Sea, 5000 km from the Sahara Desert, may represent the background distribution of the mineral component of the atmospheric aerosol.

14.4 CONCLUSIONS

The transport model offered the possibility to study some important features of Sahara dust transport within the NE-tradewind zone. From the calculations it clearly turns out that a very different behaviour for various particle size ranges must be expected when moving westward from the source. Submicron particles, which cause atmospheric turbidity, remain within the plume over large transport distances. Their removal from the plume, which mainly takes place by turbulent diffusion, is strongly affected by the tradewind inversion. This layer suppresses vertical exchange and leads to a delayed downward flux of these particles. Giant particles, however, remain unaffected by the inversion layer due to their high sedimentation velocities. Thus they are rapidly removed from the plume within relatively short distances from the source. Most of this particle fraction, which represents a considerable part of the mineral aerosol mass, will be deposited within the first 1500 km of transport.

A comparison between measured and calculated mass concentrations at sea level suggests $\nu^* = 2$ as a reasonable value for the number size distribution in the form of a power function within the range $r \geqslant 0.1$ μm. This indicates a relatively large fraction of giant particles produced at the source and confirms earlier findings by the author in the Sahara desert. Furthermore this agrees well with data obtained by Gillette *et al.* (1972) for airborne soil particles over Texas and Nebraska and might indicate that the value of $\nu^* = 2$ is significant for soil-derived aerosols close to the source.

Finally it can be pointed out that this model allows further applications, as is shown by Jaenicke (1979). His source strength estimates of the Sahara Desert and the budget studies of transported and deposited mineral mass over the Atlantic are based on this model.

REFERENCES

Carlson, T. N., and Prospero, J. M. (1972). The large-scale movement of Saharan air outbreaks over the northen equatorial Atlantic. *J. Appl. Meteorol.*, **11**, 283–297.

Delany, A. C., Delany, Audrey Claire, Parking, D. W., Griffin, J. J., Goldberg, E. D., and Reimann, B. E. F. (1967). Airborne dust collected at Barbados. *Geochim. Cosmo-Chim. Acta*, **31**, 885–909.

Dubief, J. (1953). Les vents de sable au Sahara français. *Colloq. Int. CNRS*, **35**, 45–70.

Gillette, D. A., Blifford, I. H. Jr., and Fenster, C. R. (1972). Measurements of aerosol size distribution and vertical fluxes of aerosols on land subject to wind erosion. *J. Appl. Meteorol.*, **11**, 977–987.

Jaenicke, R. (1979). *Monitoring and Critical Review of the Estimated Source Strength of Mineral Dust from the Sahara.* In this report.

Jaenicke, R., and Schütz, L. (1977). A comprehensive study of physical and chemical properties of the surface aerosols in the Cape Verde Islands region. Submitted to *J. Geophys. Res.*

Prospero, J. M., and Carlson, T. N. (1970). Radon – 222 in the North Atlantic trade winds: Its relationship to dust transport from Africa. *Science*, **167**, 974–977.

Prospero, J. M., and Carlson, T. N. (1972). Vertical and areal distribution of Saharan dust over the western equatorial North Atlantic Ocean. *J. Geophys. Res.*, **77**, 5255–5265.

Schütz, L. (1977). Die Saharastaub-Komponente über dem subtropischen Nord-Atlantik. *Ph.D. Thesis*, Univ. Mainz, FRG., 153 pp.

Schütz, L., and Jaenicke, R. (1974). Particle number and mass distributions above 10^{-4} cm radius in sand and aerosol of the Sahara desert. *J. Appl. Meteorol.*, **13**, 863–870.

Semmelhack, W. (1934). Die Staubfälle im Nordwestafrikanischen Gebiet des atlantischen Ozeans. *Ann. Hydrol.*, **62**, 273–277.

SMN. (1968). *Servicio Meteorologico National, Climatologia Aeronautica de Portugal.* Aeroporto do Sal, Lisboa.

CHAPTER 15

A Possible Method for the Sampling of Saharan Dust

B. STEEN

ABSTRACT

Most sampling methods for suspended particles in ambient air used today are designed for the urban environment and its possible related health effects. They are designed for measuring concentrations between 10 and 200 $\mu g/m^3$ particle sizes below 20 μg, moderate wind speed and a time scale of 0.5 to 24 hours. The data obtained with most of these methods are useful if the conditions do not differ too much from those described. In the case of the problematic erosion of Saharan dust, most of the critera mentioned above would probably not be fulfilled: The concentrations are likely to be much higher where the erosion occurs, the particle sizes up to several hundreds of microns, the wind speed high and the time scale longer (at least for practical reasons). If one wants to determine the real mass of particles suspended in air in this case or collect a representative sample, no other solution immediately presents itself than isokinetic sampling – at least for the larger particles.

This paper will describe such a method which has been developed at IVL.*

Basically the sampler consists of a small cylinder which is open at both ends. During sampling its axis is oriented parallel to the flow lines. The fluid is allowed to pass isokinetically through a nozzle at the front end of the cylinder by means of its own inertia. The particles are charged and deposited on the cylinder wall. The cylinder acts as a collecting electrode in an electrostatic precipitator and its weight is determined before and after the sampling. The collection efficiency was found to be above 90% for most particles. The sampler is battery-operated and its total weight is 2.3 kg.

15.1 INTRODUCTION

Aerosols emitted to or present in ambient air are mostly characterized by their concentrations. Consequently, most techniques for sampling and measurements of particles are designed to determine concentrations. In many cases, however, concentrations are not of prime interest. When determining emissions from stacks,

*Swedish Water and Air Pollution Research Laboratory.

ventilators or 'diffuse' sources such as stockyards, roads, etc., the prime interest is the flux of particles, i.e. the amount of particles transported through a certain area during a certain period of time. In such cases the mere measurement of the flux may be advantageous compared to conventional techniques that use instruments for the determination of both concentration and flow velocity.

15.2 GENERAL DESCRIPTION

The sampler consists basically of a nozzle, a cylindrical collecting electrode, an emission electrode, a flow regulator, a shaft, and a high voltage generator. (Figures 15.1 and 15.2).

During sampling, the axis of the cylinder is oriented parallel to the flow lines by a vane and the fluid is allowed to pass isokinetically through the nozzle by means of its own inertia. The particles are charged and collected on the inner cylinder wall due to electrostatic forces. The cylinder is weighed before and after the sampling to determine the particle mass. The particles may also be analysed by chemical or physical means.

15.2.1 Nozzle

Several different inlets may be used.

First, the cylinder was used without any nozzle. This makes it possible to measure a very low particle flux, but the collection efficiency may decrease significantly with increasing flow velocity.

Second, three similar nozzles with different inlet diameters were made of teflon (Figure 15.2). The diameters chosen were 5.0, 8.2, and 11.2 mm, giving approximately 25, 10, and 5 times reduction in flow velocity respectively from nozzle inlet to collection area.

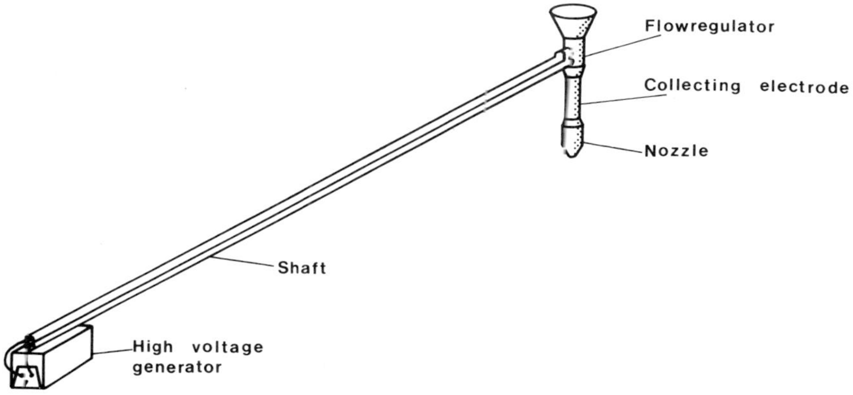

Figure 15.1 Sketch of particle sampler

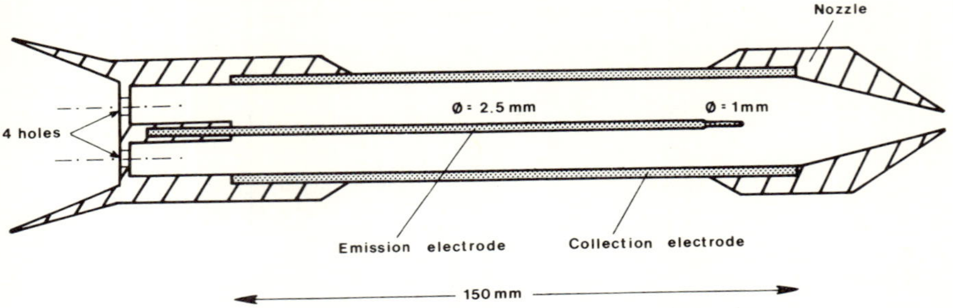

Figure 15.2 Cross-section of sampler

Third, a special 5 mm 'expansion-contraction' nozzle was constructed to be used in those cases when the collection efficiency obtained with the ordinary 5 mm nozzle is significantly reduced due to turbulence. The nozzle gives initially a reduction in flow velocity of 50 times just as the other nozzles but it also consists of a contraction section with an area ratio of 2 : 1 where the turbulence is smoothed out before the sample enters the cylindrical part.

Most of the tests and measurements described in this paper were carried out with the normal 5 mm nozzle.

15.2.2 Collecting electrode

The collecting electrode consists of a 150 mm acid-proof steel cylinder with an inner diameter of 25 mm and a weight of 145 g. Inside it may be covered with an aluminium foil. The foils used weigh around 3 g and may easily be folded and stored with the particles collected on it.

15.2.3 Emission electrode

The tip of the emission electrode is located 20 mm from the front end of the cylinder as shown in Figure 15.2.

15.2.4 Flow regulator

A cross section of the flow regulator is shown in Figure 15.2. It is made of teflon to give a good insulation and resistance to temperature. Its rear end is expanded in order to increase the flow velocity on the outside and provide an extra vacuum on the inside. This is necessary in order to achieve isokinetic sampling through the nozzle inlet. Without this expansion and a nozzle the friction losses on the inside would lower the flow rate through the cylinder resulting in sub-isokinetic sampling. When nozzles are used the sample flow is above that of isokinetic sampling if no extra flow restriction is made. Therefore an exchangeable plate with four holes is

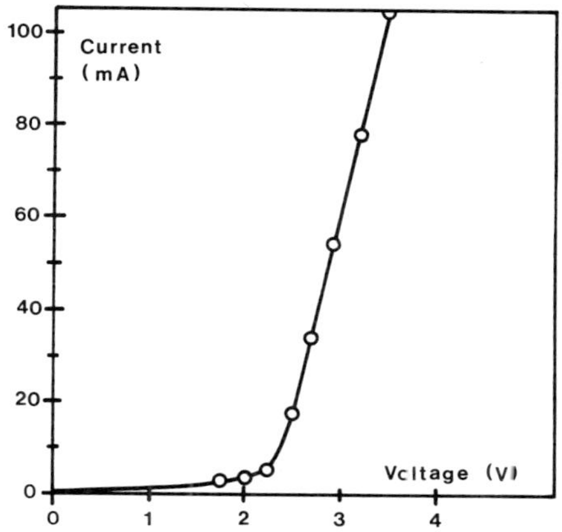

Figure 15.3 Battery current–voltage characteristics for high voltage supply working on sampler

placed just before the expansion at the rear end. The diameters of the holes are matched with the respective nozzles in order to obtain isokinetic sampling.

15.2.5 High voltage generator

The high voltage generator is battery-operated and consists of an oscillator, a transformer and a voltage multiplier. The output of the generator is around 12 kV at 4.5 V input voltage. The battery current as a function of battery voltage is shown in Figure 15.3 and indicates the formation of a stable corona discharge in the sampler.

15.3 PERFORMANCE

15.3.1 Flow characteristics

The flow velocity in the nozzle tip and outside the sampler was determined by means of a hot-wire anemometer. The sampler was placed in a wind tunnel and velocity measurements were made at a wind speed varying from 0.7 to 11.2 m/s.

The results given in Table 15.1 show that the deviations from isokinetic sampling conditions generally are small but ought to be considered when sampling is carried out at low flow rates or without a nozzle.

15.3.2 Collection efficiency

The collection efficiency was determined experimentally as a function of various

TABLE 15.1 Ratio of Flow Velocity Between Sample at Nozzle Inlet and Main Flow

Main flow velocity (m/s)	Relative Sample Flow Velocity			
	5 mm nozzle	8.2 mm nozzle	11.2 mm nozzle	Without nozzle
0.7	1.02	0.96	0.95	0.79
1.1	1.01	1.00	1.01	0.85
2.0	1.03	0.99	1.01	0.90
3.1	1.00	0.99	1.02	0.90
4.3	1.03	1.01	1.02	0.95
11.2	1.00	0.99	1.03	1.04

parameters. The experimental set-up consisted either of:

(a) the sampler, a filter, a dry gas meter and a pump, or
(b) the sampler, a Royco 220 light scattering particle counter and a pump.

15.3.2.1 Particle size

Table 15.2 shows the results obtained by using a Royco particle counter when determining the concentration of different particle size classes present in a room atmosphere before and after the sampler. The sampler was used without a nozzle and with a flow velocity of 1 m/s. The particle number concentrations were low, around 1000/1 in the 0.3–0.4 μm range and 10–20/1 in the 3.4–6.6 μm range.

TABLE 15.2 Collection Efficiency of Sampler for Different Particle Size Classes at Different Battery Currents

Latex equivalent particle diameter (μm)	Sampling efficiency (%)				
	Battery current (mA)				
	6	8	10	50	196
0.3–0.4	73.7	83.2	93.7	97.6	99.0
0.4–0.6	72.9	83.8	94.5	97.5	98.7
0.6–1.0	75.6	86.1	96.0	98.0	99.2
1.0–1.8	77.6	92.2	96.3	99.3	100
1.8–3.4	78.7	92.4	99.7	100	100
3.4–6.6	78.2	92.9	98.7	100	100
6.6–13	88.3	(80.0)	100	100	100

15.3.2.2 Concentration

Approximately 35 mg of a talc powder, with a mass median particle diameter of 2.5 μm ('ITX talc powder'), was dispersed in an air stream and fed into the 5 mm nozzle inlet at concentrations varying from 23 mg/m^3 to 15 g/m^3. The flow velocity was 25 m/s and the battery current was 50 mA.

The collection efficiency was determined by the filter method. No correlation between concentration and collection efficiency was found. The results showed a collection efficiency between 93.8 and 97.9%.

15.3.2.3 Flow velocity

Using a battery current of 50 mA and 5 mm nozzle, the collection efficiency was determined with ITX talc powder dispersed in a wind tunnel. The collection efficiency was 88.7, 91.8, 99.3% at 22.9, 16.6, and 10.4 m/s, respectively.

15.3.2.4 Amount of collected material

Varying the sampling time, and keeping the concentration constant the collection efficiency was determined as a function of the amount of ITX talc powder on the collecting electrode. The results, given in Table 15.3 indicate a decreasing collection efficiency with an increasing amount of talc. As talc is one of the most easily dispersed powders redispersion of talc particles may be the major cause of the decreasing collection efficiency. There are two things that indicate this. First the collection efficiency increases when an 'expansion–contraction' nozzle is used to suppress turbulence (Table 15.3). Second, when sampling more sticky dust, such as the fumes from an electric arc steel converter, a much higher collection efficiency will be obtained (Table 15.4).

TABLE 15.3 Influence of Electrode Deposits on Collection Efficiency

Amount of collected talc (mg)	Battery current (mA)	Nozzle type	Fluid velocity (m/s)	Collection efficiency (%)
5.2	50	normal, 5 mm	25	96.3
18.9	50	normal, 5 mm	25	95.5
40.6	50	normal, 5 mm	25	88.6
79.6	50	normal, 5 mm	25	72.9
27.7	250	normal, 5 mm	25	97.9
69.0	250	normal, 5 mm	25	78.4
84.0	400	normal, 5 mm	25	80.8
33.7	50	expansion–contraction, 5 mm	25	98.2
103.7	50	expansion–contraction, 5 mm	25	94.0
255.7	50	expansion–contraction, 5 mm	11.7	96.7

TABLE 15.4 Collection Efficiency Obtained from Measurements of Different Types of Particles

Gas velocity at nozzle (m/s)	Nozzle diameter (mm)	Battery current (mA)	Type of particles	Amount of collected dust (mg)	Collection efficiency (%)
0.8	25	50	Redispersed soot from oil combustion	27.8	99
	25	50	Redispersed Na_2SO_4 from craft pulp mill recovery furnace	8.7	98
0.8	25 5	50	ITX-talc Emissions from electric arc steel furnace	23.6	98
0.8	25	50	Fe_2O_3	8.4	100
0.8	25	50	Particles in ambient air	5.3	91
0.8	25	50	Sulphate in particles in ambient air	0.201	87

The sampler was later tested for redispersion by the filter method. The flow velocity was 17.0 m/s at the inlet of the 5 mm nozzle. The amount of initially collected ITX-talc was 38.8 mg. During 2 hours of extra sampling of dustfree air the redispersion was 0.9 mg corresponding to a decrease in the collection efficiency of 2.4%. This proves that a redispersion takes place, but the amount of redispersed talc is less than that indicated by Table 15.3. A possible cause is the presence and absence respectively of simultaneous collection of particles.

15.3.2.5 Types of aerosols

The collection efficiency obtained by the filter method for different types of aerosol is shown in Table 15.4.

15.3.2.6 Battery current

The battery current is the most simple electrical parameter to determine. According to the manufacturer of the high voltage generator, a 6 volt DC input should give a 15 kV DC output. The battery current as a function of the battery voltage is shown in Figure 15.3, which indicates that an input voltage of more than 2.25 V is needed to form a corona in the sampler. An example of the influence of the battery current on the collection efficiency is shown in Table 15.2.

15.3.3 Battery life

Battery life depends mainly upon the current output and the battery temperature. At temperatures above 23°C the 1.5 V Berec alcaline MN1300 batteries which were used in this study could run during approximately 75 hours before the battery voltage dropped from 4.5 V to 3 V.

15.3.4 Practical use

In order to get a general picture of the performance of the sampler in practical field operations, it was used to determine the total particulate emission from a ventilator system above two electric arc steel furnaces. 10 samplers were built and used simultaneously on different ventilator drums. 5 mm nozzles and an aluminium foil were used. The battery current was constantly above 140 mA and the flow velocity constantly below 15 m/s.

More than 35 samples were taken. Positive experience was gained from the measurements. The aluminium foils made the work handy as they could be folded and easily stored after sampling. Scattered rain storms did not seem to influence the operation although the high-voltage source and the shaft was getting wet. The sampling efficiency was 97.5% as determined with the filter method, the amount of collected particles being as high as 389 mg. Visually, it could be checked that the sampling efficiency was normal by the sight of the particle deposits on each sample. In such a case the rear end of the foil ought to be relatively free from particles and the front end, especially the first centimetres, well covered. Although there was rather intense vibrations from the fans, the deposits on the foils seemed to be intact.

CHAPTER 16

Techniques for Measuring Dry Deposition. Summary of WMO Expert Meeting on Dry Deposition, April 18—22, 1977, Gothenburg

B. STEEN

There are, in principle, three ways of determining dry deposition. They may be referred to as (1) direct accumulation, (2) flux-gradient, and (3) eddy correlation techniques.

Direct accumulation techniques use some kind of sampling surface – either natural or artificial – and determine the amount of material on the surface, or an adjacent bulk, before and after exposure. As they are easy to make, many of these samplers are home-made and there are therefore many different designs. However, only two (HASL and HARWELL methods) were described that had been used to any greater extent for dry deposition. Most were used for the sum of wet and dry deposition. The samplers discussed and their principal design are listed in Table 16.1.

TABLE 16.1 Summary of Accumulation Techniques for Measuring Dry Deposition of Particles

Type of sampler	Specifications	Principal sketch of sampler	Tests of performance
Bucket	HASL: aperture diameter = 300 mm (Volchok, 1977)		Sr^{90} against grass (Volchock, 1977)
Bucket	NILU: aperture diameter = 200 mm		Correlations with deposition on moss (Skärby, 1977)
Bucket	BERGERHOFF: aperture diameter = 90 mm (VDI, 1971a). Turbulence shielded		

TABLE 16.1 (continued)

Type of sampler	Specifications	Principal sketch of sampler	Tests of performance
Bucket	HIBERNIA: aperture diameter = 250 mm (VDI, 1971b)		compared with each other (Köhler and Fleck, 1963)
Bucket	BRITISH STANDARD DEPOSITE GAUGE: aperture diameter = 300mm (BSI, 1969). Turbulence shielded		BS tested in wind tunnel (Ralph and Barrett, 1976)
Bucket	LÖBNER: aperture diameter = 290 mm (VDI, 1971b)		
Vertical gauge	CERL: tube diameter = 75 mm, opening width = 45 mm (BSI, 1972)		Wind tunnel (Ralph and Barrett, 1976)
Plane surface	HARWELL METHOD: filter under cover (Pattenden, 1977)		
Natural surfaces	Growing moss (Rühling and Tyler, 1971), mossbags (Goodman *et al.*, 1979), snow (Dovland and Eliassen, 1976), water, trees		

The flux-gradient technique requires determination of the concentration gradient and the turbulence.

For the eddy correlation technique, continuous determination of the concentration and the vertical component of wind speed is needed. The last two techniques have been used almost exclusively for gases.

REFERENCES

British Standards Institution (1969). *Methods for the Measurement of Air Pollution.* BS 1747, Part 1. London, 17 pp.

British Standards Institution (1972). *Methods for the Measurement of Air Pollution.* BS 1747, Part 5. London, 18 pp.

Dovland, H., and Eliassen, A. (1976). Dry deposition on a snow surface. *Atmos. Environ.*, **10**, 783–785.

Goodman, G. T., Inskip, M. J., Smith, S., Parry, G. D. R., and Burton, M. A. S. (1979). *The Use of Moss-bags in Aerosol Monitoring.* In this report.

Köhler, A., and Fleck, U. W. (1963). Untersuchungen zur Festlegung eines Standardmessgerätes für Staubniederschlag. T. H. Darmstadt. *Inst. Meteorol. Abschlussbericht No. J.* (84) 2.

Pattenden, J. (1977). *Pers. Comm. AERE*, Bldg 364, Harwell, Oxon, U.K.

Ralph, M. O., and Barrett, C. F. (1976). A wind-tunnel study of the efficiency of deposit gauges – interim report. *Warren Spring Lab. Rep. LR 235 (AP).* Stevenage, Herts, 17 pp.

Rühling, Å., and Tyler, G. (1971). Moss analysis – a method for surveying heavy metal deposition. *Proc. Sec. Int. Clean Air Congr. Wash., D.C.*, Acad. Press, London, New York, 129–132.

Skärby, L. (1977). Correlation of moss analysis to direct measurement by deposit gauge of cadmium, lead, and copper. *Swed. Water Air Pollut. Res. Lab.*, publ. B377a, 24 pp.

Verein Deutscher Ingenieure (1971a). *Bestimmung des Partikelförmigen Niederschlags mit dem Bergerhoffgerät (Standardverfaren).* VDI 2119, Blatt 2, VDI-Verlag GmbH, Düsseldorf, FRG, 4 pp.

Verein Deutscher Ingenieure (1971b). *Bestimmung des Partikelförmigen Niederschlags mit dem Hibernia – und Löbner-Liesegang – Gerät (Standardverfaren).* VDI 2119, Blatt 3, VDI-Verlag GmbH, Düsseldorf, FRG, 4 pp.

Volchok, H. (1977). *Pers. Comm.* HASL-ERDA, New York, N.Y. 10014.

Index

SCOPE 1: Global Environmental Monitoring 1971, 68pp (out of print)

SCOPE 2: Man-Made Lakes as Modified Ecosystems, 1972, 76pp

SCOPE 3: Global Environmental Monitoring System (GEMS): Action Plan for Phase 1, 1973, 132pp

SCOPE 4: Environmental Sciences in Developing Countries, 1974, 72pp

Environment and Development, proceedings of SCOPE/UNEP Symposium on Environmental Sciences in Developing Countries, Nairobi, February 11–23, 1974, 418pp

SCOPE 5: Environmental Impact Assessment: Principles and Procedures, 1975, 160pp

SCOPE 6: Environmental Pollutants: Selected Analytical Methods, 1975, 277pp (available from Butterworth & Co. (Publishers) Ltd., Sevenoaks, Kent, England)

SCOPE 7: Nitrogen, Phosphorus, and Sulphur: Global Cycles, 1975, 192pp (available from Dr Thomas Rosswall, Swedish Natural Science Research Council, Stockholm, Sweden)

SCOPE 8: Risk Assessment of Environmental Hazard, 1978, 132pp

SCOPE 9: Simulation Modelling of Environmental Problems, 1978, 128pp

SCOPE 10: Environmental Issues, 1977, 242pp

SCOPE 11: Shelter Provision in Developing Countries, 1978, 112pp

SCOPE 12: Principles of Ecotoxicology, 1979, 372pp

SCOPE 13: The Global Carbon Cycle, 1979, 480pp (approx.)

SCOPE 14: Saharan Dust: Mobilization, Transport, Deposition, 1979, 320pp

Scientists have recently estimated that each summer during the early 1970's no less than 60–200 million tons of soil dust were blown out of the Sahara and surrounding areas over the Atlantic Ocean. This enormous transport of dust coincided with the years of drought and famine in the Sahel of West Africa.

What processes are responsible for the transport of this dust and what does it mean in terms of the loss of productive soil by wind erosion in Africa, the pollution of the air and possible impact on climate, and the supply of nutrients and soil particles to the ocean and land areas under the dust trajectories?

In order to highlight these questions a Workshop on Saharan Dust was arranged in Gothenberg, Sweden, in April 1977, with the participation of scientists of several different disciplines including meteorology, ecology, sedimentology and pedology. The outcome of this workshop is published in the present SCOPE Report 14 which contains the Review and Recommendations from the workshop together with sixteen papers given there.

JOHN WILEY & SONS
Chichester · New York · Brisbane · Toronto

ISBN 0 471 99680 7